Patterns, Defects and Materials Instabilities

NATO ASI Series

Advanced Science Institutes Series

A Series presenting the results of activities sponsored by the NATO Science Committee, which aims at the dissemination of advanced scientific and technological knowledge, with a view to strengthening links between scientific communities.

The Series is published by an international board of publishers in conjunction with the NATO Scientific Affairs Division

A Life Sciences **B Physics**	Plenum Publishing Corporation London and New York
C Mathematical and Physical Sciences **D Behavioural and Social Sciences** **E Applied Sciences**	Kluwer Academic Publishers Dordrecht, Boston and London
F Computer and Systems Sciences **G Ecological Sciences** **H Cell Biology**	Springer-Verlag Berlin, Heidelberg, New York, London, Paris and Tokyo

Series E: Applied Sciences - Vol. 183

Patterns, Defects and Materials Instabilities

edited by

D. Walgraef
Service de Chimie Physique,
Université Libre de Bruxelles,
Brussels, Belgium

and

N. M. Ghoniem
Mechanical, Aerospace and Nuclear Engineering Department,
University of California,
Los Angeles, U.S.A.

Kluwer Academic Publishers

Dordrecht / Boston / London

Published in cooperation with NATO Scientific Affairs Division

Proceedings of the NATO Advanced Study Institute on
Patterns, Defects and Materials Instabilities
Cargèse, France
September 4–15, 1989

Library of Congress Cataloging in Publication Data

```
NATO Advanced Study Institute on Patterns, Defects, and Materials
   Instabilities (1989 : Cargèse, France)
    Pattern, defects and materials instabilities / edited by D.
 Walgraef and N.M. Ghoniem.
       p.  cm. -- (NATO ASI series. Applied sciences ; 183)
    "Proceedings of the NATO Advanced Study Institute on Patterns,
 Defects, and Materials Instabilities, Cargese, France, September
 4-15,1989."
    "Published in cooperation with NATO Scientific Affairs Division."
    ISBN 0-7923-0753-4 (U.S. : alk. paper)
    1. Chaotic behavior in systems--Congresses.  2. Space and time-
 -Congresses.  3. Fluctuations (Physics)--Congresses.  4. Materials-
 -Congresses.  I. Walgraef, D. (David) II. Ghoniem, N. M. (Nars
 Mostafa)  III. North Atlantic Treaty Organization. Scientific
 Affairs Division.  IV. Title.  V. Series: NATO ASI series. Series
 E, Applied sciences ; no. 183.
 Q172.5.C45N383  1989
 530.4'1--dc20                                            90-34912
```

ISBN 0–7923–0753–4

Published by Kluwer Academic Publishers,
P.O. Box 17, 3300 AA Dordrecht, The Netherlands.

Kluwer Academic Publishers incorporates the publishing programmes of
D. Reidel, Martinus Nijhoff, Dr W. Junk and MTP Press.

Sold and distributed in the U.S.A. and Canada
by Kluwer Academic Publishers,
101 Philip Drive, Norwell, MA 02061, U.S.A.

In all other countries, sold and distributed
by Kluwer Academic Publishers Group,
P.O. Box 322, 3300 AH Dordrecht, The Netherlands.

Printed on acid-free paper

All Rights Reserved
© 1990 by Kluwer Academic Publishers
No part of the material protected by this copyright notice may be reproduced or utilized in any form or by any means, electronic or mechanical, including photocopying, recording or by any information storage and retrieval system, without written permission from the copyright owner.

Printed in the Netherlands

CONTENTS

PREFACE VII

ONE-DIMENSIONAL CELLULAR PATTERNS
P. Coullet and G. Iooss 1

DEFECTS AND DEFECT-MEDIATED TURBULENCE
J. Lega 7

PHASE DYNAMICS - THE CONCEPT AND SOME RECENT DEVELOPMENTS
H.R. Brand 25

TRANSIENT PATTERN DYNAMICS : GENERAL CONCEPTS AND THE
FREEDERICKSZ TRANSITION IN NEMATICS
M. San Miguel and F. Sagues 35

LOCALIZED STRUCTURES IN REACTION-DIFFUSION SYSTEMS
G. Dewel and P. Borckmans 63

KINETIC MODELS FOR DEFECT POPULATIONS IN DRIVEN MATERIALS
D. Walgraef 73

EXTERNAL NOISE AND PATTERN SELECTION IN CONVECTIVELY
UNSTABLE SYSTEMS
R J. Deissler 83

SECONDARY INSTABILITY OF TRAVELING INCLINED ROLLS IN
TAYLOR-DEAN SYSTEM
I. Mutabazi 89

EXPERIMENTS ON THE FORMATION OF STATIONARY SPATIAL
STRUCTURES ON A NETWORK OF COUPLED OSCILLATORS
T. Dirksmeyer, R. Schmeling, J. Berkemeier, H.-G. Purwins 91

STUDIES ON INSTABILITIES AND PATTERNS IN EVAPORATING LIQUIDS
AT REDUCED PRESSURE AND/OR MICROWAVE IRRADIATION
G. Bertrand, M. Lallemant, A. Steinchen, P. Gillon,
P. Courville and D. Stuerga 109

DIRECTIONAL SOLIDIFICATION : THEORETICAL METHODS AND
CURRENT UNDERSTANDING
H. Levine 123

NEW INSTABILITIES IN DIRECTIONAL SOLIDIFICATION
OF SUCCINONITRILE
P.E. Cladis, J.T. Gleeson and P.L. Finn 135

STATIONNARY CELLS IN DIRECTIONAL SOLIDIFICATION
M. Mashaal and M. Ben Amar 147

RECENT PROGRESS IN THE THEORY OF THE GROWTH
OF NEEDLE CRYSTALS
M. Ben Amar and Y. Pomeau 159

STRUCTURAL INVARIANTS AND THE DESCRIPTION OF THE LOCAL
STRUCTURE OF CONDENSED MATTER
A.C. Mitus and A.Z. Patashinskii 185

STRUCTURAL ASPECTS OF DOMAIN PATTERNS IN CERAMICS
AND ALLOYS
Ch. Leroux, G. Van Tendeloo and J. Van Landuyt 195

SELF-ORGANIZATION IN FAR-FROM-EQUILIBRIUM REACTIVE
POROUS MEDIA SUBJECT TO REACTION FRONT FINGERING
P. Ortoleva and W. Chen 203

NONLINEARITY AND SELFORGANIZATION IN PLASTICITY
AND FRACTURE
E.C. Aifantis 221

PLASTIC INSTABILITIES AND THE DEFORMATION OF METALS
H. Neuhäuser 241

DISLOCATION PATTERNS AND PLASTIC INSTABILITIES
L.P. Kubin, Y. Estrin and G. Canova 277

NUMERICAL SIMULATION OF DISLOCATION PATTERNS DURING
PLASTIC DEFORMATION
N.M. Ghoniem and R.J. Amodeo 303

PATTERN FORMATION DURING CW LASER MELTING OF SILICON
K. Dworschak, J.S. Preston, J.E. Sipe and H.M. Van Driel 331

IRRADIATION-INDUCED CAVITY LATTICE FORMATION IN METALS
J.H. Evans 347

THE FORMATION OF CLUSTERS OF CAVITIES DURING IRRADIATION
S.M. Murphy 371

A MESOSCOPIC THEORY OF IRRADIATION-INDUCED VOID-LATTICE
FORMATION
P. Hähner and W. Frank 381

INDEX 383

PARTICIPANTS 389

PREFACE

Understanding the origin of spatio-temporal order in open systems far from thermal equilibrium and the selection mechanisms of spatial structures and their symmetries is a major theme of present day research into the structures of continuous matter. The development of methods for producing spatially ordered microstructures in solids by non-equilibrium methods opens the door to many technological applications. It is also believed that the key to laminar/turbulence transitions in fluids lies in the achievement of spatio-temporal order.

Let us also emphasize the fact that the idea of self-organization in itself is at the origin of a reconceptualisation of science. Indeed, the appearance of order which usually has been associated with equilibrium phase transitions appears to be characteristic of systems far from thermal equilibrium. This phenomenon which was considered exceptional at first now appears to be the rule in driven systems. The chemical oscillations obtained in the Belousov-Zhabotinskii reaction were initially considered to be thermodynamically impossible and were rejected by a large number of chemists. Now these oscillations and related phenomena (waves, chaos, etc.) are the subject of intensive research and new classes of chemical oscillators have been recently discovered. Even living organisms have long been considered as the result of chance rather than necessity. Such points of view are now abandoned under the overwhelming influence of spatio-temporal organization phenomena in various domains ranging from physics to biology via chemistry, nonlinear optics, and materials science .

Today, materials science is undergoing a complete revolution. Indeed, by the use of new technologies (laser and particle irradiation, ion implantation, ultrafast quenches, etc.) it is possible to escape from the tyranny of the phase diagram and to process new materials with unusual properties. In order to describe and understand such materials, dynamical concepts related to nonequilibrium phenomena, irreversible thermodynamics, nonlinear dynamics, and bifurcation theory, are required.

The development of a theoretical framework to describe and interpret self-organization phenomena was made easier by the progress of thermodynamics of irreversible processes and by the introduction of the concept of dissipative structure. In this context it is clear that the nonlinearities of the dynamics and the distance from thermal equilibrium are at the origin of to spatio-temporal organization. Similar phenomena appear in very different systems : spiral waves in chemical systems (but also in the cortex or cardiac activity), the aggregation of micro-organisms, and convective rolls associated with hydrodynamical instabilities in normal fluids and liquid crystals. These varied appearances show that these phenomena are not induced by the microscopic properties of the systems but are triggered by collective effects including a large number of individuals (atoms, molecules, cells, etc.).

The role of fluctuations is also very important in such circumstances. Effectively, near instability points, the space and time scales are so large that the structures are particularly sensitive to even small fluctuations. When different states are simultaneously stable beyond an instability, such fluctuations or small external fields may affect the pattern selection mechanisms. Furthermore, in the case of spatial patterns, the position and orientation of the structure which are described by phase variables are usually fixed by the boundary conditions in small systems. This is of course not the case in large systems where phase fluctuations may trigger the nucleation of defects analogous to dislocations and disclinations. These effects show the importance of a stochastic description of self-organization phenomena far from equilibrium.

In pioneering fields, such as hydrodynamics or nonlinear chemistry, the comparison between theoretical predictions and experimental observations has long been qualitative but has reached the quantitative level recently. This is because of new experimental methods using laser and computer technology and of theoretical progress based on the theory of dynamical systems, on bifurcation calculus, and on the development of supercomputers which make numerical simulations feasible.

While quantitative and systematic experimental analysis followed theoretical analysis in the case of nonlinear chemistry, the evolution has been quite different in the field of hydrodynamics. Despite the fact that convective instabilities and turbulence have been studied for more than a century , definite progress in understanding pattern formation, selection and stability, and the origin of chaotic behavior were achieved only recently It is worth noting that these problems present severe difficulties. From the experimental point of view, the absence of any operational definition of turbulence, the lack of sensitivity of traditional measurement techniques to the temporal behavior of hydrodynamical flows, and a poor resolution of boundary effects limited the progress until the last decade. From the theoretical viewpoint, a major difficulty has been finding analytic solutions because of the complexity of the Navier-Stokes equations.

Significant progress have been achieved in the experimental analysis of instabilities and hydrodynamical flows because of new techniques (laser velocimetry, cryogenic techniques, image processing, etc.) , the systematic use of computer science in data processing and experiment control, and the linkage with new theoretical approaches based on instability and bifurcation theory. On the other hand, the study of the succession of instabilities obtained by increasing the bifurcation parameter requires nonlinear analysis which extends far beyond the classical studies in the field. Hence a few relatively simple systems (Rayleigh-Benard, Taylor-Couette, Benard-

Marangoni, etc.) became very popular as prototypes of complex behavior where nonlinear theories of pattern formation may easily be tested.

Although the Rayleigh-Benard type of instabilities have been discussed at length in the literature and are still providing new challenges for theorists and experimentalists, some of their basic aspects bear reviewing. When a thin horizontal layer of fluid is heated from below or cooled from above, a temperature gradient is generated across the sample. For small gradients, the fluid remains in a conductive state but, on increasing the temperature difference between the horizontal fluid boundaries, the gradient may reach a threshold where this conductive state becomes unstable. Beyond this threshold (instability or bifurcation point), convection sets as cellular structures associated with periodic spatial variations of the hydrodynamic fluid velocity field and of the temperature field. Several types of structures may be obtained according to the working conditions: rolls, hexagons, squares, traveling or standing waves. On increasing further the bifurcation parameter, these patterns may in turn become unstable causing successive bifurcations to occur driving the system to chaos.

From the theoretical point of view, while the first bifurcation may easily be determined from the Navier-Stokes equation, it is a formidable task to determine the behavior of the system beyond the hydrodynamic instabilities with these equations. Fortunately, the derivation of amplitude equations for the patterns led to definite progress in the study of their formation, selection and stability properties. These equations which are usually of the Ginzburg-Landau type, correspond to reduced versions of the complete dynamics which contain all the symmetries of the problem. They may be solved more easily and describe correctly the dynamics of the system on long space-time scales close to the bifurcation point.

Because of the permanent interactions between theory, experiment and numerical analysis, significant progress have been made during the past 20 years on the mathematical methods of nonlinear dynamics and in the understanding of simple fluids instabilities. It has become quite clear that such instabilities manifest themselves in the form of various patterns which vary from the simple to the complex. More recently, the growth in the body of knowledge of liquid-crystal hydrodynamics furnished an exciting ground for further experimental observations on the nature of transitions from one pattern structure to another. Nonlinear interactions at the micro level can explain, to a large degree, the onset and propagation of instabilities at the macro level. Instabilities are saturated through the formation of what is now commonly known as "dissipative" structures. The geometry and properties of these structures can be well explained by a competition between local and nonlocal transport reactions. This framework appears to be quite general, at least conceptually, and can be seen in many physical phenomena (e.g. laser-material interactions, energetic particle-material interactions, magnetic fluids, plastic instabilities, plasma and electric systems). General observations of these vastly diverse physical systems show striking similarities in the nature and occurrence of patterns as manifestations of instabilities.

Physicists and mathematicians have already observed that beyond the onset of instability, patterns which form to "dissipate" the instability are rarely perfect. Imperfections, or defects, can be shown to develop in all pattern-forming instabilities. In some simplified models, one can mathematically find conditions for defect creation in otherwise regular structures. On the other hand, it is already known to materials scientists that defects play an important role in determining material properties. Point defects play a major role in all macroscopic material properties which are related to atomic diffusion mechanisms, and to electronic properties in semiconductors. Line defects, or dislocations, are unquestionably recognized as the basic elements which lead to metal plasticity and fracture. As a consequence, the study of the <u>individual</u> properties of solid state defects is at an advanced level. However, studies of the collective behavior of line defects are still elementary. At the present time, it is important to note that the collective behavior of point defects is well described within the rate theory framework, in analogy to the concepts developed earlier for chemical kinetics. Theoretical description of point-defect interactions can be described as reaction-diffusion equations. On the other hand, major theoretical challenges are encountered in the development of statistically based models of the collective behavior of line defects. Nonetheless, significant progress has been made in the field of dislocation dynamics and plastic instabilities over the past several years.

Physical systems which comprise many interacting entities (e.g. fluids, solids, plasmas, etc.) have been described at different levels. (1) The most fundamental and detailed level is a description based on the equations of motion (EOMs) for the individual entities. Hence, the framework of Newtonian or quantum mechanics is appropriate. (2) A higher level in the hierarchy is statistical mechanics, where the concern is with distribution functions in phase space, rather than the individual EOMs. Thus we are able at this level of description to discuss collective properties such as diffusion, conduction, viscosity, permeability, etc.. (3) The description of continuum mechanics is more appropriate for studying macroscopic length and time scale resolutions. Navier-Stokes, continuity and compatibility equations provide the primary vehicles for a macroscopic description of the continuum. At this level, constitutive relations are needed to complete the framework. It is hoped that such constitutive relations are derivable at the statistical mechanics level. However, this is not usually the case. More often, a phenomenological model is used to obtain the needed constitutive relations. (4) Continuous media deform generally homogeneously if the externally perturbing field induces small deviations from equilibrium. Critical levels of non-equilibrium perturbing field lead to dynamical instabilities in the continuous distribution of matter and to the eventual emergence of patterns or dissipative structures. In past few decades astonishingly rapid progress has been made in our understanding of the nature of pattern-forming instabilities. The appropriate framework at this level of analysis is nonlinear dynamics, where the dynamical equations which describe the bifurcations and instabilities are obtained from appropriate con-

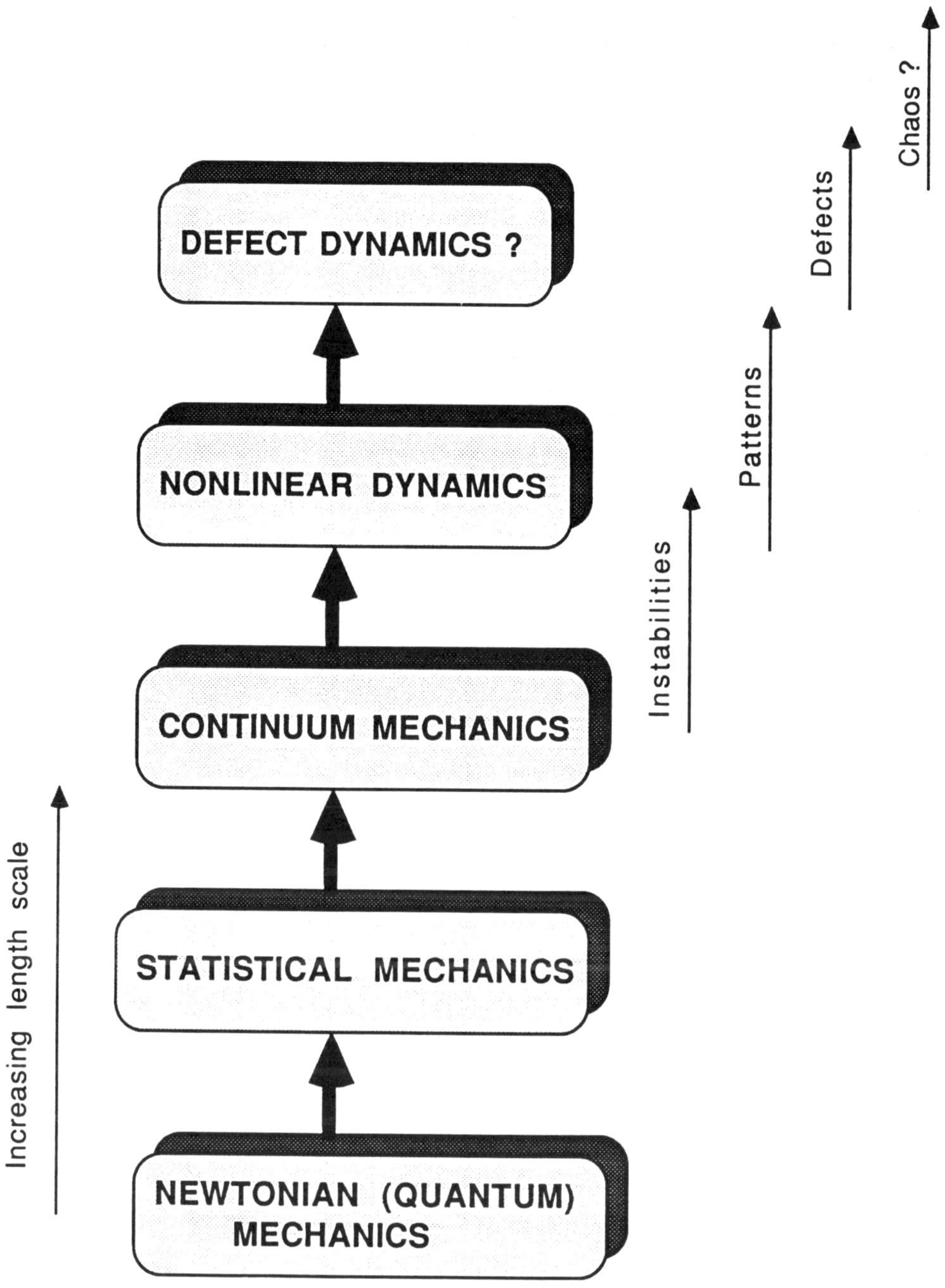

Fig. 1: Schematic representation of a hierarchical framework for describing material instabilities, patterns, and defects.

tinuum equations. Thus from Navier-Stokes or reaction-diffusion equations, for instance, one is able to develop dynamical equations of the Ginzburg-Landau type, which are capable of describing the instabilities in a manner reminiscent of phase transitions in thermodynamic systems. (5) Solutions of these dynamical equations lead to conditions where "defects" are obtained in otherwise periodically perfect structures. This has, in fact, already been observed in several systems, most notably in liquid crystals. At this level of description, the system characteristics are manifest in the dynamics of interaction between such defects, and one may be able to develop the framework of "defect dynamics". This ascension of levels is shown schematically in Fig.1.

Physicists, chemists, material scientists, and mathematicians have the goal of developing a unified framework for explaining pattern-forming instabilities and defects. To this end, the proceedings of this meeting represent a combination of lectures and contributed presentations on patterns, defects, and material instabilities and we hope that this ASI has advanced this goal within the community. We are grateful to the NATO Scientific Affairs Division, to the Directorate General for Science, Research and Development of the Commission of the European Communities, to the CNRS (France) and to the International Solvay Institutes for their generous financial support. Special thanks are also due to the secretarial staff in Brussels and Cargèse, in particular Ms M.F.Hanseler and N.Sardo, for its efficiency.

N. M. Ghoniem   D. Walgraef

January 1990

ONE-DIMENSIONAL CELLULAR PATTERNS

P. COULLET and G. IOOSS
Institut de Mathématiques et de Sciences Physiques
Université de Nice - Parc Valrose
06034 Nice cedex
France

ABSTRACT. A classification of the generic instabilities that one-dimensional cellular patterns can suffer is presented.

1. Introduction

Stationary cellular patterns are frequently observed in nature. Recently a classification of the various bifurcations of one-dimensional periodic patterns has been proposed [1]. In this paper we summarize these bifurcations on the basis of symmetry considerations.

Our initial hypothesis is the existence of a one-dimensional stationary cellular pattern which can be described by a solution $U_0(x)$ of an evolution equation

$$\partial_t U = f(U) \qquad (1)$$

where $U = (U_1, ..., U_N)$ are, for the sake of simplicity chosen to be scalar quantities. This evolution equation is assumed to describes one-dimensional physical systems which are invariant under the following symmetries

$$T_\theta \quad : \quad t \to t + \theta \qquad (2)$$
$$T_\sigma \quad : \quad x \to x + \sigma \qquad (3)$$
$$P \quad : \quad x \to -x \qquad (4)$$

Since U is a scalar $PU(x) \equiv U(-x)$. The solution U_0 which describes the periodic cellular pattern is such that

$$T_\theta U_0 = U_0 \qquad (5)$$
$$T_a U_0 = U_0 \qquad (6)$$
$$P U_0 = U_0 \qquad (7)$$

In words, U_0 is stationary, periodic in space with a period a, and can be chosen even by an appropriate coordinate change.

2. Normal modes of a perturbation

The stability analysis of U_0 proceeds as follows. Let

$$U(x,t) = U_0(x) + u(x,t) \tag{8}$$

where $u(x,t)$ is a small perturbation. At the first order in $u(x,t)$ the equation for the perturbation reads

$$\Lambda u = 0 \tag{9}$$

where $\Lambda = \partial_t - \mathcal{L}(x)$ and $\mathcal{L}(x) = \partial f(U)/\partial U|_{U=U_0(x)}$ is the Jacobian operator. Λ has the following properties:

$$[\Lambda, \mathcal{T}_\theta] = 0 \tag{10.a}$$

$$[\Lambda, \mathcal{T}_a] = 0 \tag{10.b}$$

$$[\Lambda, P] = 0 \tag{10.c}$$

where $[A, B] = AB - BA$. These properties are now used to solve Eq. (9). Since Λ commutes with $\mathcal{T}_\theta$, $\mathcal{T}_a$ and P, they have a common spectral decomposition,

$$u = \sum_j u_j e_j(x,t) \tag{11}$$

where, in the space of bounded functions, a typical element of this basis

$$e_j(x,t) = B_j(x) exp(s_j t) \tag{12}$$

is such that

$$\mathcal{T}_\theta e_j(x,t) = \lambda_\theta e_j(x,t) \tag{13.a}$$

with $\lambda_\theta = exp(s_j \theta)$ and

$$\mathcal{T}_a e_j(x,t) = \lambda_a e_j(x,t) \tag{13.b}$$

with $|\lambda_a| = 1$. The eigenfunction of the discrete translation $\mathcal{T}_a$ are Bloch functions. They have the general form $B(x) = \eta(x) \exp(ikx)$, with $\mathcal{T}_a \eta(x) = \eta(x)$. The condition of boundness ($|\lambda_a| = 1$) is satisfied by three different type of eigenvalues (spatial analogs of Floquet multipliers)

(a) $\lambda_a = 1$
(b) $\lambda_a = \exp(i\phi), \exp(-i\phi)$
(c) $\lambda_a = -1$

In case (b), in general, $\phi/2\pi \neq n/m$. The corresponding Bloch functions are

(a) $B(x) = \eta(x)$ $(k=0)$
(b) $B(x) = \eta(x)\exp(ik_0 x), \bar{\eta}(x)\exp(-ik_0 x)$ $(k=k_0)$
(c) $B(x) = \eta(x)\exp(i\pi x/a) \equiv \hat{\eta}(x)$ $(k=\pi/a)$

In case (a), the eigenvector has the same period as the basic pattern U_0 and the corresponding eigenvalue is generically simple. In case (b) it is bi-periodic (modulation with a period $2\pi/k_0$ generally irrationaly related to a). The corresponding eigenvalue is generically double. In case (c), the eigenvalue is generically simple and the eigenvector has twice the period of the basic pattern. The spectral decomposition can be pushed further when one takes into account the parity. In case (a) the eigenspace naturally splits into two components which correspond to the two eigenspaces of P, namely the even functions of x ($P\eta_E(x) = \eta_E(x)$) and the odd function of x ($P\eta_O(x) = -\eta_O(x)$). The same property is true for case (c), where $\hat{\eta}(x)$ splits into even and odd parts ($\hat{\eta}_E(x)$ and $\hat{\eta}_O(x)$ respectively). In case (b), the eigenspace has already a dimension two, no further splitting occurs since even and odd functions of x can be generated in this space. The spectral decomposition is summarized in Table I (Case I and Case II respectively correspond to real and complex s).

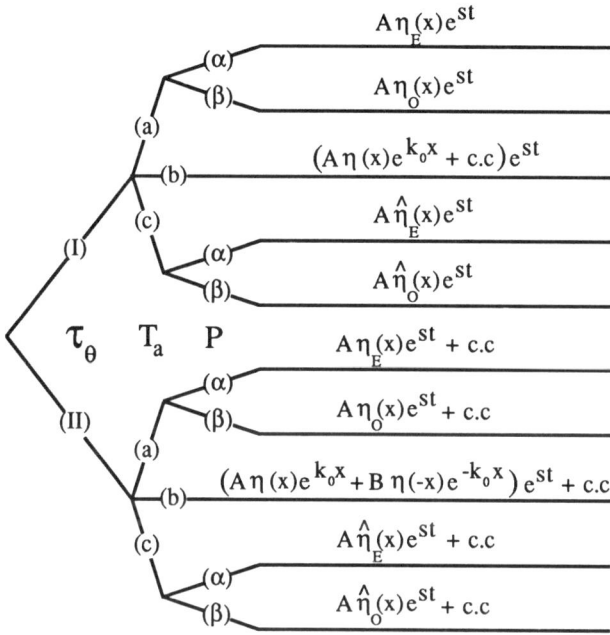

Table I

3. Instabilities

A bifurcation occurs whenever $\mathcal{R}e(s) = 0$. Replacing s by 0 or $\pm i\omega_0$, the previous table turns into the classification of the ten generic instabilities that one-dimensional periodic patterns can suffer [1]. Case (I) corresponds to stationary bifurcations. The associated marginal eigenvector does not break time translations. Case (II) corresponds to oscillatory or Hopf bifurcations. The marginal eigenfunctions do break time translations. In case (a) the discrete translational symmetry T_a is not broken. Case (b) and case (c) do break it. Case (b) is associated with a spatial modulation of the pattern and case (c) corresponds to a spatial period doubling bifurcation. In case (α) the parity is not broken, while it is in case (β). In case (b) the parity is also generically broken.

4. Normal forms

As usual, the next step in the bifurcation analysis consists in establishing the equations for the amplitudes of the weakly off marginal modes. We first remark that $u = \partial_x U_0(x)$ is a solution of Eq. (9) with $s = 0$. It is a straightforward consequence of the translational symmetry (T_σ) which has been broken by the cellular pattern $U_0(x)$

$$\partial_x f(U_0(x)) = \mathcal{L}(x)\partial_x U_0(x) = 0 \tag{14}$$

A simple way to account explicitely for this degree of freedom (phase order parameter) consists in replacing Eq.(8) by

$$U(x,t) = U_0(x+\phi) + u(x+\phi,t) \tag{15}$$

together with an orthogonality condition

$$\langle u, \partial_x U_0 \rangle = 0$$

where $\langle V, W \rangle = \int_{-\infty}^{+\infty} \sum_j V_j(x)\bar{W}_j(x)dx$. In order to get the equations for the phase and amplitude perturbations, we now insert (14) into Eq. (1) and keep only first order terms in ϕ and u

$$\partial_t \phi = \langle \mathcal{L}u, \partial_x U_0 \rangle / |\partial_x U_0|^2 \tag{16.a}$$

$$\partial_t u = L(x)u \tag{16.b}$$

where $L(x)u = \mathcal{L}(x)u - \partial_x U_0 \langle \mathcal{L}u, \partial_x U_0 \rangle / |\partial_x U_0|^2$. These equations describe, in the linear theory, the coupling between the phase order parameter ϕ and the amplitude modes. Near an instability, in the linear approximation, u becomes one of the eigenmodes given in Table I, where $s = 0$ or $s = \pm i\omega_0$. The amplitudes of these marginal modes turn into order parameters associated with the instability. Center manifold theorem allows to express, near the bifurcation threshold, the amplitudes of all other modes in terms of the order parameters. The equations for ϕ and for these order parameters are the normal forms associated with the bifurcation. Simple symmetry arguments can be used in order to establish the form of these equations.

5. An example of bifurcation: the parity breaking instability.

Let us now illustrate this method for finding the amplitude equations. Case (I.a.β) corresponds to a stationary instability characterized by a mode which has the same period than the basic cellular pattern, but with a different parity. From Table I one

learns that $u = A\eta_0(x)$, with A real, $\eta_0(x+a) = \eta_0(x)$ and $\eta_0(-x) = -\eta_0(x)$. Near the parity breaking instability threshold a solution of Eq. (1) is looked in the form
$$U(x,t) = U_0(x+\phi) + A\eta_0(x+\phi) + ... \tag{17}$$
where the dots stand for center manifold corrections. Solvability conditions give us equations for ϕ and A which are supposed to be slowly varying in time.
$$\partial_t \phi = \Phi(\phi, A) \tag{18.a}$$
$$\partial_t A = \mathcal{A}(\phi, A) \tag{18.b}$$
Since $U(x,t)$ is assumed to be a solution of Eq. (1), $T_\theta U(x,t)$, $T_\sigma U(x,t)$, $T_a U(x,t)$ and $PU(x,t)$ must also be solutions of Eq. (1)
$$T_\theta U(x,t) = U_0(x + \phi(t+\theta)) + A(t+\theta)\eta_0(x + \phi(t+\theta)) + ... \tag{19.a}$$
$$T_\sigma U(x,t) = U_0(x + \phi(t) + \sigma) + A(t)\eta_0(x + \phi(t) + \sigma) + ... \tag{19.b}$$
$$T_a U(x,t) = U_0(x + \phi(t)) + A(t)\eta_0(x + \phi(t)) + ... \tag{19.c}$$
$$PU(x,t) = U_0(x + \phi(t)) - A(t)\eta_0(x + \phi(t)) + ... \tag{19.d}$$

From (19.a) one learns that Eqs. (18) must be invariant under the change $t \to t + \theta$. From (19.b) one deduces that the change $\phi \to \phi + \sigma$ must leave invariant Eqs. (18). We learn nothing from Eq. (19. c) since the discrete symmetry T_a is not broken in this case. We finally learn from (19.d) that Eqs (18) must be invariant under the transformations $\phi \to -\phi$, $A \to -A$. These equivariance properties allow us to write the general form of the normal form for the parity breaking bifurcation as
$$\partial_t \phi = \Phi(A) \tag{20.a}$$
$$\partial_t A = \mathcal{A}(A) \tag{20.b}$$
where $\Phi(-A) = -\Phi(A)$ and $\mathcal{A}(-A) = \mathcal{A}(A)$. At the leading order in the order parameter, after appropriate scalings, these equations become
$$\partial_t \phi = A \tag{21.a}$$
$$\partial_t A = \mu A \pm A^3 \tag{21.b}$$
where μ is the distance from instability threshold. Parity breaking bifurcation is thus a pitchfork bifurcation. In the supercritical case (minus sign in Eq. (21.b)), the parity order parameter A eventually converges towards one of the two solutions $A^*_\pm = \pm\sqrt{\mu}$. One can then solve Eq. (21.a): $\phi = A^*_\pm t$. Back to the expression for the actual solution
$$U(x,t) = U_0(x + A^*_\pm t) + A^*_\pm \eta_0(x + A^*_\pm t) + ... \tag{22}$$
one discovers that the parity breaking bifurcation generically implies a translation of the pattern at a constant speed which scales as the parity order parameter.

6. Conclusion

The domain of validity of normal forms in extended physical systems is generally small. A better description consists in writing envelope equations for order parameters which vary slowly both in space and time. Symmetry arguments similar to those presented above also apply. In the case of the parity breaking bifurcation one gets the equations

$$\partial_t \phi = A + \gamma \phi_{xx} + \delta A_{xx}... \qquad (23.a)$$

$$\partial_t A = \mu A \pm A^3 + A_{xx} + \alpha A_x A + \beta \phi_x A + \eta \phi_{xx}... \qquad (23.b)$$

where the subscript x stands for the derivative with respect to x and α, β and δ are real constants.

The analysis presented here and in Ref [1] can be extended to stationary cellular patterns in 2 dimensions. The phenomenology associated with these envelope equations is likely to be very rich. It includes defects, phase and defects turbulence and pattern transitions. Some relevant references related to this work are found in Ref.[1]

Acknowlegment. One of the author (P.C.) author wants to thank A.Newell for his warm hospitality and useful discussions while this work was completed. He also gratefully acknowledge the research support from Arizona center for Mathematical Sciences, sponsored by AFSOR Contract F49620-86-C0130 with the university Research Initiative Program at the University of Arizona. J. Lega is acknowledged for discussions and a careful reading of this paper. Part of this work have been supported by the DRET under contract (312/88).

Reference.

1. P. Coullet and G. Iooss, *Instability of one-dimensional periodic patterns*, to appear in Phys. Rev. Lett. (1990).

DEFECTS AND DEFECT-MEDIATED TURBULENCE

J. LEGA
Laboratoire de Physique Théorique
Université de Nice - Parc Valrose
06034 Nice cedex
France

ABSTRACT. We give a description of defects of macroscopic structures by means of numerical simulations of Ginzbug-Landau equations, and describe a mechanism for spontaneous appearance of defects in a far from equilibrium system.

1. Introduction

These few pages are intended to give a characterization of some defects of two dimensional patterns, and to emphasize their role in the disorganization of the structure in which they have appeared. The description is made by means of Ginzburg-Landau equations, which rule the behavior of the slowly varying envelope of a macroscopic structure near threshold. Instead of going into great details, we shall restrict ourselves to an outline of the main ideas, referring to pictures and to a list of references, a more detailed study being published elsewhere [1]. After an introduction to amplitude equations and a short description of some of their defect solutions, we shall focus on a mechanism responsible for the spontaneous nucleation of defects in the system, which gives rise to a turbulent state.

2. Amplitude equations description of a macroscopic structure

2.1. STATIONARY PERIODIC PATTERNS

When driven far from equilibrium by an external parameter, many physical systems display macroscopic structures (see for instance [2]). The most classical example is that of the Rayleigh-Bénard experiment, which consists in heating from below a thin horizontal layer of fluid contained between two plates. Above a critical value of the difference of temperature between the two plates, convection sets in (for a review on convection see [3]; for a revue about hydrodynamic instabilities, see [4]), and pairs of parallel convective rolls appear in the system (see for instance [5] for Rayleigh-Bénard convection). Hence, a bifurcation towards a spatial periodic pattern has occurred.

Close to this instability threshold, any physical quantity, such as the temperature of the fluid, can be described by means of a complex amplitude A which depends only on horizontal space coordinates (x and y) and on time (t):

$$T = T_{cd}(z) + [A(x,y,t)\exp(ik_0 x) + c.c. + \ldots]\Phi(z) \tag{1}$$

where $T_{cd}(z)$ is the conductive profile of temperature, A is assumed to vary slowly in space and time, k_0 is the critical wavevector at threshold, $\Phi(z)$ is a structure function which describes the behavior of the temperature along the vertical direction, and the dots stand for higher order terms. Such a description assumes an unique direction of the rolls in the whole system, which is not the general situation in experiments. It is at least valid in parts of the experimental system with same roll orientation (for a description of the case of various orientations of the rolls, see [6] and [7]). Near threshold, the amplitude A is shown to obey a complex Ginzburg-Landau equation [8-11], which reads:

$$\frac{\partial A}{\partial t} = \mu A + \left(\frac{\partial}{\partial x} - \frac{i}{2k_0}\frac{\partial^2}{\partial y^2}\right)^2 A - |A|^2 A. \tag{2}$$

The first term of the right hand side of this equation, where μ measures the distance from instability threshold, describes linear instability of the conductive solution ($A = 0$). The second one allows spatial modulations of the complex order parameter A, and the third stands for nonlinear saturation. Indeed, $A = \sqrt{\mu}$ is a stable solution of Eq. (2) and corresponds to a perfect convective pattern, since T is then given by:

$$T = T_{cd}(z) + [2\sqrt{\mu}\cos(k_0 x) + \ldots]\Phi(z).$$

Finally, let us mention that this equation is left unchanged by the transformation

$$A \to A\exp(i\varphi),$$

which reflects the invariance of the physical system under space translations along x, and by the transformation

$$x \to -x \quad y \to -y \quad A \to \bar{A}$$

which is a consequence of parity invariance of the system, and leads to the fact that Eq. (2) has real coefficients.

2.2. TEMPORAL PATTERNS

The same kind of description can be given in the case of a temporal periodic pattern [12], as those occuring for instance in chemical reactors. A relevant quantity C is then described near the instability (Hopf bifurcation) threshold by:

$$C = C_0 + A\exp(i\omega_0 t) + c.c. + \ldots \tag{3}$$

where ω_0 is the critical frequency, A is assumed to vary slowly in space and time and obeys the following complex Ginzburg-Landau equation [12]:

$$\frac{\partial A}{\partial t} = \mu A + (1 + i\alpha)\Delta A - (1 + i\beta)|A|^2 A. \tag{4}$$

Such an equation is left unchanged by the transformation $A \to A\exp(i\varphi)$ which reflects the invariance of the physical system under time translations. Its main difference with Eq. (2) is that its coefficients are complex numbers, and that the diffusive term is isotropic.

2.3. WAVE PATTERNS

Our last example is the case of the occurrence of a normal roll wave pattern in an anisotropic, parity invariant medium. Assuming the wave vector $\vec{k}_0$ parallel to the x direction and ω_0 denoting the temporal frequency, any relevant quantity U is described by:

$$U = U_0 + A\exp[i(-k_0 x + \omega_0 t)] + B\exp[i(k_0 x + \omega_0 t)] + c.c. + \ldots \quad (5)$$

where the complex amplitudes A and B of right and left traveling waves vary slowly in space and time and obey the two coupled Ginzburg-Landau equations:

$$\frac{\partial A}{\partial t} = \mu A + (1 + i\alpha_x)\frac{\partial^2 A}{\partial x^2} + (1 + i\alpha_y)\frac{\partial^2 A}{\partial y^2} - c\frac{\partial A}{\partial x}$$
$$- (1 + i\beta)|A|^2 A - (\gamma + i\delta)|B|^2 A \quad (6.a)$$

$$\frac{\partial B}{\partial t} = \mu B + (1 + i\alpha_x)\frac{\partial^2 B}{\partial x^2} + (1 + i\alpha_y)\frac{\partial^2 B}{\partial y^2} + c\frac{\partial B}{\partial x}$$
$$- (1 + i\beta)|B|^2 B - (\gamma + i\delta)|A|^2 B, \quad (6.b)$$

where c is related to the group velocity of the wave pattern, and γ turns out to rule the competition between traveling (one of the two order parameters is zero) and standing (A and B are both finite) waves.

The transformations which leave unchanged these equations are:
- $A \to B$, $B \to A$, $x \to -x$, and $y \to -y$ which reflects parity invariance,
- $A \to A\exp(i\varphi)$, $B \to B\exp(-i\varphi)$ which reflects invariance under space translations,
- $A \to A\exp(i\theta)$, $B \to B\exp(i\theta)$ which reflects invariance under time translations.

For more detail about the derivation and the solutions of these equations, see for instance [13] (see also [14]). The one dimensional version of Eqs. (6) can be found in [15-17], and the case of a 2D isotropic medium is given for instance in [18] (see also [19]).

Hence, the occurrence of spatial, temporal, or spatio-temporal patterns in a system driven far from equilibrium by an external parameter can be described by means of Ginzburg-Landau equations. Their form is independant of the very nature of the physical system under interest and is only related to the kind of bifurcation which occurs and to the symmetries of the problem.

3. Some topological defects of two dimensional patterns

Macroscopic patterns observed in experiments often exhibit defects (many examples can be found in [2] and [20]). The core of the defect is the place where a change of

SINK OF TRAVELING WAVES

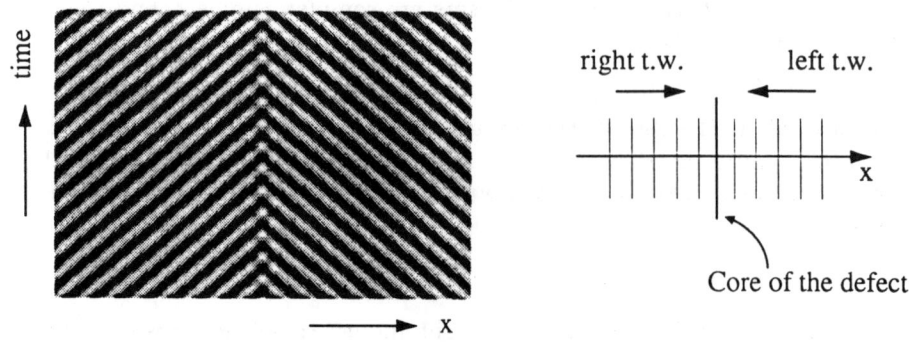

SOURCE OF TRAVELING WAVES

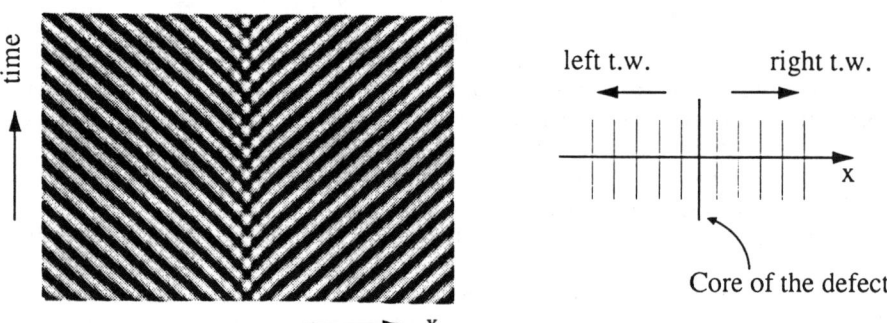

Figure 1: Numerical simulation ($\mu = 1$, $c = 0.5$, $\alpha_x = 1$, $\beta = 1$, $\gamma = 1.5$, $\delta = -1$, $L_x = 200$=box length, $N_x = 300$=number of collocation points) of the one dimensional version of Eqs. (6) showing (x,t) diagrams of a sink and a source of traveling waves.

PAIR OF DISLOCATIONS OF A STATIONARY ROLL PATTERN

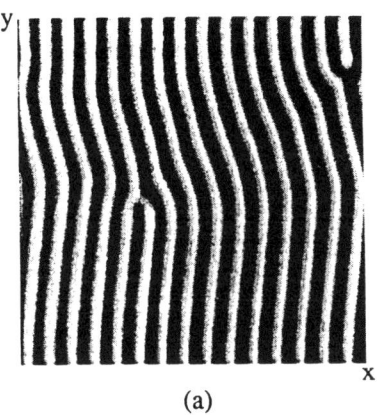

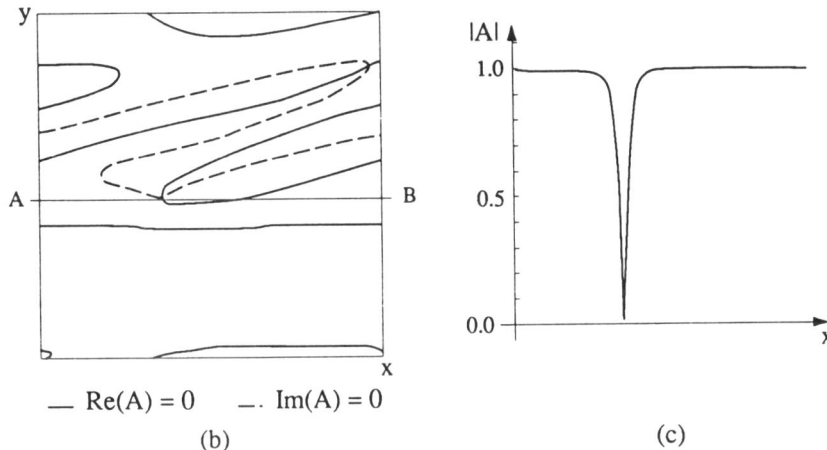

Figure 2: Numerical simulation ($\mu = 1$, $k_0 = 0.6\pi$, $L_x = L_y = 50$, $N_x = N_y = 80$) of Eq. (2) showing a pair of dislocations of a stationary roll pattern. (a) Behavior of T given by Eq. (1) as a function of x and y. Minima are in white, maxima in black. (b) Lines where $\Re e(A) = 0$ and $\Im m(A) = 0$. (c) Cross section of the amplitude $|A|$ along the line $A - B$ of (b).

structure occurs, and corresponds in two dimensions either to a line or to a point. Defects can be characterized by means of topological considerations on the set of stable solutions of Ginzburg-Landau equations (see [18] and for more detail [21]). Such arguments will not be developed here, and we shall only give a description of defects by means of numerical simulations of amplitude equations.

3.1. LINE DEFECTS

In stationary patterns, line defects are grain boundaries between regions with different roll orientations. In wave patterns, they are also [18] sinks and sources of traveling waves (see for instance experiments in [22-25]) and zipper states (experiment in [26]). A sink or a source of traveling waves is a line towards or from which waves converge or emanate. Figure 1 shows (x,t) diagrams associated with such defects. Here x is both the direction transverse to the core of the defect and the direction of propagation of waves.

3.2. POINT DEFECTS

3.2.1. Stationary periodic patterns. As shown in Fig. 2, the point defect of a stationary roll pattern is a dislocation, which corresponds to the insertion of an extra pair of rolls. Round the core of the defect, the phase of the amplitude A turns by 2π, that is

$$\oint \vec{\nabla}\varphi . \vec{dl} = \pm 2\pi \qquad \text{where } A = |A|\exp(i\varphi),$$

and as a consequence, A goes to zero at the core of the defect (see Fig. 2c). As shown in Fig. 2b, the lines where $\Re e(A) = 0$ and $\Im m(A) = 0$ meet at the core of the defect.

3.2.2. Temporal patterns. The analog in a temporal pattern of the dislocation of a stationary periodic pattern is a spiral wave [12]. As shown in Fig. 3b, the lines where $\Re e(A) = 0$ and $\Im m(A) = 0$ have a spiral shape around the core of the defect where they meet. Since the whole pattern oscillates periodically in time, the spiral seems to wind or unwind round its core, whence the name of spiral wave.

3.2.3. Traveling wave patterns. A traveling wave corresponds to a solution of Eqs. (6) where one of the two order parameters, say B, is equal to zero. Hence, Eqs. (6) reduce, modulo a change of referential, to

$$\frac{\partial A}{\partial t} = \mu A + (1+i\alpha_x)\frac{\partial^2 A}{\partial x^2} + (1+i\alpha_y)\frac{\partial^2 A}{\partial y^2} - (1+i\beta)|A|^2 A,$$

which is analog to Eq. (4). Thus, the point defect of such a pattern is also a spiral wave for the amplitude A (see Fig. 4b). Nevertheless, the relevant physical quantity U given by Eq. (5) exhibits (Fig. 4a) a dislocation whose core is located at the same place as that of the spiral for the amplitude A. The defects move with a velocity close to the group velocity c, which is different from the phase velocity ω_0/k_0.

SPIRAL WAVE

(a)

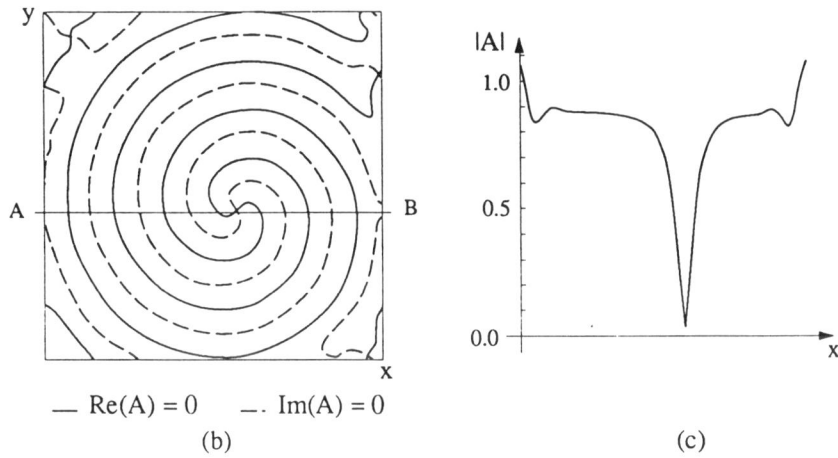

— Re(A) = 0   _. Im(A) = 0

(b)   (c)

Figure 3: Numerical simulation ($\mu = 1$, $\alpha = 1$, $\beta = -0.6$, $L_x = L_y = 50$, $N_x = N_y = 80$) of Eq. (4) showing a spiral defect. (a) Behavior of C given by Eq. (3) as a function of x and y. Minima are in white, maxima in black. (b) Lines where $\Re e(A) = 0$ and $\Im m(A) = 0$. (c) Cross section of the amplitude $|A|$ along the line $A - B$ of (b).

PAIR OF DISLOCATIONS OF A TRAVELING WAVE PATTERN

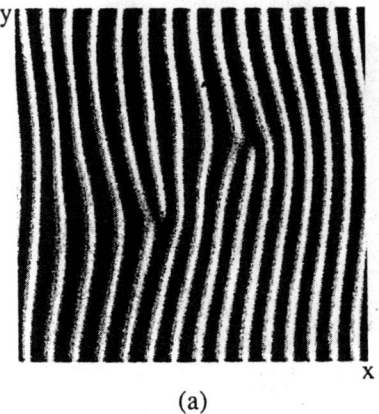

(a)

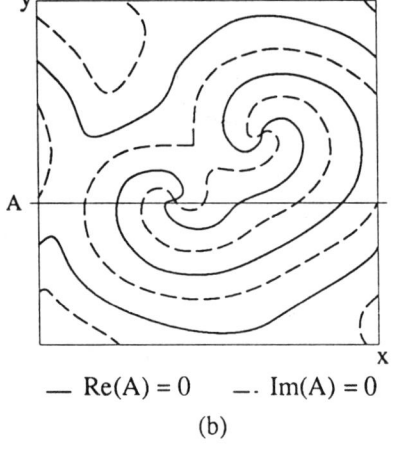

— Re(A) = 0   — · Im(A) = 0

(b)

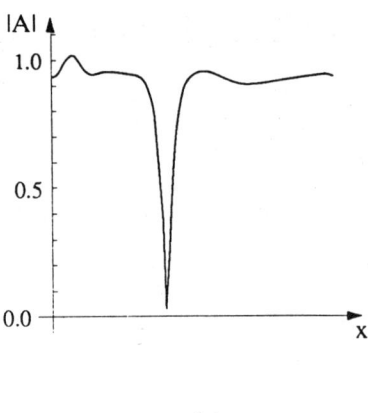

(c)

Figure 4: Numerical simulation ($\mu = 1$, $c = 0.1$, $\alpha_x = 0.5$, $\alpha_y = -1$, $\beta = -1.8$, $\gamma = 1.5$, $\delta = -1$, $L_x = L_y = 50$, $N_x = N_y = 80$) of Eqs. (6) showing a pair of dislocations of a traveling wave pattern. (a) Behavior of U given by Eq. (5) as a function of x and y. Minima are in white, maxima in black. (b) Lines where $\Re e(A) = 0$ and $\Im m(A) = 0$. (c) Cross section of the amplitude $|A|$ along the line $A - B$ of (b).

DISLOCATION OF A STANDING WAVE PATTERN

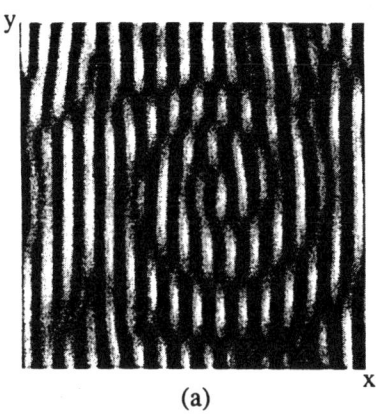

(a)

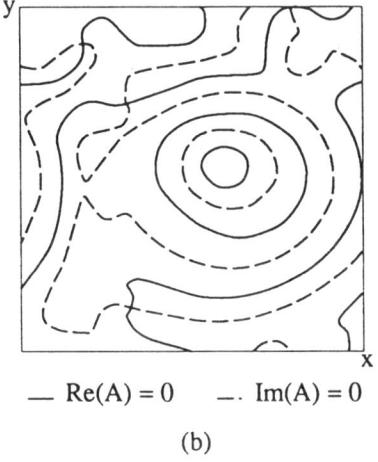

— Re(A) = 0   —. Im(A) = 0

(b)

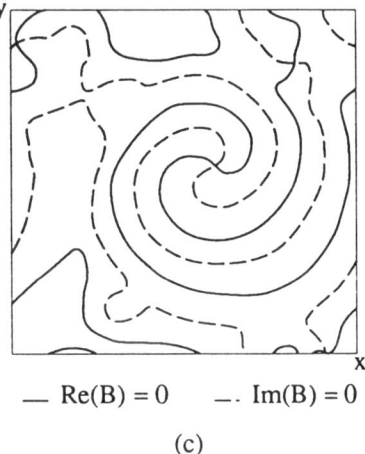

— Re(B) = 0   —. Im(B) = 0

(c)

Figure 5: Numerical simulation ($\mu = 1$, $c = 0.1$, $\alpha_x = 0.6$, $\alpha_y = 0.3$, $\beta = -0.6$, $\gamma = 0.5$, $\delta = -1$, $L_x = L_y = 50$, $N_x = N_y = 80$) of Eqs. (6) showing a dislocation of a standing wave pattern. (a) Behavior of U given by Eq. (5) as a function of x and y. Minima are in white, maxima in black. (b) Lines where $\Re e(A) = 0$ and $\Im m(A) = 0$. (c) Lines where $\Re e(B) = 0$ and $\Im m(B) = 0$.

3.2.4. Standing wave patterns. The elementary defect of a standing wave pattern [27] corresponds to a dislocation carried by one of the two waves, say that associated with B. Here again, as shown in Fig. 5c, the lines where $\mathfrak{Re}(B) = 0$ and $\mathfrak{Im}(B) = 0$ have a spiral shape around the core of the defect. But the surprising feature is the presence of a target pattern on A centered at the core of the spiral of B (see Fig. 5b), which leads to the existence of a spiral-shaped line where the phase of the standing rolls is inverted (see Fig. 5a). So far, standing waves and their defects have not been clearly identified experimentally in systems driven far from equilibrium.

4. Defect-mediated turbulence

The description of defects we have just given shows that such objects correspond to localized amplitude perturbations and that they break, at least locally, the order of the macroscopic structure. In particular, experimental results have shown that defects may change the stability of the structure in which they have appeared [28], and conversely, that an instability of this structure may lead to the spontaneous nucleation of defects [29-30]. Moreover, many experiments have revealed the presence of defects spontaneously appeared in the system during transitions towards turbulent states [31-35]. Some authors have developed the analogy [31,36-37] between such phenomena and two dimensional melting, and more recently, a mechanism of spontaneous nucleation of defects has been given (see [38-39] and for a review [1]). The latter leads to a turbulent state which has been called defect-mediated turbulence [38].

The main idea is that a large scale instability, namely a phase instability of the pattern, will lead to the spontaneous nucleation of defects in the system. A first step in understanding this phenomenon is to study the linear stability of the perfect pattern (which corresponds to the homogeneous solution of the amplitude equation) with respect to small perturbations. This study is made in the case of Eq. (4), which is the simplest equation which can lead to such a behavior.

4.1. LINEAR STABILITY ANALYSIS

The homogeneous solution of Eq. (4) is $A_0 = \sqrt{\mu}\exp(-i\beta\mu t)$. In order to study its linear stability, we look for a solution of Eq. (4) in the form $A = (\sqrt{\mu} + a)\exp(-i\beta\mu t)$, and compute the eigenvalues of the linearized system for a and $\bar{a}$. They read:

$$\sigma_R = -2\mu - (1 - \alpha\beta)k^2 + \ldots$$

$$\sigma_\varphi = -(1 + \alpha\beta)k^2 - \frac{\alpha^2(1 + \beta^2)}{2\mu}k^4 + \ldots$$

As shown in Fig. 6, σ_R is always negative, but σ_φ may become positive for small values of $|k|$ if $1 + \alpha\beta < 0$. Hence, a large scale instability [40-43], which is also called phase instability since σ_φ is associated with phase perturbations, may occur.

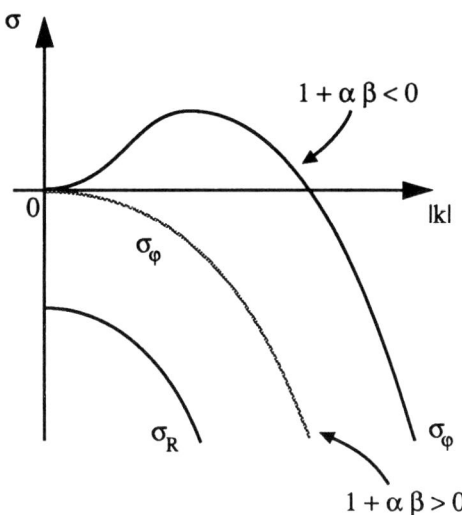

Figure 6: Schematic behavior of the eigenvalues of the linearized system associated with Eq. (4) around the homogeneous solution A_0.

4.2. DEFECT-MEDIATED TUBULENCE

When $1 + \alpha\beta < 0$, the unstable pattern evolves towards a turbulent state where defects appear spontaneously, move through, and annihilate in the system [38]. The appearance of defects is related to a strong decrease of the correlation function of the field [38,1], given by

$$\mathcal{C}(r) = \frac{\mathcal{F}(r)}{\mathcal{F}(0)}, \qquad \mathcal{F}(r) = < \sum_{x^2+y^2=r^2} \Re e[A(x_0, y_0, t)\bar{A}(x, y, t)] >,$$

where $< . >$ denotes time averaging. Figure 7 shows the behavior of $\mathcal{C}$ when there are defects in the system and where there are not. The correlation length is of the same order as the mean distance between a defect and its nearest neighbor.

The number N of defects in the system oscillates (see Fig. 8) in time around a mean value which seems roughly proportional to the number of unstable modes, and which scales linearly with the area of the box. Figure 9 shows the distribution of N over a sample of 20000 time iterations. This distribution (see Fig. 9, dashed line) and its characteristic features can be reproduced [44] by a simple probabilistic model, assuming a constant rate of creation of pairs of defects, and a rate of annihilation proportional to the square of the number of pairs. Moreover, this theoretical distribution fits the experimental results very well [35].

The parameters α and β play of course different roles [38], β being more crucial than α. Namely, the former rules the order of magnitude of the number of defects in the system and the value of the correlation length, while changes of the latter

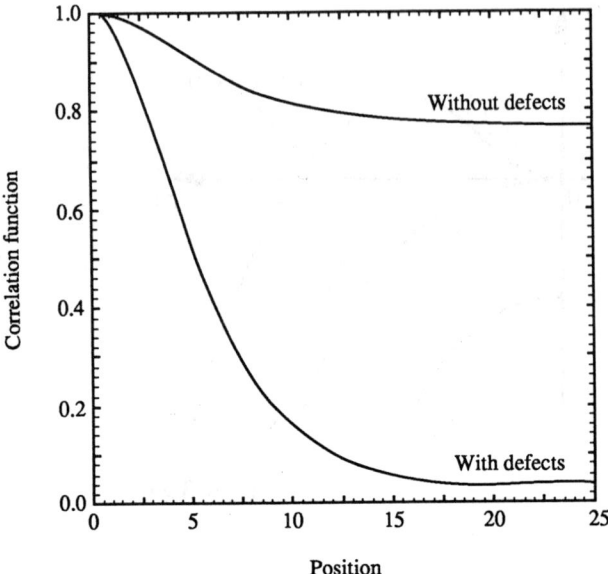

Figure 7: Correlation function of the field when there are defects in the system and where there are not (numerical simulation of Eq. (4), $\mu = 1$, $\alpha = 2$, $\beta = -0.82$).

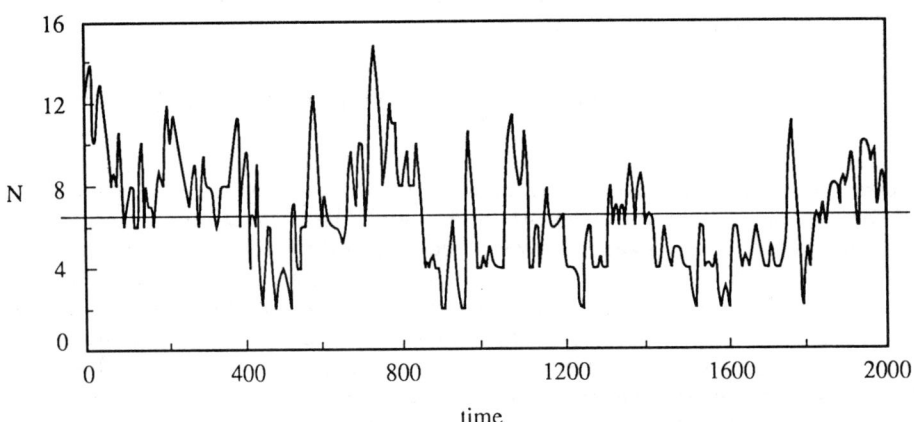

Figure 8: Number N of defects in the system as a function of time (numerical simulation of Eq. (4), $\mu = 1$, $\alpha = 2$, $\beta = -0.85$). Each plotted value of N corresponds to an average over 4 units of time. The number of defects fluctuates around a mean value, which is about 6.5.

lead to small deviations of these values. When β is increased, defect-mediated turbulence merges progressively into amplitude turbulence [12,45-46] characterized by a large number of excited amplitude modes.

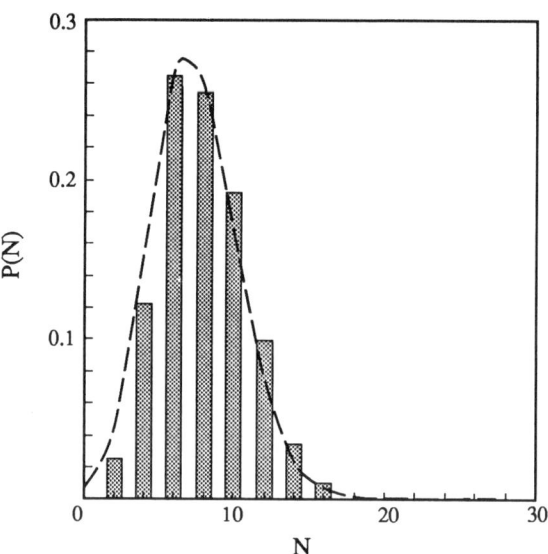

Figure 9: Distribution of the number N of defects in the box (grey bars), for $\mu = 1$, $\alpha = 2$, and $\beta = -0.85$. Since defects appear by pairs, N is always even. The dashed line corresponds to the theoretical distribution given in [44].

4.3. PHASE TURBULENCE AND DEFECT-MEDIATED TUBULENCE

As far as the homogeneous solution of Eq. (4) $A_0 = \sqrt{\mu}\exp(-i\beta\mu t)$ is stable, and since homeogeneous phase perturbations are marginal while other phase perturbations and amplitude perturbations are damped, one can reduce the dynamics of Eq. (4) to that of the phase of the order parameter A when the latter is close to A_0 [47]. This leads, in the limit $1 + \alpha\beta$ small, to the two-dimensional analog of the Kuramoto-Sivashinsky [47-48] equation:

$$\frac{\partial\varphi}{\partial t} = (1+\alpha\beta)\Delta\varphi + (\beta-\alpha)(\nabla\varphi)^2 - \frac{\alpha^2(1+\beta^2)}{2\mu}\Delta^2\varphi. \tag{7}$$

In one dimension, this equation is known (for a review see [49]) to produce a "soft" turbulence, called phase turbulence [47,41], when $1 + \alpha\beta < 0$. In two dimensions, numerical simulations of Eq. (7) show [1] that it also displays turbulent behaviors. Moreover, it turns out that the maximum value of the modulus of the phase gradient $\vec{\nabla}\varphi$ in the system oscillates as a function of time around a mean value, which scales linearly with the size of the system (see Fig. 10). Since the derivation of the phase equation (7) fails when phase gradients are of order $\sqrt{\mu}$, one can conclude that, at least in sufficiently large systems, phase approximation cannot hold, in other words that phase perturbations should destabilize slaved amplitude modes, and may lead to the spontaneous appearance of topological defects in the system.

In order to check this, we have to compute the behavior of the gradient of the phase of the order parameter A given by Eq. (4) while defect-mediated turbulence sets in. Better than the phase, one should look at the eigenvector Γ associated with phase perturbations of the homogeneous solution A_0. As shown in Fig. 11, the amplitude starts varying through the system while $|\Gamma|$ grows, and when it becomes high enough, the minimum value of the amplitude in the system goes to zero, that is defects have appeared.

Thus, the mechanism giving rise to defect-mediated turbulence is quite simple. Namely, phase perturbations destabilize slaved amplitude modes and are responsible for the spontaneous creation of defects in the system.

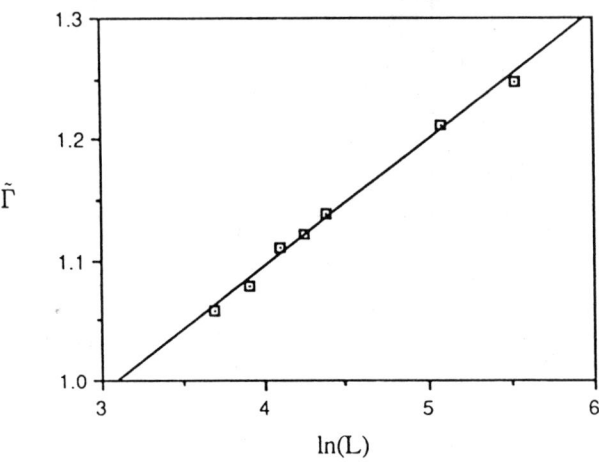

Figure 10: Numerical simulation ($\mu = 1$, $\alpha = 3$, $\beta = -2$) of Eq. (7) showing the behavior of $\tilde{\Gamma} =< \max_{(x,y)}[|\vec{\nabla}\varphi|] >_t$ as a function of the logarithm of the size of the box.

4.4. DEFECT-MEDIATED TUBULENCE IN WAVE PATTERNS

Defect-mediated turbulence also occurs in wave patterns of either isotropic or anisotropic [39] systems. Figure 12 shows a numerical simulation of Eqs. (6) where one can see the destabilization of a wave pattern, leading to the spontaneous nucleation of dislocations in the system. At $t = 300$ (Fig. 12.b), there is compression and dilatation of the rolls. At $t = 400$ (Fig. 12.c), the first pairs of dislocations have appeared. Figure 12.d shows defect-mediated turbulence.

5. Conclusion

The aim of this paper was twofold, first to give a description of topological defects of macroscopic structures by means of Ginzburg-Landau equations, second to point out a mechanism of spontaneous nucleation of topological defects in two dimensional

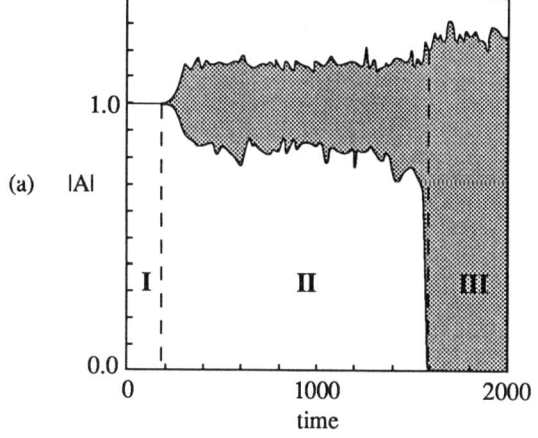

(a) |A|

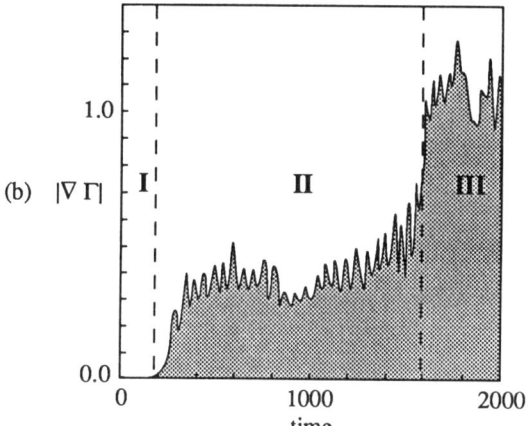

(b) |∇Γ|

I. The system is close to the unstable solution. The solution A is nearly homogeneous, i.e. there exist phase variations but no amplitude variations.

II. Phase gradients become higher. The amplitude is no more homogeneous.

III. Phase gradients have exceeded a critical value, and as a consequence, defects have appeared in the system.

Figure 11: Numerical simulation ($\mu = 1$, $\alpha = 2$, $\beta = -0.82$) of Eq. (4) showing how defect-mediated turbulence sets in. (a) Minima and maxima of $|A|$ in the system as a function of time. (b) Same plot for $|\vec{\nabla}\Gamma|$.

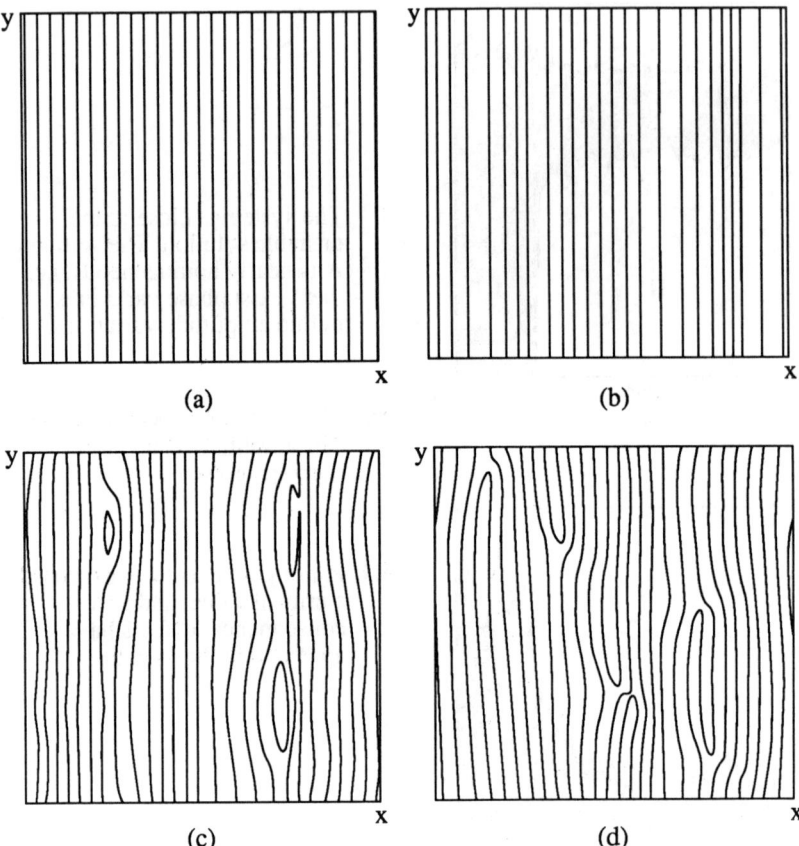

Figure 12: Numerical simulation ($\mu = 1$, $\alpha_x = 1$, $\alpha_y = -2$, $\beta = -1.8$, $\gamma = 1.5$, $c = 5$) of Eqs. (6) showing the destabilization of a traveling wave pattern, giving rise to dislocations. The lines where $\mathcal{R}e(A \exp i(-k_0 x + \omega_0 t)) = 0$ have been plotted in the (x, y) plane, at times (a) 100, (b) 300, (c) 400, and (d) 500. The area between two lines corresponds to a traveling roll.

patterns. The idea was to show through pictures the main phenomena rather than to give details of calculations, which can be found elsewhere. The mechanism we have given is quite generic, and seems to have experimental relevance. Namely, some experiments show disordered regimes associated with defects (dislocations of wave patterns [30,35] as well as spiral waves [50]) which may be related to defect-mediated turbulence. A large part of this work relies on numerical simulations and one may wonder whether a regime of defect-mediated turbulence could be described analytically. The only possible way seems to derive from the Ginzburg-Landau equation a set of coupled equations describing defects as isolated objects interacting between them and evolving in the phase field [51-52]. Recent work [53], giving such a description of spiral defects of Eq. (4), may lead to a better understanding of defect-mediated turbulent regimes, at least in the approximation of a dilute gas of defects.

Acknowledgements

The author would like to acknowledge P. Coullet, L. Gil, and J. L. Meunier with whom large parts of the work presented here have been done. I also thank the CCVR (Centre de Calcul Vectoriel pour la Recherche) and the CICNT (Centre Inter-universitaire de Calcul de Nice et Toulon) where the numerical computations have been performed, the NCAR (National Center for Atmospheric Research) for the use of its graphic library, and the EEC twinning project *Chaos spatio-temporel dans les systèmes étendus* (n° SC1-0035-C (CD)) for a financial support.

References

1. P. Coullet, L. Gil, and J. Lega, Physica **D 37**, 91 (1989).
2. *Cellular structures in instabilities*, J. E. Wesfreid, and S. Zaleski, Eds., Lecture Notes in Physics **210**, Springer-Verlag, Berlin (1984).
3. C. Normand, Y. Pomeau, and M. Velarde, Rev. Mod. Phys. **49**, 581 (1977).
4. S. Chandrasekhar, *Hydrodynamic and hydromagnetic stability*, Clarendon, Oxford University Press (1961).
5. P. Bergé and M. Dubois, Contemp. Phys. **25**, 535 (1984).
6. M. C. Cross, and A. C. Newell, Physica **10D**, 299 (1984).
7. A. C. Newell, T. Passot, and M. Souli, *The phase diffusion and mean drift equations for convection at finite Rayleigh numbers in large containers I.*, preprint (1989).
8. A. C. Newell, and J. A. Whitehead, J. Fluid Mech. **38**, 279 (1969).
9. L. A. Segel, J. Fluid Mech. **38**, 203 (1969).
10. E. D. Siggia, and A. Zippelius, Phys. Rev. Lett. **47**, 835 (1981).
11. E. D. Siggia, and A. Zippelius, Phys. Rev. **A 24**, 1036 (1981).
12. Y. Kuramoto, *Chemical oscillations, waves and turbulence*, Springer Series in Synergetics, Vol. **19**, Springer-Verlag (1984).
13. W. Pesch, and L. Kramer, Z. Phys. **B 63**, 121 (1986).
14. J. Lega, Le Journal de Physique Colloque C3, **50**, C3-193 (1989).
15. C. S. Bretherton, and E. A. Spiegel, Phys. Lett. **A 96**, 152 (1983).
16. P. Chossat, and G. Iooss, Jpn. J. Appl. Math. **2**, 37 (1985).

17. P. Coullet, S. Fauve, and E. Tirapegui, J. Phys. (Paris) Lett. **46**, 787 (1985).
18. P. Coullet, C. Elphick, L. Gil, and J. Lega, Phys. Rev. Lett. **59**, 884 (1987).
19. H. R. Brand, P. S. Lomdahl, and A. C. Newell, Phys. Lett. **A 118**, 67 (1986).
20. *Propagation in systems far from equilibrium*, J. E. Wesfreid, H. R. Brand, P. Manneville, G. Albinet, and N. Boccara, Eds., Springer-Verlag, Berlin, Heidelberg (1988).
21. L. Gil, and J. Lega, in [20], p. 164.
22. S. Ciliberto, and M. A. Rubio, Phys. Rev. Lett. **25**, 2652 (1987).
23. E. Moses, J. Fineberg, and V. Steinberg, Phys. Rev. **A 35**, 2757 (1987).
24. A. Chiffaudel, B. Perrin, and S. Fauve, *Spatiotemporal dynamics of oscillatory convection at low Prandtl number: waves and defects*, preprint (1988).
25. A. Joets, and R. Ribotta, *Defects and transition to disorder in space-time patterns of non-linear waves*, to appear in *New trends in non-linear dynamics and pattern forming phenomena: the geometry of non-equilibrium*, P. Coullet and P. Huerre, Eds., NATO ASI Series, Plenum (1990).
26. R. W. Walden, P. Kolodner, A. Passner, and C. M. Surko, Phys. Rev. Lett. **55**, 496 (1985).
27. J. Lega, C. R. Acad. Sci. Paris, **II 309**, 1401 (1989).
28. R. Ribotta, and A. Joets, in [2], p. 249.
29. E. Guazzelli, C. R. Acad. Sci. Paris, **B 291**, 9 (1980).
30. X. D. Yang, A. Joets, and R. Ribotta, in [20], p. 194.
31. J. M. Dreyfus, and E. Guyon, J. Physique **42**, 283 (1981).
32. J. Occelli, E. Guazzelli, and J. Pantaloni, J. Phys. (Paris) Lett. **44**, 567 (1983).
33. A. Joets, and R. Ribotta, in [2], p. 294.
34. A. Pocheau, V. Croquette, and P. Le Gal, Phys. Rev. Lett. **55**, 1094 (1985).
35. I. Rehberg, S. Rasenat, and V. Steinberg, Phys. Rev. Lett. **62**, 756 (1989).
36. J. Toner, and D. R. Nelson, Phys. Rev. **B 23**, 316 (1981).
37. D. Walgraef, G. Dewel, and P. Borckmans, Z. Phys. **B 48**, 167 (1982).
38. P. Coullet, L. Gil, and J. Lega, Phys. Rev. Lett. **62**, 1619 (1989).
39. P. Coullet, and J. Lega, Europhys. Lett. **7**, 511 (1988).
40. Y. Kuramoto, Suppl. Prog. Theor. Phys. **64**, 346 (1978).
41. Y. Pomeau, and P. Manneville, J. Phys. Lett. **40**, 609 (1979).
42. T. B. Benjamin, and J. E. Feir, J. Fluid Mech. **27**, 417 (1967).
43. A. C. Newell in Lectures in Applied Mathematics, Vol. **15**, 157, Am. Math. Society, Providence (1974).
44. L. Gil, J. Lega, and J. L. Meunier, *Statistical properties of defect-mediated turbulence*, to appear in Phys. Rev. A.
45. Y. Kuramoto, Prog. Theor. Phys. **71**, 1182 (1984).
46. H. R. Brand, P. S. Lomdahl, and A. C. Newell, Physica **23D**, 345 (1986).
47. Y. Kuramoto, and T. Tsuzuki, Prog. Theor. Phys. **55**, 356 (1976).
48. G. I. Sivashinsky, Acta Astronautica 4, 1177 (1977).
49. J. M. Hyman, B. Nicolaenko, and S. Zaleski, Physica **D 23**, 265 (1986).
50. S. Nasuno, M. Sano, and Y. Sawada, *Phase wave propagation in rectangular convective structure of nematic liquid crystal*, submitted to J. Phys. Soc. Jpn.
51. K. Kawasaki, and T. Ohta, Physica **116A**, 573 (1983).
52. K. Kawasaki, Prog. Theor. Phys. Suppl. **79**, 161 (1984).
53. S. Rica and E. Tirapegui, *Interaction of defects in two dimensional systems*, submitted to Phys. Rev. Lett.

PHASE DYNAMICS –
THE CONCEPT AND SOME RECENT DEVELOPMENTS

HELMUT R. BRAND
FB 7, Physik
Universität Essen
D 43 Essen 1
West Germany
  and
Theoretische Physik
Universität Bayreuth
D 8580 Bayreuth
West Germany

0. Introductory Remarks

In this note we will introduce in the first section the concept of phase dynamics, followed by an outline of some recent developments associated with the properties of nonlinear phase equations, and we will conclude with an extensive list of references on phase dynamics and closely related subjects including envelope equations, the motion of defects, and the coupling between amplitude and phase equations.

For a short review of the field up to about summer 1987 we refer to our contribution to the proceedings of the Les Houches conference 'Propagation in Systems Far From Equilibrium', which took place in spring 1987 (H.R. Brand, 1988). A detailed and extensive review is at present under way (H.R. Brand, 1990 II).

1. Phase Dynamics – The Concept

One of the most successful macroscopic concepts in condensed matter physics over the last few decades has been the derivation and exploitation of hydrodynamic equations, that is of balance equations for the hydrodynamic variables. Hydrodynamic variables come in two groups:
i) conserved macroscopic quantities of a given condensed system including for example density, density of linear momentum, energy density, the concentration of one component in nonreactive, miscible binary mixtures or the magnetization in a magnetic system
ii) variables associated with a spontaneously broken continuous symmetry such as the superfluid velocity in the superfluid phases of ^{4}He and ^{3}He (due to broken gauge

invariance), the director(s) in uniaxial (and biaxial) nematic liquid crystals (broken rotational symmetry), and the displacement vector in crystals (broken translational invariance).

After identification of the hydrodynamic variables one uses irreversible thermodynamics and general symmetry arguments (including invariance under rotations and translations, Galilean invariance, behaviour under time reversal and spatial parity etc.) to write down dynamic equations. Key assumptions entering this derivation are the applicability of a continuum approximation (that is one must have many atoms or molecules to average over), the absence of long range forces - e.g. Coulomb forces -, and the existence of an expansion in the spatial gradients. Furthermore one is restricting oneself to the 'hydrodynamic regime': sufficiently small frequencies (small compared to all microscopic frequencies, e.g. collision frequencies between atoms) and long wavelengths (length scales large compared to all microscopic lengths such as interatomic distances etc.).

As an output one obtains from such a hydrodynamic approach in the linearized domain the normal modes, which are either diffusive (heat diffusion, spin diffusion, vortex diffusion etc.) or propagative (first sound, second sound, spin waves, etc.). In the nonlinear domain one finds localized and wall-like solutions, and for sufficiently high values of an external stress parameter (temperature gradient, shear rate etc.) the onset of spatially and/or temporally periodic and eventually disordered patterns signalling the onset of turbulent motion.

The major drawback of this rigorous hydrodynamic approach is the occurence of phenomenological parameters such as static susceptibilities (specific heat, compressibility, etc.) and transport parameters (viscosities, heat conductivities etc.), which must be either determined experimentally or calculated approximately by a more microscopic technique including for example Boltzmann- and Dyson equations.

Since the close analogy and the similarity between phase transitions in equilibrium systems and nonequilibrium phase transitions has been recognized about two decades ago (Graham and Haken, deGiorgio and Scully, 1970) and has been exploited ever since, it seems quite natural to investigate to what extent one can carry over other concepts well established close to equilibrium to systems driven far from equilibrium. And one approach which one would like to apply to nonequilibrium systems as well, since it has been successful close to equilibrium, is that of a macroscopic dynamics valid for low frequencies and long wavelengths. That is, one is seeking dynamic equations for a small number of collective variables, which describe the behaviour of a pattern forming nonequilbrium system in the hydrodynamic regime. To describe such a system in a continuum approximation, one must clarify first what the analogue of the atoms and molecules of ordinary hydrodynamics would have to be. Analyzing the experimental results obtained on spatially extended pattern forming nonequilbrium systems including for example Rayleigh-Bénard convection (a thin layer of a simple fluid is heated from below) and the Taylor instability (the fluid in the gap between two concentric cylinders is subjected to an external torque by rotating the inner or both cylinders), it emerges that the analogue of the atoms in hydrodynamics close to equilibrium are the unit cells (rolls in Bénard convection, vortices in the Taylor instability, hexagons in surface tension driven convection etc.) in spatially periodic patterns in nonequilibrium systems. From this observation it follows immediately that one must investigate large aspect ratio pattern forming nonequilbrium systems to test the analogue of hydrodynamics in such a situation, since one must have a sufficiently large number of unit cells to justify a continuum approximation.

The next question which arises is the one concerning the analogue of the hydrodynamic variables. The first important step in this direction has been done about a decade ago, when Pomeau and Manneville pointed out that a dynamic equation for the slow spatial and temporal variations of the phase Φ, which characterizes the position of the rolls in Rayleigh-Bénard convection, is applicable even well above the onset of the instability as long as there is no change to a different pattern. This is just analogous to a hydrodynamic equation close to equilibrium, which is also only applicable as long as one is sufficiently far away from a phase transition. For the variations of this phase Φ Pomeau and Manneville derived the phase diffusion equation

$$\Phi_t = D_1 \Phi_{xx} + D_2 \Phi_{yy} \tag{1.1}$$

where x denotes the direction parallel to the normal of the convective rolls and y the direction along the crest of the rolls, from the consideration of an underlying microscopic model in the limit of long wavelengths. Because of this approach they were able to express the phase diffusion coefficients D_1 and D_2 in terms of the parameters of a microscopic model, just in the same spirit as one can use a Boltzmann equation to evaluate transport parameters in a hydrodynamic equation - compare the discussion sketched above. The structure of eq.(1.1) can also be obtained directly from symmetry considerations in the limit of a large system size taking into account that the system is invariant under the transformations $x \to -x$ and $y \to -y$ separately and by assuming that its dynamics is purely dissipative.

Triggered by the work of Pomeau and Manneville, Wesfreid and Croquette demonstrated one year later (1980) the applicability of a phase diffusion equation experimentally for Rayleigh-Bénard convection in a high Prandtl number silicone oil and since then, the Santa Barbara group (Ahlers et al., 1986) also showed its validity for the Taylor vortex flow state (a set of azimuthally symmetric vortices along the cylindrical axis) of the Taylor instability. Very recently (Ning et al., 1989) it also emerged that the phase diffusion equation can be used to give a quantitative description well above the onset of a pattern provided one calculates the phase diffusion coefficient D_1 accurately from the underlying hydrodynamic equations (Riecke and Paap, 1987).

For the case of Rayleigh-Bénard convection with spatial variations in two dimensions it has become clear in the meantime that vertical vorticity effects are important for small Prandtl numbers (Siggia and Zippelius, 1981, Cross, 1983) and that one therefore needs to take into account the influence of a mean flow field.

In 1984 we (Brand, 1984) generalized - pursuing the hydrodynamic spirit - the linear phase diffusion equation into the nonlinear domain considering spatial variations in one dimension and we obtained to lowest order in the higher order derivative terms for a stationary pattern

$$\psi_t = D\psi_{xx} + E\psi_x\psi_{xx} + F\psi_x^2\psi_{xx} - G\psi_{xxxx} \tag{1.2}$$

or, rewritten as a diffusion equation for an 'effective' diffusion coefficient

$$\psi_t = (D + E\psi_x + F\psi_x^2 - G\partial_{xx})\psi_{xx} \tag{1.3}$$

In the next section we shall briefly review some of the progress achieved recently on the analysis of the properties of eqs.(1.2) and (1.3).

Once the existence of a diffusive mode had been established theoretically and experimentally, the next question was whether it would be possible to have the analogue of a propagating mode (a sound wave) in a large aspect ratio pattern forming system. The first pattern which was identified as a candidate for such a mode was the wavy mode state in the Taylor instability (Brand and Cross, 1983), a nonequilibrium state that can be characterized as a combination of Taylor vortices along the axis of the cylinders and waves traveling in azimuthal direction on top of the vortices. One can then associate a phase with each of these directions (axial and azimuthal) and we (Brand and Cross, 1983) have shown that the coupled dynamic equations for these two phases can give rise to a propagating mode for sufficiently large wavelengths of the perturbation.

In the sequel a propagating phase motion has been predicted to occur for several other nonequilibrium states, but up to now no experiment designed to detect such a propagative phase motion has been reported. This would be a crucial experiment to perform in order to demonstrate experimentally that there is for the linearized macroscopic equations a very close correspondence between ordinary hydrodynamics and phase dynamics, the analogue of hydrodynamics for large aspect ratio pattern forming nonequilibrium systems.

2. Some Recent Developments

Over the last year or so there have been mainly two developments in phase dynamics both of which have been triggered by experimental results.
Stimulated by the observation of localized excitations propagating across the interface in cellular solidification in nematic liquid crystals (Simon et al., 1988) and in mixtures close to the eutectic point (Faivre et al., 1989), Coullet and collaborators (Coullet et al., 1989 II, III) investigated the effects of the coupling of a phase equation to an envelope equation describing a secondary pattern. This approach will be discussed in detail in these proceedings in the chapter by Pierre Coullet.

A stationary wavelength distribution of the unit cells of a pattern has been observed for the case of the Taylor instability between co-rotating cylinders (Baxter and Andereck, 1986, Andereck and Baxter, 1988) and more recently for the case of slot convection (Dubois et al. 1989, 1990), for which the thickness of the convective cell is less than the height. For the former the interpretation is made complicated by the fact that the vortices with the larger wavelength show turbulent behaviour in azimuthal direction, whereas in the latter class of observations a quasi one-dimensional (1D) situation seems to prevail. Triggered by these observations of a stationary wavelength distribution in a stationary pattern, we (Brand and Deissler, 1989 II, IV, Brand, 1990 I, Deissler et al., 1990) have studied in some detail the nonlinear phase equation for a stationary pattern (eqs.(1.2), (1.3)) and also the dynamic equation for the local wavevector $q = \psi_x$

$$\dot{q} = (D + Eq + Fq^2)\, q_{xx} + (E + 2Fq)(q_x)^2 - Gq_{xxxx} \tag{2.1}$$

In the following we briefly review some of the main results of our analysis.

Firstly one observes that eq.(1.3) can be derived from a Liapunov functional

$$\dot{\psi} = -\frac{\delta V}{\delta \psi} \tag{2.2}$$

with

$$V(\{\psi\}) = \int dx \left[\frac{D}{2}(\psi_x)^2 + \frac{E}{6}(\psi_x)^3 + \frac{F}{12}(\psi_x)^4 + \frac{G}{2}(\psi_{xx})^2 \right] \tag{2.3}$$

Provided F and G are positive, $V(\{\psi\})$ guarantees global stability. To give eq.(2.3) a more familiar form, we rewrite it in terms of the local wavevector

$$V(\{q\}) = \int dx \left[\frac{D}{2}q^2 + \frac{E}{6}q^3 + \frac{F}{12}q^4 + \frac{G}{2}(q_x)^2 \right] \tag{2.4}$$

Inspection of eq.(2.4) shows the close analogy to a generalized Ginzburg-Landau free energy for a weakly first order phase transition. It is also possible to show that eq.(2.4) is a Liapunov functional for the nonlinear equation for the local wavevector q. Thus we can conclude that the local wavevector for a stationary 1D pattern of the type discussed here can be interpreted as an order parameter. In passing we note that we have also shown, that $V(\{\psi\})$ and $V(\{q\})$ in eqs.(2.3) and (2.4) can be used as generalized thermodynamic potentials if one incorporates fluctuations which are delta correlated in space and time in eqs.(1.3) and (2.1).

To emphasize further the analogy of phase dynamics with hydrodynamics, we point out that the nonlinear phase equation (eq.(1.3)) for a 1D stationary pattern can be brought into the form of a conservation law

$$\dot{\psi} + \partial_x j^\psi = 0 \tag{2.5}$$

with the phase current j^ψ

$$j^\psi = -[D\psi_x + \frac{E}{2}\psi_x^2 + \frac{F}{3}\psi_x^3 - G\psi_{xxx}] \tag{2.6}$$

Combining this observation with the fact that all phase equations for 1D patterns of the form

$$\psi_t = \Gamma(\{\psi_x\}) \tag{2.7}$$

can be rewritten for the local wavevector $q = \psi_x$ as

$$q_t = \partial_x \Gamma(\{q\}) \tag{2.8}$$

we obtain from eqs.(2.5) - (2.8) the following equation for the local wavevector of a 1D stationary pattern

$$\dot{q} = \partial_{xx}[\frac{\delta V}{\delta q}]$$
$$= \partial_{xx}[Dq + \frac{E}{2}q^2 + \frac{F}{3}q^3 - Gq_{xx}] \quad (2.9)$$

with $V(\{q\})$ as in eq.(2.4). As can be seen by close inspection of eq.(2.9) it can be brought into the form of the dynamic equation for a conserved order parameter by a shift of q by a constant value: $q = q_0 - E/(2F)$ and then shows close similarity with the equation for the concentration in spinodal decomposition.

We remark that it is crucial for the derivation of the structure of eq.(2.9) that the phase satisfies a conservation law, which is not the case for the other prototype of a nonlinear phase equation, the Kuramoto-Sivashinsky equation, as it arises for the phase variable associated with a propagative motion

$$\dot{\phi} = D\phi_{xx} - \tilde{D}\phi_{xxxx} + \frac{E}{2}\phi_x^2 \quad (2.10)$$

It is well known that this equation can show chaotic behaviour (Kuramoto, 1980). As a phase equation it is no longer valid as soon as defects become important and it is then necessary to go back to a description keeping amplitude and phase.

The existence of a conservation law for the phase associated with 1D stationary patterns opens an interesting possibility. By pinning the phase at the boundaries it is possible to obtain a distribution in the wavelengths of the unit cells, a localized state in phase dynamics, for which there are only wavelength and no amplitude variations. Due to the conservation law, phase is not generated or annihilated in the bulk in contrast e.g. to the case of the Kuramoto-Sivashinsky equation. If one allows the phase to relax freely, e.g. by implementing a ramp (Kramer et al., 1982), a state with a constant wavelength of the unit cells (for example rolls in Bénard convection) will result.

We close by noting that the nonlinear equations for the phase ψ and the wavevector also have interesting analytically accessible localized and wall-type solutions. Shifting q as described above, we find for example for q_0 a localized solution of the form

$$q_0^2 = \frac{3H}{F}(1 - \text{sech}^2[\sqrt{\frac{H}{2G}}(x - x_0)]) \quad (2.11)$$

where $H = -D + E^2/(4F)$. Further details on the nonlinear solutions will be given elsewhere.

Acknowledgements
It is a pleasure to thank Guenter Ahlers, David Andereck, Bob Deissler, and Monique Dubois for stimulating discussions.
Support of this work by the Deutsche Forschungsgemeinschaft is gratefully acknowledged.

3. References on Phase Dynamics and Related Problems

G. Ahlers, D.S. Cannell and M.A. Dominguez-Lerma,
   *Phys.Rev.Lett.***48**, 368 (1982)
G. Ahlers, D.S. Cannell and M.A. Dominguez-Lerma,
   *Phys.Rev.***A27**, 1225 (1983)
G. Ahlers, D.S. Cannell and V. Steinberg, *Phys.Rev.Lett.***54**, 1373 (1985)
G. Ahlers, D.S. Cannell, M.A. Dominguez-Lerma and R. Heinrichs,
   Physica D - Nonlinear Phenomena **23**, 202 (1986)
C.D. Andereck, R. Dickman and H.L. Swinney, *Phys.Fluids* **26**, 1395 (1983)
C.D. Andereck, S.S. Liu and H.L. Swinney, *J.Fluid Mech.* **164**, 155 (1986)
C.D. Andereck and G.W. Baxter, p.315 ff, in
   Propagation in Systems far from Equilibrium, J.E. Wesfreid et al., Eds.,
   Springer Series in Synergetics, Springer, Heidelberg, 1988
M.A. Azouni, *J.Cryst.Growth* **42**, 405 (1977)
P. Bak, *Rep.Prog.Phys.* **45**, 587 (1981)
G.W. Baxter and C.D. Andereck, *Phys.Rev.Lett.* **57**, 3046 (1986)
D. Bensimon, P. Kolodner, C.M. Surko, H. Williams and V. Croquette, preprint,
   March 1989
H.R. Brand, *Prog.Theor.Phys.* **71**, 1096 (1984)
H.R. Brand, *Phys.Rev.* **A31**, 3454 (1985)
H.R. Brand, *Phys.Rev.* **A32**, 3551 (1985)
H.R. Brand, *Phys.Rev.Lett.* **57**, 2768 (1985)
H.R. Brand, p. 206, in *Propagation in Systems far from Equilibrium*,
   J.E. Wesfreid et al., Eds., Springer Series in Synergetics,
   Springer, Heidelberg, 1988
H.R. Brand, p. xxx, in the Proceedings of the NATO Advanced Research
   Workshop in Streitberg, Sept.1989, F.H. Busse and L. Kramer, Eds., Plenum,
   New York, 1990 I
H.R. Brand, *Phys.Rep.*, to be published (1990 II)
H. Brand and H. Pleiner, *J.Phys.(Paris)* **41**, 553 (1980)
H. Brand and P. Bak, *Phys.Rev.* **A27**, 1062 (1983)
H. Brand and M.C. Cross, *Phys.Rev.* **A27**, 1237 (1983)
H.R. Brand and H. Pleiner, *J.Phys.(Paris)* **45**, 563 (1984)
H.R. Brand, P.C. Hohenberg and V. Steinberg, *Phys.Rev.* **A30**, 2548 (1984)
H.R. Brand and K. Kawasaki, *J.Phys.***A17**, L905 (1984)
H.R. Brand, J.E. Wesfreid, M.A. Azouni and S. Kai, *Chem.Phys.Lett.*
   126, 447 (1986)
H.R. Brand, P.S. Lomdahl and A.C. Newell, *Physica D - Nonlinear Phenomena*
   23, 345 (1986)
H.R. Brand, P.S. Lomdahl and A.C. Newell, *Phys.Lett.***A118**, 67 (1986)
H.R. Brand and B.J.A. Zielinska, *Phys.Rev.Lett.* **57**, 2789 (1986)
H.R. Brand and R.J. Deissler, *Phys.Rev.* **A39**, 462 (1989 I)
H.R. Brand and J.E. Wesfreid, *Phys.Rev.* **A39**, 6319 (1989)
H.R. Brand and R.J. Deissler, *Phys.Rev.Lett.* **63**, 508 (1989 II)
H.R. Brand and R.J. Deissler, *Phys.Rev.Lett.* **63**, 2501 (1989 III)

H.R. Brand and R.J. Deissler, *Phys.Rev.* **A41**, xxxx (1990)
K. Bühler, *Z.Angew.Math.Mech.* **64**, T180 (1984)
H. Chaté and P. Manneville, *Phys.Rev.Lett.* **58**, 112 (1987)
P. Coullet, *Phys.Rev.Lett.***56**, 724 (1986)
P. Coullet, S. Fauve and E. Tirapegui, *J.Phys.(Paris)Lett.***46**, 787 (1985)
P. Coullet and S. Fauve, p.290, in *Lecture Notes in Physics* **230** (1985)
P. Coullet and S. Fauve, *Phys.Rev.Lett.***55**, 2857 (1985)
P. Coullet and P. Huerre, *Physica D - Nonlinear Phenomena* **23**, 27 (1986)
P. Coullet, L. Gil, and J. Lega, *Phys.Rev.Lett.* **62**, 1619 (1989 I)
P. Coullet and G. Iooss, private communication 1989 II
P. Coullet, R.E. Goldstein, and G.H. Gunuratne, *Phys.Rev.Lett.*
   63, 1954 (1989 III)
V. Croquette and J.E. Wesfreid, p.399, in *Symmetries and Broken Symmetries in Condensed Matter Physics*, N. Boccara, Ed., IDSET, Paris, 1981
V. Croquette, P. Le Gal, A. Pocheau and R. Guglielmetti, *Europhys. Lett.*
   1, 393 (1986)
M.C. Cross, *Phys.Rev.***A27**, 490 (1983)
M.C. Cross and A.C. Newell, *Physica D - Nonlinear Phenomena* **10**, 299 (1984)
R.J. Deissler, Y.C. Lee, and H.R. Brand, to be published
M. Dubois, R. DaSilva, F. Daviaud, P. Bergé, and A. Petrov, *Europhys.Lett.*
   8, 135 (1989)
M. Dubois et al. in *The Geometry of Nonequilibrium* ,
   eds. P. Huerre and P. Coullet, Plenum, New York, 1990, to appear
   and private communication
M. Dubois et al., to be published
G. Faivre, S. de Cheveigné, C. Guthmann, and P. Kurowski, *Europhys.Lett.*
   9, 779 (1989)
D. Forster, *Hydrodynamic Fluctuations, Broken Symmetries
   and Correlation Functions*, Benjamin, Reading, Mass. 1974
V. deGiorgio and M.O. Scully, *Phys.Rev.***A2**, 1170 (1970)
M. Gorman and H.L. Swinney, *J.Fluid Mech.***117**, 123 (1982)
R. Graham, *Phys.Rev.Lett.* **31**, 1479 (1973)
R. Graham, *Phys.Rev.* **A10**, 1762 (1974)
R. Graham and H. Haken, *Z.Phys.***237**, 31 (1970)
H. Haken, *Rev.Mod.Phys.* **47**, 67 (1975)
P. Huerre, *Nucl.Phys.Suppl.***B2**, 159 (1987)
G. Iooss, Y. Demay and P.H. Coullet, preprint, 1986,
K. Kawasaki, *Prog.Theor.Phys.Suppl.***79**, 161 (1984)
K. Kawasaki and H.R. Brand, *Ann.Phys.(N.Y.)* **160**, 420 (1985)
G.P. King and H.L. Swinney, *Phys.Rev.***A27**, 1240 (1983)
P. Kolodner, D. Bensimon and C.M. Surko, *Phys.Rev.Lett.* **60**, 1723 (1988)
L. Kramer, E. Ben-Jacob, H. Brand and M.C. Cross, *Phys.Rev.Lett.*
   49, 1891 (1982)
L. Kramer and H. Riecke, *Z.Phys.***B59**, 245 (1985)
Y. Kuramoto, *Prog.Theor.Phys.Suppl.* **64**, 346 (1980)
Y. Kuramoto, *Prog.Theor.Phys.***71**, 1183 (1984)

Y. Kuramoto, *Chemical Oscillations, Waves and Turbulence*,
  Springer, N.Y., 1984
Y. Kuramoto and T. Tsuzuki, *Prog.Theor.Phys.***54**, 687 (1975)
P. Lallemand, S. Zaleski and P. Tabeling, *Phys.Rev.* **A32**, 655 (1984)
J.S. Langer, *Ann.Phys.(N.Y.)* **65**, 53 (1971)
J.S. Langer, *Physica* **73**, 61 (1974)
M. Lowe and J.P. Gollub, *Phys.Rev.***A31**, 3895 (1985)
M. Lowe, B.S. Albert and J.P. Gollub, *J.Fluid Mech.***173**, 253 (1986)
T.C. Lubensky, *Phys.Rev.***A6**, 452 (1972)
J.M. Luijkx, J.K. Platten and J.C. Legros, *Int.J.Heat Mass Transfer*
  24, 1287 (1981)
P. Manneville, p.265, in J.E. Wesfreid et al., Eds.
  Propagation in Systems far from Equilibrium,
  Springer Series in Synergetics, Springer, Heidelberg, 1988
P. Manneville and J.M. Piquemal, *Phys.Rev.***A28**, 1774 (1983)
P.C. Martin, O. Parodi and P.S. Pershan, *Phys.Rev.***A6**, 2401 (1972)
S.C. Müller, S. Kai and J. Ross, *Science* **216**, 635 (1982)
A.C. Newell, p.157, in *Lectures in Applied Mathematics*, Vol.15, M. Kac, Ed.,
  American Mathematical Society, RhI, 1974
A.C. Newell, p.244, in *Synergetics*, H. Haken, Ed., Springer, N.Y., 1979
A.C. Newell, *Solitons in Mathematics and Physics*, Society for
  Industrial and Applied Mathematics, Philadelphia, PA 19103, 1985
A.C. Newell, p. 122 in *Propagation in Systems far from Equilibrium*,
  J.E. Wesfreid et al., Eds., Springer Series in Synergetics,
  Springer, Heidelberg, 1988
A.C. Newell and J.A. Whitehead, *J.Fluid Mech.***38**, 279 (1969)
A.C. Newell and J.A. Whitehead, IUTAM Symposium in Herrenalb,
  H. Leipholz, Ed., Springer, N.Y. 1970
J. Niemela, G. Ahlers and D.S. Cannell, *Bull.Am.Phys.Soc.* **33**, 2261 (1988),
  and preprint Nov.89
L. Ning, G. Ahlers, and D.S. Cannell, preprint, October 1989
C. Nouar, R. Devienne, G. Cognet and M. Lebouche, Abstracts of the
  4th Taylor Vortex Flow Working Party, Karlsruhe, West Germany, 1985,
  K. Bühler, M. Wimmer and J. Zierep, Eds.
P. Ortoleva and J. Ross, *J.Chem.Phys.* **58**, 5673 (1972)
R.L. Pfeffer and W.W. Fowlis, *J.Atm.Sci.* **25**, 361 (1968)
A. Pocheau, *J.Phys.(Paris)* **49**, 1127 (1988)
A. Pocheau, *J.Phys.(Paris)* **50**, 2059 (1989)
A. Pocheau, V. Croquette, P. Le Gal, and C. Poitou, *Europhys. Lett.*
  3, 915 (1987)
Y. Pomeau and P. Manneville, *J.Phys.(Paris) Lett.***40**, 609 (1979)
H. Riecke, *Europhys.Lett.***2**, 1 (1986)
H. Riecke, preprint
H. Riecke and H.G. Paap, *Phys.Rev.***A33**, 547 (1986)
H. Riecke and H.G. Paap, *Phys.Rev.Lett.***59**, 2570 (1987)
W. van Saarloos and P.C. Hohenberg, preprint, Dec. 1989

L.A. Segel, *J. Fluid Mech.***38**, 203 (1969)
B.I. Shraiman, *Phys.Rev.Lett.***57**, 325 (1986)
E.D. Siggia and A. Zippelius, *Phys.Rev.Lett.***47**, 835 (1981)
A.J. Simon, J. Bechhoefer, and A. Libchaber, *Phys.Rev.Lett.* **61**, 2574 (1988)
H.A. Snyder, *Phys.Fluids* **11**, 728 (1968)
P. Tabeling, *J.Phys.(Paris) Lett.***44**, 665 (1983)
P. Tabeling and C. Trakas, in *Cellular Structures in Instabilities*,
  J.E. Wesfreid and S. Zaleski, Eds., Springer, N.Y., 1984, p.285
O. Thual and S. Fauve, *J.Phys.(Paris)* **49**, 1829 (1988)
D. Walgraef, *Phys.Rev.* **A34**, 3270 (1986)
D. Walgraef, *Europhys.Lett.* **7**, 485 (1988)
J.E. Wesfreid, Y. Pomeau, M. Dubois, C. Normand and P. Bergé,
  J.Phys.(Paris) **39**,725 (1978)
J.E. Wesfreid and V. Croquette, *Phys.Rev.Lett.* **45**, 634 (1980)
J.E. Wesfreid, H.R. Brand, P. Manneville, G. Albinet and N. Boccara,
  Eds. *Propagation in Systems Far From Equilibrium*, Springer, N.Y. 1988
M. Wimmer, *Z.Angew.Math.Mech.***63**, T299 (1983)
M. Wimmer, *Z.Angew.Math.Mech.***65**, T255 (1985)
S. Zaleski, Thèse de troisième cycle, Université de Paris, 1981

TRANSIENT PATTERN DYNAMICS: GENERAL CONCEPTS AND THE FREEDERICKSZ TRANSITION IN NEMATICS

M. SAN MIGUEL
Departament de Física
Universitat de les Illes Balears
E-07071 Palma de Mallorca
Spain

F. SAGUES
Departament de Química-Física
Universitat de Barcelona
Diagonal 647, E-08028 Barcelona
Spain

ABSTRACT. The problem of transient patterns is generally addressed. Some general concepts and methods used in the problem of spinodal decomposition are discussed as a framework of reference. The formation of transient patterns in the Fréedericksz transition in nematics is considered using a Ginzburg-Landau formulation of stochastic nematodynamics. The stages of pattern formation are described by the dynamics of a structure factor and the late stages of pattern decay by domain wall dynamics.

1. Introduction

Among the many problems considered in the context of pattern formation studies[1], perhaps the most typical is the one in which given a system in an homogeneous steady state, one asks for which critical value λ_c of an appropriate control parameter the homogeneous state becomes unstable and a stable spatial pattern emerges. The interest is then to describe the pattern by quantities like a characteristic wavelength, and to check its stability and possible further bifurcations. These questions are concerned with steady state properties. A different, and less popular problem, is the description of the transient dynamical evolution to a stable steady state pattern from an initial homogeneous state, following a change of the control parameter from $\lambda_i < \lambda_c$ to $\lambda_f > \lambda_c$. One is then dealing with the decay from an unstable state, which is typically accompanied by large transient fluctuations. Examples of this problem are the transient evolution to a periodic pattern[2] in a one-dimensional version of the Swift-Hohenberg model of a convective instability[3], and the dynamics of pattern formation in the Turing Optical instability[4]. A third different problem, which will be the subject of these lectures is the one of transient pattern formation during the temporal evolution between two homogeneous steady states. As in the previous case, one is here interested in transient dynamics, but now the emerging pattern disappears in the final state. Relevant questions in this type of problems are the conditions for the occurrence of the transient pattern, onset time of appearance of the pattern, time scales for pattern development and decay, mechanisms of pattern growth and decay, etc...

The problem of transient patterns is rather old. A prototype example with a long tradition of studies because of its importance in Materials Science is that of spinodal decomposition. From a fundamental point of view, renovated interest in this problem appears associated with general studies of nonlinear and nonequilibrium phenomena. New methods and ideas have been developed in this context[5-7] which give a useful framework of reference to discuss transient pattern dynamics. Here we first summarize some basic ideas and methods involved in the description of transient patterns in spinodal decomposition. From this general perspective the problem of transient patterns[8-10] in the magnetic Fréedericksz transition[11] in nematic liquid crystals will be considered.

Our discussion of transient patterns is limited here to problems in which a potential exists. This means that the final steady state, in the presence of fluctuations, is described in terms of a free energy F by a known probability distribution $P_{st} \sim \exp(-\beta F)$, $\beta=(k_B T)^{-1}$. The consequence is that the final steady state is obtained from a minimizing principle associated with F. However, it should be emphasized that the transient dynamics is not determined by F alone, because of two other possible ingredients. First, is the occurrence of wavenumber dependent kinetic coefficients. A second ingredient is the possibility of nonrelaxational dynamics. This means that nondissipative terms, which do not modify the final steady state, might occur in the equation of motion. The role of these two ingredients in the transient dynamics will be considered throughout the paper. Other general aspect of our discussion is the consideration of different time scales of evolution. The onset time for the emergence of the pattern is determined by fluctuations, so that a stochastic description is essential in the early stages of development. We will further argue that linear theory has a well defined time scale of validity fixed by a Mean First Passage Time. Such time scale is fixed by the parameters of the system, being rather large in the case of the Fréedericksz transition. The final stage of pattern decay is governed by defect dynamics.

The outline of the paper is as follows. In Section 2 we briefly recall basic aspects of spinodal decomposition emphasizing some specific points particularly relevant for the discussion of transient patterns. In Section 3 we consider the coupled equations for the director field and velocity flow in a nematic state as needed to study the Fréedericksz transition. We formulate stochastic nematodynamic equations which have the general form of a Time Dependent Ginzburg Ladau model (TDGL). This puts the general problem of pattern formation in nematics on a similar formal basis than the one of spinodal decomposition. Section 4 considers pattern formation in the Fréedericksz transition. We describe the dynamical stages of emergence of the pattern. Section 5 considers pattern decay through the analysis of the dynamics of Fréedericksz domain walls.

2. Spinodal Decomposition: Some Aspects Relevant for Pattern Formation

If a binary alloy is prepared at a high temperature in a disordered state and then rapidly quenched to a low temperature, it orders kinetically. Small domains of the ordered phase form and then grow to macroscopic size as time increases. The early stages of phase transformation can occur via the mechanisms of nucleation or spinodal decomposition. Nucleation is an activated mechanism which occurs when the system is quenched to a point of the phase diagram close to the coexistence curve. On the other hand, if the alloy is quenched in the center of the miscibility gap, the system is unstable with respect to arbitrarily small long wavelength fluctuations. The decay process is associated with the decay of an unstable state. The early stage of such decay process is called spinodal decomposition[5]. The transformation is caused by the temporal development of a spatial fluctuation which takes the initially homogeneous state through a sequence of nonequilibrium states. These states exhibit a transient pattern which displays a characteristic interconnected structure. The dynamics of the process involves a conservation law related to the conservation of the number of particles of a given species. The same qualitative evolution occurs, in different time scales, in other systems which phase separate, as for example binary fluids or polymer mixtures. The name of spinodal decomposition is also often used referring to the early stages of phase transformations triggered by long wavelength fluctuations in which there is no conservation law for the corresponding order parameter. This implies an essentially different dynamics for the formation and evolution of the transient pattern. The spinodal decomposition process can be visualized, for example by Transmission Electron Microscope (TEM) techniques for binary alloys[12], or by optical microscope for polymer mixtures[13]. What it is typically seen is a first stage of evolution in which an initially homogeneous phase develops a number of interfaces. These are first very diffuse. In a later stage the interfaces become sharp and separate domains in local equilibrium. The sample has then a well developed pattern. The late stages of the phase transformation are associated with domain growth and the dynamical evolution proceeds through the motion of the interfaces. In situations in which more than two final phases are possible, more complicated defects occur. The late stage dynamics is determined by the motion of such defects.

A quantitative description of pattern formation and evolution during spinodal decomposition and subsequent domain growth can be given in terms of the time dependent structure factor for the fluctuations, $\Delta\psi = \psi(\mathbf{r},t) - \psi_0$, of the local relevant order parameter

$$S(\mathbf{q},t) = \int d(\mathbf{r}-\mathbf{r'}) \exp[i\mathbf{q}(\mathbf{r}-\mathbf{r'})] \langle \Delta\psi(\mathbf{r},t) \Delta\psi(\mathbf{r'},t) \rangle \quad . \tag{2.1}$$

Scattering experiments permit the measurement of $S(\mathbf{q},t)$. For a phase separation process it is typically observed that, starting from a nearly $\mathbf{q}$-independent function, the scattering intensity develops a growing peak at a value $Q(t)$ which moves to smaller values of $q=|\mathbf{q}|$. The growth of the peak is associated with order development and the motion of the peak with domain growth. The characteristic peak is due to the existence of a forbidden mode at $q=0$ where growth never occurs due to the conservation law involved. The existence of the peak reveals a dynamical mechanism of wavenumber selection such that a pattern with a characteristic length emerges. For an order-disorder transformation the scattering intensity has its growing peak at $q=0$ and domain growth can then be associated with the time dependence of the width of the structure factor. In this case the form of the transient pattern depends on the fluctuations in the very early stages of evolution. From the structure factor it is possible to calculate an average domain size $R(t)$ whose growth is usually characterized in terms of power laws

$$R(t) \approx At^m \quad . \tag{2.2}$$

Typical values of m are m=1/2,1/3,1. The value m=1/2 (Cahn-Allen law[5]) appears in cases with nonconserved order parameter. It is associated with curvature driven motion of antiphase boundaries. The case m=1/3 (Lifhitz-Slyozov law) appears in the presence of a conservation law[5,14]. Finally m=1 appears in diffusionless first order transitions and also in binary fluids[5,15]. A question which has attracted a large attention of theoreticians, simulators and experimentalists is the observation of a dynamical scaling law for the structure factor[5,7],

$$S(q,t) = R^d(t) F(qR(t)) \quad ; \quad t > t_0 \quad . \tag{2.3}$$

where d is the dimensionality of the system and F the scaling function. This scaling law implies the existence of a single function for the structure factor which permits to collapse into a single curve measurements made at different times by appropriate rescaling of axis. Physically it means that the evolution of the system is invariant under changes of length scales, provided that a corresponding change in time scale is done. The relationship between the two rescalings is of course the time dependent domain size. Different values of the exponent m and different scaling functions give a classification of nonequilibrium universality classes. Examples of experimental observations of the scaling law are those of Ref. 16 for phase separation and Ref. 17 for a case with a nonconserved order parameter. Growth laws and scaling behavior have not been considered at length in other contexts of pattern formation problems.

The mathematical modeling of spinodal decomposition and domain growth falls in general into three large groups. A first group of models are Kinetic Ising models[5,6]. These are lattice models easily amenable to Monte Carlo simulation. Those models are a first well known example of possible different dynamics with the same equilibrium properties, as determined by the Ising Hamiltonian. Two of such models are the Spin Flip Kinetic Ising (SFKI) associated with the name of Glauber and the Spin Exchange Kinetic Ising (SEKI) associated with the name of Kawasaki. The SEKI involves a conservation law during the transient dynamics while the SFKI models transformations with a nonconserved order parameter. Monte Carlo simulations of these models have been very useful to exhibit the transient patterns formed and to estabilish growth and dynamical scaling laws. An example of an extensive simulation of this type is given in Ref. 18. A second group of models, discussed below, is formed by field theory models of the Time Dependent Ginzburg Landau (TDGL) form. Finally a third class of models considers evolution equations for the defects dominating domain growth[19,20]. Renormalization group techniques have also been used in the description of domain growth[7,21].

2.1. TDGL MODELS: POTENTIAL AND KINETICS

TDGL models are field theory dynamical models based on a Ginzburg-Landau free energy $F[\psi(r,t)]$ for the order parameter ψ. The free energy describes the equilibrium properties through the Boltzmann factor $\exp(-\beta F)$ which gives the statistical weight associated with a given configuration $\psi(r,t)$. The main idea of these models is to incorporate different dynamical processes in the evolution of ψ such that in any case the equilibrium solution is given by $\exp(-\beta F)$. TDGL models can be justified by a generalization of irreversible thermodynamics to a nonlinear regime. Appropriate generalized Onsager coefficients are considered and noise sources satisfying fluctuation-dissipation relations are introduced. Such models have been extensively used in studies of critical dynamics[22] and kinetics of phase transitions[5]. A classification of a number of these models is given in Ref. 22. The simplest form of a TDGL model for a scalar order parameter takes the form of a relaxation dynamics:

$$\partial_t \psi(r,t) = - L\left[\delta F[\psi] / \delta\psi(r,t)\right] + \xi(r,t) \quad . \tag{2.4}$$

L is a kinetic coefficient and $\xi(r,t)$ models fluctuations as Gaussian white noise of zero mean satisfying a fluctuation-dissipation relation with L

$$\langle \xi(r,t)\, \xi(r',t') \rangle = 2 k_B T L\, \delta(r - r')\, \delta(t - t') \quad . \tag{2.5}$$

The Fokker-Planck equation, associated with (2.4) for the probability density $P(\psi,t)$ of a configuration $\psi(r,t)$ can be written as a continuity equation,

$$\partial_t P(\psi,t) = - \int dr\; \delta J / \delta\psi(r,t) \quad , \tag{2.6}$$

with probability current

$$J[\psi] = - L\left[(\delta F / \delta\psi) + k_B T(\delta / \delta\psi)\right] P(\psi,t) \quad . \tag{2.7}$$

The stationary solution of (2.6) is of the desired form $\exp(-\beta F)$ whatever the form of F and of the kinetic coefficient L. The free energy F acts as a potential and different relaxational dynamics are associated with different forms of L for the same F. Two simple models known as models A and B are of the form above with F given by

$$F = \int dr\, \left[(k/2)\, (\nabla\psi)^2 + f(\psi)\right] \quad , \tag{2.8}$$

$$f(\psi) = - (r / 2)\, \psi^2 + (u / 4)\, \psi^4 \quad . \tag{2.9}$$

Model A is characterized by a constant kinetic coefficient L=M, while $L=-M\nabla^2$ for model B. Thus, model B includes a conservation law for the dynamics. It is the continuous version of the SEKI, while model A is associated with the SFKI. A number of numerical simulations of models A and B have been reported[23]. These simulations show the same qualitative features than Monte Carlo simulations of KI models. They exhibit the different transient pattern dynamics of both models, and growth and dynamical scaling laws are obtained. An example of a numerical simulation of model B showing the evolution of the transient pattern is seen in Fig.1. Related models defined by discretized maps[24] for the dynamical evolution have also been studied numerically showing the same general features.

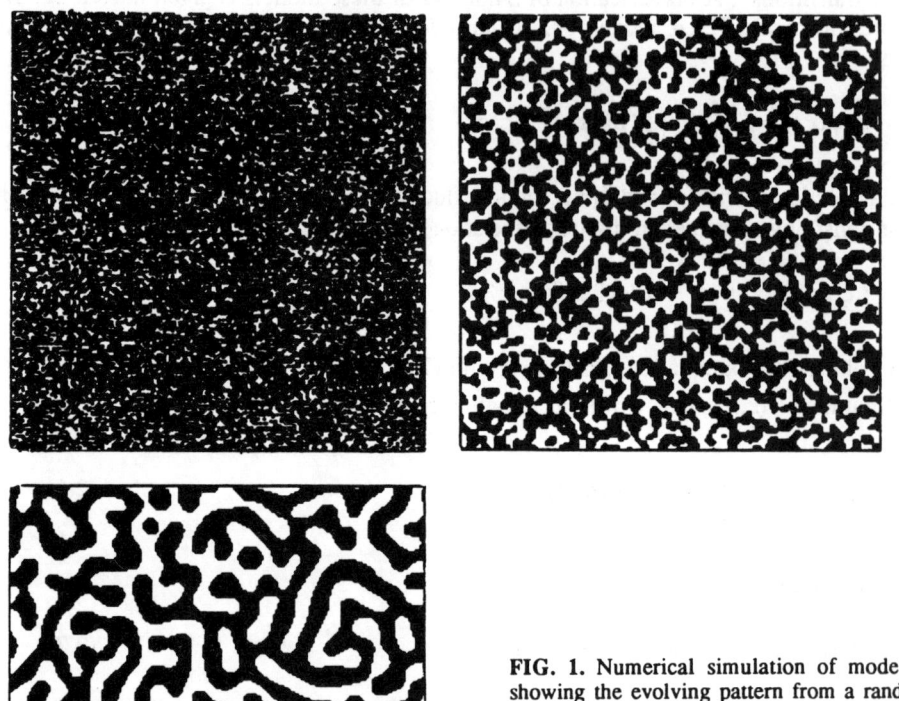

FIG. 1. Numerical simulation of model B showing the evolving pattern from a random initial configuration. Typical configurations at times t=0.2,1,80 are shown. (Parameter values k=r=u=1).

A general form of a TDGL model for a vectorial order parameter ψ is

$$\partial_t \psi_i(\mathbf{r},t) = V_i(\psi) - \mathcal{L}_{ij}(V,\psi)\left[\delta F[\psi] / \delta \psi_j(\mathbf{r},t)\right] + \xi_i(\mathbf{r},t) \quad . \tag{2.10}$$

In the right hand side of (2.10) one now distinguishes a relaxational part given by a general free energy F and kinetic coefficients $\mathcal{L}_{ij}$ and a nonrelaxational term $V_i(\psi)$. The fluctuation-dissipation relation is now the obvious generalization of (2.5)

$$\langle \xi_i(\mathbf{r},t)\, \xi_j(\mathbf{r}',t') \rangle = 2k_B T\, \mathcal{L}_{ij}(V,\psi)\, \delta(\mathbf{r}-\mathbf{r}')\, \delta(t-t') \quad . \tag{2.11}$$

The associated Fokker-Planck equation is in many cases of the form (2.6) with a probability current

$$J_i[\psi,t] = V_i(\psi) - \mathcal{L}_{ij}(\nabla,\psi)\left[(\delta F/\delta\psi_j) + k_BT(\delta/\delta\psi_j)\right] P(\psi,t) \quad . \tag{2.12}$$

The stationary solution of the Fokker-Planck equation is still of the form $P_{st}(\psi) \sim \exp(-\beta F)$ whatever the form of F and $\mathcal{L}_{ij}(\nabla,\psi)$ provided that the nonrelaxational part satisfies the requirement

$$\int dr \ (\delta/\delta\psi_i) \left(V_i \exp(-\beta F)\right) = 0 \quad . \tag{2.13}$$

This is fulfilled if

$$\int dr \ \delta V_i/\delta\psi_i = 0 \quad ; \quad \int dr \ V_i\left(\delta F/\delta\psi_i\right) = 0 \quad . \tag{2.14}$$

The first condition in (2.14) indicates a divergence-free dynamics associated with V_i. The second one indicates that V_i does not contribute to dissipation, so that F acts as a potential. For the general case (2.10) and for a given F, different dynamics leading to the same stationary distribution are obtained by considering different kinetic coefficients $\mathcal{L}_{ij}$ and different nonrelaxational contributions.

2.2. TDGL MODELS AND PATTERN FORMATION

We address here the question of the origin of transient pattern formation in the context of the TDGL models. To be specific we consider models with a scalar order parameter. It is instructive to compare models A and B above with a version of the Swift-Hohenberg[3] model often used in studies of formation of stationary patterns. This model can be written in the TDGL form (2.4) with L=M and a free energy

$$F_{SH} = (1/2) \int dr \left[-r\psi^2 + ((\nabla^2 + q_0^2)\psi)^2\right] + (u/4) \int dr \ \psi^4 \quad . \tag{2.15}$$

Models A and B have the same static properties but different dynamics due to the different kinetic coefficient. On the other hand a comparison of model A with the SH model shows different statics due to a different free energy. Different dynamics are also due just to the difference in free energy. These simple facts identify the existence of a stationary pattern in the SH model absent in models A and B: The free energy F is minimized by configurations of zero wavenumber, while F_{SH} selects $|q|=q_0$. Common wisdom in the analysis of instabilities leading to pattern formation is to consider the linearized form of the equation around the homogeneous initial state in a Fourier-space representation,

$$\partial_t \psi_q(t) = \omega(\mathbf{q}, r) \psi_q(t) + \xi_q(t) \tag{2.16}$$

where r is the control parameter of the instability.

The usual statement, in the case considered here of no complex eigenvalues, is that a pattern appears when ω(q,r) becomes first zero at the instability (r=0) at a wavenumber $q_0 \neq 0$. This is the case of the SH model (see Fig.2). For models A and B, ω(q=0,r=0) = 0. In spite of this, transient patterns do occur in models A and B.

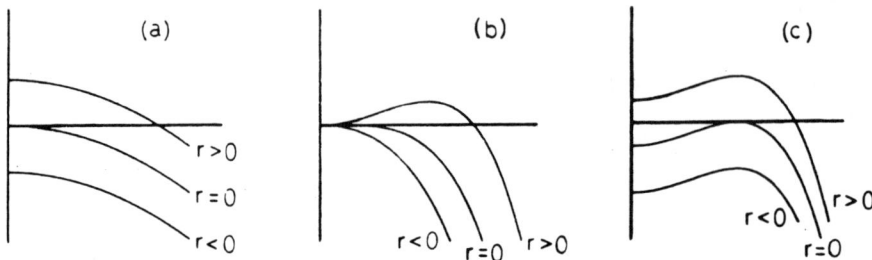

FIG. 2. Linear factor ω(q,r) vs. q for different values of r; (a), (b) and (c) correspond respectively to models A, B and SH. For models A and B: M=k=1; for model SH: M=1 and q_0=1.

The difference between the models A and B is that beyond the instability point (r>0), ω(q,r) has its maximun at **q**=0 for model A but it develops a peak at **q**≠ 0 for model B. This has important consequences when studying transient evolution following a change of r from an initial r_i<0 to a final value r_f>0. For model B a dynamical mechanism of selection of the wavenumber of fastest growth exists, while in model A the emergence of the pattern is due to the equilibrium fluctuations in the initial state at r=r_i. In any case transient patterns can occur in both models.

2.3. VALIDITY OF LINEAR THEORIES

Following the discussion above, the natural question arises of the validity of linear analysis. We are here concerned with transient evolution and since the instability starts linearly, the meaningful question is the time interval during which a linear approximation remains valid. This question is easily answered for the zero dimensional version (no spatial dependence) of the model defined by (2.4), (2.8) and (2.9). One then considers the relaxation of a variable ψ from the unstable state ψ=0 in a double well potential. This relaxation is triggered by noise. The linear approximation given by

$$\partial_t \psi = r\psi + \xi(t) \quad , \quad r > 0 \quad , \quad \langle \xi(t)\xi(t') \rangle = 2\varepsilon\delta(t-t') \quad , \tag{2.17}$$

consists in approximating the double well potential by an inverted parabola centered at ψ=0. This is obviously correct while ψ remains close to ψ=0. The solution of (2.17) can be written as,

$$\psi^2(t) = h^2(t) \exp(2rt) \quad , \qquad (2.18)$$

where h(t) is a stochastic Gaussian process,

$$h(t) = \int_0^t ds \exp(-rs) \, \xi(s) \quad , \qquad (2.19)$$

playing the role of a random effective initial condition. We are interested in the time at which ψ reaches a prescribed value ψ_0 outside the vecinity of $\psi=0$. This time is a random quantity whose average, known as the Mean First Passage Time (MFPT), sets the time domain of validity of the linear theory. The relation (2.18) can be inverted for times of interest rt>>1. In this limit h(t) becomes a time independent random Gaussian variable h(∞) so that

$$t = (1/2r) \ln\left[\psi_0^2 / h^2(\infty)\right] \quad . \qquad (2.20)$$

The statistical properties of t are then determined by those of h(∞). The MFPT T≡< t > is weakly dependent on ψ_0. In the limit of ε<<1 one obtains

$$T \simeq [1/2\omega(q,r)] \ln \varepsilon^{-1} \quad . \qquad (2.21)$$

This result justifies the assumption rt>>1. It gives a time scale determined by the noise intensity and practically independent of ψ_0. The inclusion of nonlinear terms in the calculation of T does not modify these conclusions[25]. A similar analysis[25] including space dependence gives the analogous result,

$$T \simeq [1/2\omega(q,r)] \ln \varepsilon^{-1} \quad , \qquad (2.22)$$

where $\omega(q,r)$ is the coefficient in (2.16) and ε measures the intensity of the noise ξ_q (t).

It is of interest to address this question in the specific context of model B defined above. The explicit equation of motion is

$$\partial_t \psi(r,t) = M\nabla^2 (-k\nabla^2\psi - r\psi + u\psi^3) + \xi (r,t) \quad . \qquad (2.23)$$

The equation can be reduced to a single parameter problem, chosen as the noise intensity, by the following rescaling x=(r/k)r, τ=(2Mr²/k)t, ϕ=(u/r)$^{1/2}\psi$:

$$\partial_\tau \phi(x,\tau) = (1/2) \, \nabla^2 (-\nabla^2\phi - \phi + \phi^3) + \varepsilon^{1/2}\eta (x,t) \quad , \qquad (2.24)$$

$$\langle \eta(x,\tau) \eta(x',\tau') \rangle = -\nabla^2\delta(x - x') \, \delta(\tau - \tau') \quad , \qquad (2.25)$$

where

$$\varepsilon = (k_B T / r^2) \, u \, (r/k)^{d/2} \quad . \tag{2.26}$$

Writing ε in terms of mean field quantities, it is possible[26] to identify it with the Ginzburg parameter such that for $\varepsilon \ll 1$ a mean-field description of equilibrium critical properties is accurate. In particular ε is proportional to ζ_0^{-d} where ζ_0 is the range of interaction and d the spatial dimensionality. A perturbative expansion [26] in ε can be understood alternatively, as a low temperature, weak coupling or small noise expansion. The lowest order approximation in such expansion is the linear theory obtained neglecting the ϕ^3 term in (2.24). The linear theory makes then sense for small ε, in particular for systems with long range interactions[26,27]. The meaningful question is the time domain of validity of the theory. This is fixed by ε through the expression (2.22) for the MFPT. The linear theory in this context is known as the Cahn-Hilliard-Cook theory. The associated equation for the structure factor (2.1) is

$$\partial_\tau S(\mathbf{q},\tau) = \omega(q, r=1) \, S(\mathbf{q},\tau) + \varepsilon q^2 \quad , \tag{2.27}$$

$$\omega(q, r=1) = -q^2 (q^2 - q_c^2) \quad , \quad q_c^2 \equiv 1 \quad . \tag{2.28}$$

This identifies the instability discussed in Fig. 2b with a range of unstable modes $0 < q^2 < 1$ and with a fastest growing mode at $q = 2^{-1/2}$. Whenever there is a time scale in which the linear theory is of experimental relevance ($\varepsilon \ll 1$) the selection of a wave number in the emerging pattern can be monitored by looking at the maximum q_m of the structure factor solution of (2.27) as a function of time. The value of q_m is not the constant mode of fastest growth due to the noise term in (2.27) and also because of the form of the initial condition $S(q,0)$. The initial condition is chosen as determined by the equilibrium fluctuations at the initial state. Typically $q_m(\tau)$ sets off from zero at a well defined onset time and approaches asymptotically $2^{-1/2}$, being close to that value at a time τ of the order of the MFPT. Beyond this time nonlinear saturating effects come into play.

2.4. LATE STAGES: DEFECT DYNAMICS

During the transient dynamics associated with ordering kinetics, topological defects like domain walls, vortices, dislocations etc. are created. Such defects are long lived and dominate the process of pattern decay. The variables that describe the slowly varying deformations of a quasistationary evolving pattern are called phase variables. The importance of the possibility of a reduced description of the late stage dynamics in terms of defect and phase variables has been emphasized[19,20]. This possibility is based on the existence of a small parameter given by the ratio of the defect core size to the domain size. The name of pattern dynamics has been suggested by Kawasaki[19] for the set of coupled equations of motion for topological defects and phase variables.

An example of this sort of description is given[19] by the evolution equation for a domain wall in model A. The domain wall is the planar interface solution of (2.4) with (2.8)-(2.9):

$$M(z-z_0) = (r/u)^{1/2} \tanh\left[(r/2k)^{1/2}(z-z_0)\right] \quad , \tag{2.29}$$

where $\mathbf{r}=(\rho,z)$ and z is perpendicular to the interface at each point. A collective variable $f(\rho)$ which gives the local position of the interface is introduced by

$$\psi(\mathbf{r}) = M(z - f(\rho)) + \varphi(\mathbf{r}) \quad . \tag{2.30}$$

The free energy associated with the interface can be calculated, in the low temperature limit, by integrating over the variables $\varphi(\mathbf{r})$

$$F[f] = \sigma \int d\rho \left[1 + (\nabla_\rho f(\rho))^2\right]^{1/2} \quad , \tag{2.31}$$

where σ is the surface tension $\sigma = \int (\partial M/\partial z)^2 \, dz$. Equation (2.31) has the clear interpretation of (surface tension)x(area). The equation of motion for the interface can be written in terms of (2.31) by means of a reduction process from (2.4). The equation takes also the form of a relaxational TDGL model,

$$\partial_t f = -L \left(\delta F[f]/\delta f\right) + \xi \quad , \tag{2.32}$$

where $L \equiv \alpha(M/\sigma) = [1+(\nabla_\rho f)^2]^{1/2}(M/\sigma)$, and the random noise ξ satisfies the appropriate fluctuation-dissipation relation with L. The driving force is here the curvature of the interface, since the mean curvature K defined as $K = -\nabla \cdot \mathbf{n}$, being $\mathbf{n}$ the unit vector perpendicular to the interface is given by $K = -\sigma^{-1}(\delta F/\delta f)$. For the evolution of a configuration with randomly placed walls one can introduce a phase variable $u(\mathbf{r},t)$ defined by $u(\mathbf{r},t) = z - f(\rho,t)$. Under some simplifying assumptions the phase variable is seen to obey a diffusion equation from which a scaling law of the form (2.3) with m=1/2 is predicted[28]. For model B the evolution equation for domain walls is nonlocal.

A more complicated example is provided by the ordering dynamics of a Z_N-clock model when quenched to a low temperature from an initally disordered phase. In this model there are N equivalent low temperature ordered phases. Monte Carlo simulations[29] of this model show the formation of a complicated transient pattern in which late stage evolution is governed by interface and vortex motion. The interfaces occur between domains of the N-equivalent phases and vortices appear at the intersection of several interfaces. Kawasaki has also reduced in this case the dynamical description to equations for the vortices and phase variables[30]. Under some approximations, the equations can be further reduced in d=2 to a relaxational TDGL model for the motion of the vortices. The appropriate free energy turns out to be here related to the Hamiltonian of a two-dimensional Coulomb gas.

A case of special interest for pattern formation in nematics is the description of the evolution of domain walls (kinks and antikinks) in one dimensional systems. This has been studied at legth in connection with d=1 models of spinodal decomposition. A possible approach, as proposed by Langer[31], consists in analyzing the stability properties of particular stationary solutions of the deterministic version of (2.4). Square-wave like solutions, representing a certain mode of phase separation, are indeed found to be unstable with respect to eigenmodes of defined periodicity, which depends on the particular model, A or B, considered. The rate of pattern decay is then essentially extracted from this linear stability analysis. A more recent treatment has been suggested by Kawasaki at al.[20]. It amounts to reduce the original field equations to convenient equations for the wall dynamics. From this reduced description, domain growth can be conveniently studied. We postpone a detailed discussion of both schemes until Sect.5, where both will be used in relation with the pattern dynamics resulting from the Fréedericksz transition.

3. Equations of Stochastic Nematodynamics

An interesting aspect of nematic liquid crystals is the coupled dynamics of the director field **n(r)**, which identifies a preferred direction of alignment, with the velocity flow **v(r)**. A number of instabilities leading to pattern formation are known in isotropic fluids. The inclusion of a new coupled variable as the director field **n(r)** opens a richness of additional possibilities[32]. We indicate here how this coupled dynamics can be written in the form of the TDGL models discussed before[8].

The free energy associated with distortions of the director field is the Oseen-Frank free energy[33]:

$$F_{0F}[\mathbf{n}(\mathbf{r})] = (1/2) \int d\mathbf{r} \left\{ K_1 (\nabla \cdot \mathbf{n})^2 + K_2 [\mathbf{n} \cdot (\nabla \times \mathbf{n})]^2 + K_3 [\mathbf{n} \times (\nabla \times \mathbf{n})]^2 \right\} \quad , \quad (3.1)$$

where K_1, K_2 and K_3 are elastic constants associated with splay, twist and bend deformations, respectively. In the presence of a magnetic field **H** and including the hydrodynamic contributions the total free energy for the problem is

$$F = (1/2) \int d\mathbf{r} \left[K_{\alpha\beta\gamma\delta} \partial_\beta n_\alpha \partial_\delta n_\gamma - \chi_a (n_\alpha H_\alpha)^2 + \rho v^2 - 2 p(\mathbf{r}) \partial_\alpha u_\alpha \right] \quad . \quad (3.2)$$

The first term in (3.2) is a rewriting of (3.1). The second term gives the magnetic contribution, being χ_a the anisotropic part of the magnetic susceptibility. The third term gives the hydrodynamic contribution, being ρ the mean density. The last term introduces the presure $p(\mathbf{r})$ as a Lagrange multiplier for the incompressibility condition and the field $u_\alpha(\mathbf{r})$ stands for the position of a flow element.

The equations of nematodynamics are usually derived by irreversible thermodynamics arguments[11]. The dissipation -dF/dt is calculated considering the changes in F for independent variations of **n**, **v** and **u**. The dissipation can then be interpreted as a product of fluxes and forces. Fluxes are written as a combination of forces. This is done going beyond the linear regime, as in the relaxational terms of the TDGL models discussed previously. This means that forces are derivatives of a free energy which is not quadratic and that the coefficients of the combinations are not constant. In this way the nonlinear equations of nematodynamics are obtained[11]. An important point is to recognize[8] that these equations can be rewritten in the general form of the deterministic part of (2.10). When this is done thermal fluctuations leading to a stationary distribution $\exp(-\beta F)$ are introduced in a trivial way in the fully nonlinear equations.

The full set of stochastic nematodynamic equations can be written as[8]:

$$d_t n_\beta = V_{n\beta} - (1/\gamma_1)(\delta F/\delta n_\beta) + \xi_\beta(\mathbf{r},t) \quad , \tag{3.3}$$

$$d_t v_\beta = V_{v\beta} + L_{\beta\gamma}(\mathbf{n})(\delta F/\delta v_\gamma) + \partial_\alpha \Omega_{\alpha\beta}(\mathbf{r},t) \quad , \tag{3.4}$$

$$d_t u_\beta = V_{u\beta} \quad , \tag{3.5}$$

where d_t includes the convective term. The relaxational term of (3.3) gives the equation for $\mathbf{n}$ in the absence of hydrodynamic coupling. γ_1 is a viscosity coefficient and $h_\alpha = -(\delta F/\delta n_\alpha)$ is the molecular field. The noise sources $\xi_\beta(\mathbf{r},t)$ satisfy the fluctuation dissipation-relations

$$\langle \xi_\beta(\mathbf{r},t) \xi_\gamma(\mathbf{r}',t') \rangle = 2(k_B T/\gamma_1) \delta(\mathbf{r}-\mathbf{r}') \delta(t-t') \delta_{\beta\gamma} \quad , \tag{3.6}$$

The nonrelaxational part of (3.3) gives the coupling of de director with the velocity flow:

$$V_{n\beta} = [(\lambda-1)/2] n_\alpha \partial_\beta v_\alpha + [(\lambda+1)/2] n_\alpha \partial_\alpha v_\beta \quad . \tag{3.7}$$

The relaxational part of (3.4) is explicitly written as:

$$L_{\beta\gamma}(\mathbf{n})(\delta F/\delta v_\gamma) = \rho^{-1} \partial_\alpha \left[\sigma^s_{\alpha\beta} + (\lambda/2)(n_\alpha h_\beta + n_\beta h_\alpha) \right] \quad , \tag{3.8}$$

whith a kinetic coefficient,

$$L_{\beta\gamma}(\mathbf{n}) = \partial_\alpha M_{\alpha\beta\delta\gamma}(\mathbf{n}) \partial_\delta \quad , \tag{3.9}$$

$$M_{\alpha\beta\gamma\delta} = \rho^{-2} [\nu_2(\delta_{\beta\delta}\delta_{\alpha\gamma} + \delta_{\alpha\delta}\delta_{\beta\gamma}) + 2(\nu_1 + \nu_2 - 2\nu_3)n_\alpha n_\beta n_\gamma n_\delta +$$

$$+ (\nu_3 - \nu_2)(n_\alpha n_\gamma \delta_{\delta\beta} + n_\alpha n_\delta \delta_{\gamma\beta} + n_\beta n_\gamma \delta_{\delta\alpha} + n_\beta \delta_{\gamma\alpha})] \quad , \tag{3.10}$$

where the standard notation[11] is used for viscosity coefficients $\nu_1, \nu_2, \nu_3, \lambda_1, \lambda_2$, and $\lambda = -\gamma_1/\gamma_2$, and $\sigma^s_{\alpha\beta}$ is the symmetric part of $\sigma_{\alpha\beta} - \sigma^E_{\alpha\beta}$, where $\sigma_{\alpha\beta}$ is the total stress-tensor and σ_E the Ericksen stress tensor[11]. The noise sources satisfy the fluctuation-dissipation relations

$$\langle \partial_\alpha \Omega_{\alpha\beta}(r,t) [\partial_\delta \Omega_{\delta\gamma}(r',t')]^+ \rangle = -2k_B T L_{\beta\gamma} \delta(r-r') \delta(t-t') \quad . \tag{3.11}$$

We note that the part of (3.8) that does not couple to the director (first term in (3.10)) reproduces the Navier-Stokes equation except for the pressure term contained in $V_{v\beta}$. However, our stochastic treatment goes beyond the usual linearized fluctuating hydrodynamics since fluctuations are taken into account in the full nonlinear equations. The nonrelaxational part of (3.4) can be written as

$$V_{v\beta} = \rho^{-1} \partial_\alpha \left[\sigma^a_{\alpha\beta} - (\lambda/2)(n_\alpha h_\beta + n_\beta h_\alpha) \right] + \rho^{-1} \partial_\alpha \sigma^E_{\alpha\beta} \quad , \tag{3.12}$$

where $\sigma^a_{\alpha\beta}$ is the antisymmetric part of $\sigma_{\alpha\beta} - \sigma^E_{\alpha\beta}$. The quantity $V_{v\beta}$ gives and additional coupling of director and velocity. Finally (3.5) is purely nonrelaxational with $V_{u\beta}=v_\beta=\rho^{-1}$ ($\delta F/\delta v_\beta$). The fulfillment of (2.14) requires taking into account the trivial equation (3.5) to have a closed set of equations. Conditions (2.14) are easily checked when recognizing that the nonrelaxational terms in (3.3) and (3.4) can also be obtained from the free energy F and a kinetic coefficient Γ. Noting that $\partial_\alpha \sigma^E_{\alpha\beta} = -(\delta F/\delta u_\beta)$ we have

$$V_{n\beta} = \Gamma_{\beta\gamma}(n) \left(\delta F/\delta v_\gamma \right) \quad , \tag{3.13}$$

$$V_{v\beta} = -\Gamma^+_{\beta\gamma}(n) \left(\delta F/\delta n_\gamma \right) - \rho^{-1} \left(\delta F/\delta u_\beta \right) \quad , \tag{3.14}$$

where,

$$\Gamma_{\beta\gamma}(n) = (1/2\rho) \left[(\lambda+1) n_\alpha \partial_\alpha \delta_{\beta\gamma} + (\lambda-1) n_\alpha \partial_\beta \delta_{\alpha\gamma} \right] \quad . \tag{3.15}$$

Given these expressions for the nonrelaxational part, the set of the stochastic nematodynamic equations admit still a more compact form than (3.3)-(3.5). Introducing a field $\psi(r)=[n(r),v(r),u(r)]$ we have

$$d_t \psi_i = A_{ij}(\psi) \left(\delta F / \delta \psi_j \right) + \eta_i \quad ; \quad i = 1....9 \quad , \tag{3.16}$$

where $\eta \equiv (\xi, \partial_\alpha \Omega_{\alpha\beta}, 0)$ and the operator A has a dissipative A^D and a non dissipative part A^R:

$$A^D_{ij} = \begin{bmatrix} -\gamma_i^{-1} I & 0 & 0 \\ 0 & L & 0 \\ 0 & 0 & 0 \end{bmatrix} \quad ; \quad (A^D_{ij})^+ = A^D_{ij} \quad , \tag{3.17}$$

$$A^R_{ij} = \begin{bmatrix} 0 & \Gamma & 0 \\ -\Gamma^+ & 0 & -\rho^{-1}I \\ 0 & \rho^{-1}I & 0 \end{bmatrix} \quad ; \quad (A^R_{ij})^+ = -A^R_{ij} \quad . \tag{3.18}$$

Equation (3.16) is of the form (2.10) with relaxational dynamics associated with a self-adjoint operator A^D_{ij}, and nonrelaxational dynamics associated with an antiadjoint operator A^R_{ij} in the form $V_i = A^R_{ij}(\delta F/\delta \psi_j)$. This form of V_i makes trivial the fulfillment of the second condition in (2.14). The noise sources η_i satisfy fluctuation-dissipation relations with A^D_{ij}. Eq (3.16) gives a clear TDGL formulation of the complete set of nonlinear nematodynamic equations with a consistent introduction of thermal noise sources. It gives a useful starting point to study dynamical questions in these systems.

4. Pattern Formation in the Magnetically Induced Fréedericksz Transition

The magnetically induced Fréedericksz transition is essentially an instability of the state of zero distortion of a nematic liquid crystal caused by an externally homogeneous applied magnetic field, whose intensity exceeds a critical threshold. Suppose that the nematic material has been prepared with all its molecules aligned, in average, parallel to a preferred direction n^0 imposed at the boundaries. This configuration is maintained by elastic forces acting inside the material. If we apply a magnetic field H perpendicular to n^0, the anisotropic diamagnetism of the material originates a coupling that, for example for materials with $\chi_a > 0$, tend to align the molecules parallel to H. A competition is then established between elastic and magnetic torques. If the effect of the magnetic field prevails, $H > H_c$, we obtain curved deformations in the interior of the sample.

The simplest geometry we can envisage corresponds to a distortion involving essentially a twist mode. In this situation, represented in Fig.3, the sample is contained between two plates perpendicular to the z axis. The director is initially aligned along the x axis, and the magnetic field is applied along the y axis.

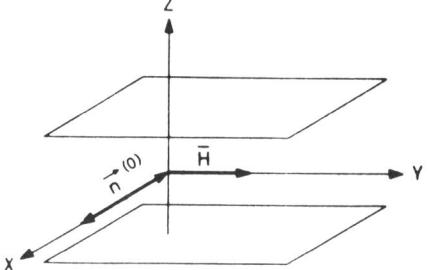

FIG. 3. Schematic representation of the geometry of the nematic sample.

The transient behavior we will describe corresponds to the switch at t=0 from an initial value $H_i < H_c$ to a final value $H > H_c$. In principle, one should expect that the simplest realization of this instability would result in a twist deformation along the z axis, being homogeneous in any plane of the sample. However, it has been experimentally shown[34] that, for the twist geometry here considered, the system may transitorily develop a more complicated structure involving bend modes along the x direction. This results in the appearance of a striped pattern of well defined periodicity, perpendicular to the initial orientation. Actually, a perpendicular striped texture was already reported for a planar geometry in the earlier work of Guyon at al.[35]. Oblique structures have also been observed[36]. Transient periodic patterns appear both for thermotropics[34-36] and lyotropics[34,36-38]. Due to the additional elastic energies involved in this reorientational mode, the texture finally decays at long times. A photomicrograph showing a series of parallel brightening stripes as observed in the experiments reported in Ref.37 is presented in Fig.4.

FIG. 4. Photomicrograph of uniform periodic structure. Field strength 7.4kG; temperature 25°C; sample thickness 50μm; spacing between stripes ≈48μm. (Reprinted from Y.W. Hui et al., Ref. 37).

The suggested explanation for this transient phenomenon involves a dynamical coupling between the director field and the velocity flows. This may give rise to a periodic arrangement of domains corresponding to opposite but equivalent reorientations. The term backflow effects has been coined to refer to this coupling. Although for the twist geometry, to what we essentially restrict here, it is well-known[11] that backflow effects do not exist at zero wavenumbers q_x, this coupling may appear at nonzero values of q_x as it is discussed below. In this section we describe the early dynamical stages during which the pattern appears and develops[8]. The dynamics is then analyzed in terms of the set of stochastic nematodynamic equations of Sect.3. Our aim is to adress typical dynamical questions as those related with the different time scales involved in this initial reorientational process, and more specifically to evaluate the onset time of appearance of the pattern and the domain of validity of a linear theory. Late dynamical stages corresponding to the decay of the structure are basically dominated by domain wall dynamics, and will be considered in Sect.5.

4.1. MINIMAL COUPLING EQUATIONS AND EFFECTIVE VISCOSITY

Let us consider the geometry of Fig.3. We assume that velocity flows only exist in the y direction. We also assume homogeneity along that direction and we make the hypothesis that the director reorientates planarly:

$$n_x(x,z) = \cos\phi(x,z)$$
$$n_y(x,z) = \sin\phi(x,z)$$
$$n_z = 0 \tag{4.1}$$

Even with these simplifications the set of nematodynamic equations (3.3)-(3.5) remain very complicated. We propose a minimal coupling approximation in which the $\mathbf{n}$ dependence of the Γ and L operators in (3.9) and (3.15) is replaced by $\mathbf{n}^0$. This procedure retains the initial coupling between $\mathbf{v}$ and $\mathbf{n}$. With this assumption the appropriate closed set of equations for ϕ and v_y are

$$d_t \begin{pmatrix} \phi \\ v_y \end{pmatrix} = \begin{pmatrix} -1/\gamma_1 & (1/2\rho)(1+\lambda)\partial_x \\ (1/2\rho)(1+\lambda)\partial_x & (1/\rho^2)(v_2\partial_z^2 + v_3\partial_x^2) \end{pmatrix} \begin{pmatrix} \delta F/\delta\phi \\ \delta F/\delta v_y \end{pmatrix} + \begin{pmatrix} \xi \\ \partial_x\Omega_{yx} + \partial_z\Omega_{yz} \end{pmatrix} \tag{4.2}$$

where d_t is the total derivative including convective terms. The usual small-amplitude expansion of the free energy leads to

$$\delta F/\delta\phi = -\left[K_2\partial_z^2\phi + K_3\partial_x^2\phi + \chi_a H^2(\phi - 2\phi^3/3) \right], \tag{4.3}$$

and the fluctuation-dissipation relations (3.6), (3.11) reduce to

$$\langle \xi(x,z,t)\,\xi(x',z',t') \rangle = 2\,(k_B T/\gamma_1 L_y)\,\delta(x-x')\,\delta(z-z')\,\delta(t-t')$$
$$\langle \Omega_{y\alpha}(x,z,t)\,\Omega_{y\beta}(x',z',t') \rangle = 2\,(k_B T/\rho^2 L_y)\,v_\alpha\,\delta_{\alpha\beta}\,\delta(x-x')\,\delta(z-z')\,\delta(t-t') \quad ; \quad \alpha,\beta=\{x,z\}, \tag{4.4}$$

with $v_x \equiv v_3$, $v_z \equiv v_2$, and L_y is the y-linear dimension of the sample. Equation (4.2) represents a particular and very clear example of utilization of the general scheme of Eq. (3.16). Diagonal and nondiagonal terms in (4.2) are respectively associated with the relaxational and nonrelaxational dynamics.

To proceed further we make the usual approximation of neglecting inertia terms assuming that the director field is the slow variable as compared to the velocity field. This permits us to obtain a closed equation for the deformation angle ϕ. This equation is more easily handled in a Fourier representation adapted to the strong anchoring boundary conditions at $z = \pm d/2$,

$$\phi(x,z,t) = \sum_m \sum_{q_x} \theta_{m,q_x}(t)\, \cos[(2m+1)\pi z/d]\, \exp(iq_x x), \tag{4.5}$$

and reads

$$(d/dt)\,\theta_{m,q_x}(t) = (1/\bar\gamma_1)\left\{ \left[\chi_a H^2 - K_2(2m+1)^2(\pi^2/d^2) - K_3 q_x^2\right]\theta_{m,q_x} - (2/3)\chi_a H^2 \times \right.$$

$$\times \sum_{q_{x_1},q_{x_2}} \sum_{n,l,p} \theta_{n,q_{x_1}} \theta_{l,q_{x_2}} \theta_{p,q_x-(q_{x_1}+q_{x_2})} \times (2/d) \int_{-d/2}^{d/2} dz \prod_{i=m,n,l,p} \cos[(2i+1)\pi z/d] \Bigg\} + \eta_{m,q_x}(t)$$

(4.6)

Notice that by eliminating the velocity we have converted a general nonrelaxational dynamics into a purely relaxational one expressed in terms of an effective q_x dependent viscosity given by

$$\bar{\gamma}_1 = \gamma_1 \left[1 - \frac{\gamma_1(1+\lambda)^2/4}{v_3 + v_2 Q^{-2} + \gamma_1(1+\lambda)^2/4} \right] \quad, \quad Q^2 \equiv \frac{q_x^2 d^2}{(2m+1)^2 \pi^2} \quad . \quad (4.7)$$

The consistency of our analysis is further checked since the thermal noise term $\eta_{m,q_x}(t)$ satisfies a fluctuation-dissipation relation in terms of $\bar{\gamma}_1$:

$$\langle \eta_{m,q_x}(t) \, \eta^*_{m,q_x'}(t') \rangle = 2 \left(k_B T / \bar{\gamma}_1 V \right) \delta_{m,n} \, \delta_{q_x,q_x'} \, \delta(t-t') \quad , \quad (4.8)$$

being V the volume of the sample.

4.2. STABILITY ANALYSIS

A simple linear analysis applied to (4.6) enables us to predict the apperance of transient patterns. Actually, unstable modes are those for which the linear amplification factor ω in (4.6), conveniently rewritten in terms of the reduced magnetic intensity :

$$h^2(m) = H^2/[(2m+1)^2 H_c^2] \quad , \quad H_c^2 \equiv K_2 \pi^2 / \chi_a d^2 \quad ,$$

$$\omega(Q^2,m,h^2(m)) = \left[K_2(2m+1)^2 \pi^2 / \bar{\gamma}_1 d^2 \right] \left[h^2(m) - 1 - K_3 Q^2/K_2 \right] \quad , \quad (4.9)$$

is positive.

For typical nematic materials $\bar{\gamma}_1$ is always positive with a maximum value, $\bar{\gamma}_1 = \gamma_1$ at $Q^2 = 0$, and a minimum $\bar{\gamma}_1 = 0.26\gamma_1$ for MBBA[11], as $Q^2 \to \infty$. This implies that the stability range does not change with respect to the situation in absence of hydrodynamical coupling, when (4.6) is still valid but with $\bar{\gamma}_1$ replaced by the pure reorientational viscosity γ_1. However due to the Q^2 dependence of $\bar{\gamma}_1$, $\omega(Q^2,m,h^2(m))$ is no longer a monotonically decreasing function of Q^2, as it is shown in Fig.5 for the most unstable twist mode m=0. The appearance of a maximum at $Q^2 \neq 0$ is here the direct signature of the pattern development. With respect to models A and B introduced in Sect.2, see Fig. 2, notice that the situation here corresponds to a somewhat intermediate case, since a maximum at $Q^2 \neq 0$ is only attained for h^2-1 finite ($h^2=1$ corresponds to the instability threshold). More precisely[8], $h^2 > 2.35$ for MBBA.

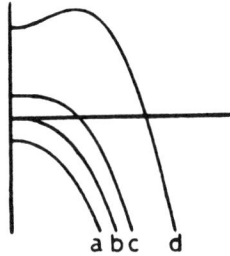

FIG. 5. Linear factor $\omega(Q^2,0,h^2(0))$ vs. q for different values of $h^2(0)$. (a), (b), (c) and (d) correspont respectively to $h^2(0)=0,1,2,5$.

4.3. STRUCTURE FACTOR DYNAMICS

Obviously, to describe the dynamics accompanying the reorientation of the director one needs to go beyond the linear analysis just presented and usually considered in the literature[34-38]. This is here done by referring to the structure factor $S(m,Q^2,t)$ introduced in Sect. 2. Focusing on the most unstable mode m=0, an equation for $S(Q^2,t) \equiv S(0,Q^2,t)$ is easily derived from (4.6)

$$(d/ds) S(Q^2,s) = (2\gamma_1/\bar{\gamma}_1)\left\{\left[h^2 - 1 - (K_3Q^2/K_2) - (3/2) h^2\left(\sum_k S(k^2,s)\right)\right]S(Q^2,s) + \varepsilon\right\}$$

(4.10)

where ε measures the strength of thermal fluctuations,

$$\varepsilon = 2k_BT / \chi_a H_c^2 V \quad , \tag{4.11}$$

and a dimensionless time variable s has been introduced according to

$$s = \tau_0^{-1}t \quad , \quad \tau_0 \equiv \gamma_1 / \chi_a H_c^2 \quad . \tag{4.12}$$

Actually, (4.12) has been written invoking a Gaussian decoupling[5,8,39] which certainly gives a more accurate description of the early stages of the relaxation from the unstable state than a simply linear theory, although it is probably unable to satisfactorily describe the late dynamics during which domain walls should play a decisive role.

Here the pattern dynamics is described in terms of the temporal evolution of the value of Q^2, noted Q^2_{max}, corresponding to the absolute maximum of $S(Q^2,s)$. A typical plot of this evolution is depicted in Fig. 6, where the linear dynamics appropiate to (4.10) is also shown for çomparison. This figure shows a first stage of evolution ending with the emergence of the periodicity at $s \approx 0.4$. This would correspond to the onset time of the pattern formation. A second stage up to $s \approx 2$, during wich linear and nonlinear descriptions agree fairly well, corresponds to the development of the pattern. The late stage decay to $Q_{max}=0$ is a typical feature of the Gaussian approximation. The validity of a linear theory extends here over a time range of experimental relevance (for typical values of $\tau_0 \approx 10$ sec,

$s\approx 2.0$ corresponds to 20 sec), and agrees with a crude estimation in terms of the MFPT (see Sect. 2.3, Eq. (2.22) with ω given in (4.9)): $T\approx 2.5$ in dimensionless units.

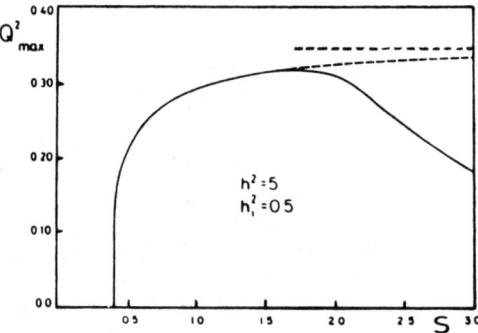

FIG. 6. Time dependence of the wavenumber correponding to the maximum of $S(Q^2,s)$. Results of the linear theory are presented in the dashed line. The assymptotic value obtained as the mode of fastest growth in a deterministic linear stability analysis is also depicted. Parameter values are those of MBBA (Ref. 11).

Finally, we note from (4.10) that in absence of hydrodynamic coupling ($\bar{\gamma}_1/\gamma_1 = 1$, $\forall Q^2$), $S(Q^2,s)$ is always peaked at the origin, so that $Q^2_{max}(s)=0$[39], although the structure factor would exhibit an amplification effect as a common expected result of the standard instability.

4.4. EXPERIMENTAL SITUATION

There are a number of experiments on the dynamics of the transient pattern formation in the twist geometry of the Fréedericksz transition. We mention here two recent representative ones[40,41]. The one in Ref. 40 is for a lyotropic sample ($\chi_a<0$). Both systems have similar time scales and critical fields with τ_o of the order 10^3 sec and H_c of the order of the kG. The work reported in Ref. 40 indicates the existence of an onset time between 170 sec and 280 sec and a clear stage of pattern development up to 400 sec in which a time independent wavevector is selected. For longer times the average dominant wavevector decreases by domain wall motion and recombination. Mc Clymer et al[40] conclude that experimental restrictions do not give enough data to show quantitative agreement with theory and suggest the study of late time dynamics of inversion walls. However their picture of time scales involved is in good qualitative agreement with the theory discussed above. In particular it is clear that the onset time can not be identified with the nonstochastic time scale defined as the inverse of the growth rate of the fastest mode which is of the order of 40 sec.

The work of Srajer et al[41] aims to test the linear theory. These authors conclude that the selected wavenumber is not the one given by the linear theory. We have indicated that the dominant wavenumber is time dependent even within linear theory. Unfortunately no experimental results are given on the time scales involved and on the time at which their observations are made. Such results are given for their numerical solution of the nonlinear nematodynamic equations with random inital conditions. They obtain an onset time of the order of 10 sec, which is much smaller than the one in Ref. 40. The reason seems to be their choice of $\varepsilon \sim 10^{-4}$ which is several orders of magnitude larger than the one given by (4.11). This precludes a comparison with experiments or with a theory in which the initial time scales are dominated by ε and which is based on the smallness of ε. In any case, the numerical results indicate an observable and well defined time scale in which pattern is developed with a selected wavenumber in agreement with linear theory. For later times a nonlinear mode seems to dominate. Comparison of the power spectrum evolution with the time evolution of θ(x,t) indicate that when this happens, dynamics and recombination of walls is already dominant. Our understanding is that their experimental observations correspond to the late stages in which the pattern had been previously well formed and it is evolving as dominated by wall dynamics.

4.5 OTHER SITUATIONS.

An interesting situation[9] is one in which the magnetic field is periodically changed in time from a value below the critical field H_c to one above H_c with a mean value $H_o<H_c$. A first interesting question concerns the location of the instability point. It turns out that for fixed amplitude and period of modulation the instability is shifted to a value of $H_o<H_c$. A second important result is the possibility of a dynamical stabilization of the otherwise transient spatial pattern. It is possible to find[9] time periodic solutions displaying pattern formation. These solutions are accompanied by very large orientational fluctuations. Also very important from the point of view of the dynamical description is that the periodic field gives a mechanism to delay the instability: When the period of modulation becomes of the order or smaller than the onset time calculated for a fixed magnetic field, the emergence of the pattern is predicted to be largely delayed[9]. This would be an interesting way of observing early stages of pattern formation and wavenumber selection during large time scales.

The splay Fréedericksz transition has also been considered in the context here analyzed. The interest in this case consists in the study of the dynamics of oblique structures which have been experimentally detected[36] starting with a planar geometry. Actually, situations of perpendicular as well as oblique field-induced stripes have been examined with explicit indications of the time scales needed for their appearance and posterior development[42].

5. Domain Wall Dynamics

As anticipated in Sect. 2.4, the late stages of the transition, during wich the transient pattern relaxes to a pure twist deformation, is likely to be dominated by the dynamics of topological defects like domain walls, disclination lines, etc. According to our description of the initial reorientation, we will here particularize our analysis[10] to the dynamics of a configuration of domains of opposite distorsion along the x axis, i.e., along the direction corresponding to the initial alignment of the director. These domains are separated by walls, commonly referred as Fréedericksz walls[11], that recombine by mutual interaction between them. Our description here will be essentially one dimensional, although we believe that a more realistic approach to the very late stages should be at least two dimensional to take into account the mechanism of nucleation of loops, resulting from the pincement of pairs of walls. This would be followed by a slow drift to a boundary along the perpendicular transverse direction y.

Let us first check the linear stability of a periodic configuration of such Fréedericksz walls. To this end we particularize (4.2) for the most unstable mode m=0. It is also convenient to introduce dimensionless space and time variables in terms of field-dependent characteristic length and time scales $\chi \equiv \zeta u$ and $t \equiv \tau s$

$$\zeta^2 \equiv 2\left(K_3 d^2 / K_2 \pi^2\right)\left(h^2 - 1\right)^{-1} \quad , \quad \tau \equiv \left(\gamma_1 d^2 / K_2 \pi^2\right)\left(h^2 - 1\right)^{-1} \quad . \quad (5.1)$$

As discussed in Sect. 4.1, by eliminating the velocity field, one is led to consider a closed purely relaxational dynamics for the distorsional angle. Instead of expressing it in a completely Fourier decomposed form, as corresponds to Eq. (4.6), we prefer here to keep implicit the dependence of the angular variable on the transverse x direction. The equation then reads

$$\partial_s \theta(u,s) = L(\partial_u) \left\{ (1/2)\partial_u^2 \theta + \theta - \left[2(1 - h^{-2})\right]^{-1} \theta^3 \right\} \quad , \quad (5.2)$$

where the kinetic operator $L(\partial_u)$ is expressed by

$$L(\partial_u) = 1 + (1/4)\,\gamma_1(1 + \lambda)^2 \left[v_3 \partial_u^2 - v_2\, 2K_3/K_2(h^2 - 1)\right]^{-1} \partial_u^2 \quad . \quad (5.3)$$

In what follows we will be essentially interested in the role of this specific kinetic operator in the domain wall dynamics. Thermal noise effects have thus been neglected since they are not essential. A Fréedericksz wall is nothing but a stationary solution of (5.2)

$$\theta_\omega(u) = \left[2(1 - h^{-2})\right]^{1/2} \tanh u \quad . \quad (5.4)$$

This solution corresponds to a profile connecting the two equivalent minima of the magneto-elastic free energy, $\pm[2(1 - h^{-2})]^{1/2}$, at $x=\pm\infty$. The width of the wall is given directly in terms of the unit length ζ wich manifestly diverges at the critical point $h^2=1$. To render the following analysis more generic we correspondingly scale the dynamical variables $\Psi^2 \equiv \theta^2/2(1-h^{-2})$, so that (5.2) converts into

$$\partial_s \psi(u,s) = L(\partial_u)\left[(1/2)\,\partial_u^2 \psi + V'(\psi)\right] \quad , \quad (5.5)$$

where the potential is given by $V(\Psi) = \Psi^2/2 + \Psi^4/4$.

It is readily checked that a solution like (5.4) is stable under a dynamics given by (5.2). However, as previously stated, a purely periodic configuration of Fréedericksz walls constitutes an unstable stationary solution of (5.5). To check this point let us write

$$\psi(u,s) - \psi_{st}(u) = \sum_n \chi_n(u) \exp(-\varepsilon_n s) \quad , \quad (5.6)$$

so that the linear problem is expressed as

$$L(\partial_u)\left[-(1/2)\partial_u^2 - 1 + 3\psi_{st}^2(u)\right]\chi_n(u) = \varepsilon_n \chi_n(u) \quad , \tag{5.7}$$

where for a particular arrangement of M walls with interspacing ρ we have

$$3\psi_{st}^2(u) - 1 = 2 - 3\sum_{j=0}^{M-1} \text{sech}^2(u - j\rho) \quad . \tag{5.8}$$

The strategy, as proposed by Langer[31], consits in transforming (5.7) into a variational problem in terms of a Scrödinger-like operator F and a set of conjugate eigenfunctions $\tilde{\chi}_n(u)$, defined respectively as:

$$F \equiv -(1/2)\partial_u^2 - 1 + 3\psi_{st}^2(u) \quad ,$$

$$L(\partial_u)\tilde{\chi}_n(u) = \chi_n(u) \quad . \tag{5.9}$$

Then, assuming periodic boundary conditions for both sets χ_n and $\tilde{\chi}_n(u)$, the variational problem is well posed and finally reduces to

$$\varepsilon_{n,\min} \leq (\chi_n, F\chi_n)/(\tilde{\chi}_n, \chi_n) \quad . \tag{5.10}$$

Selecting an appropriate set of orthogonal trial functions specified by wavenumbers q

$$\chi_q(u) = M^{-1/2} \sum_{j=0}^{M-1} \text{sech}^2(u - j\rho) \exp(iqj\rho) \quad , \quad q = (2\pi/M\rho) \times \text{integer} \quad , \tag{5.11}$$

and invoking a limit of large interspacing, $\rho \gg 1$, the final result reads

$$(\chi_q, F\chi_q) \approx -64(1 + \cos q\rho) \exp(-2\rho) \quad , \tag{5.12}$$

$$\Gamma \equiv (\tilde{\chi}_q, \chi_q) \approx 2\alpha\left[2\beta/(h^2 - 1)\right]^{1/2} + (4/3)(1 - \alpha) \quad , \tag{5.13}$$

in terms of dimensionless quantities related to the set of viscosity parameters appearing as a result of the retained hydrodynamic coupling,

$$\alpha \equiv \gamma_1(1 + \lambda)^2 / \left[4\nu_3 + \gamma_1(1 + \lambda)^2\right] \quad ,$$

$$\beta \equiv 4K_3\nu_2 / \left(K_2\left[4\nu_3 + \gamma_1(1 + \lambda)^2\right]\right) \quad . \tag{5.14}$$

According to (5.12) and (5.13), all the eigenvalues ε_q turn out to be negative, which is a direct signature of the instability of the configuration of domain walls. Note also that the most unstable mode, the one wich grows faster and leads the dynamics since we are linearizing around an unstable state, corresponds to q=0. Certainly the mode q=π/ρ is also of interest since it reflects the traslational invariance of the pattern as a whole. The corresponding eigenvalue is ε_q=0 and would be associated with a uniform drift of all the Fréedericksz walls towards a boundary. Although in terms of purely elastic energetic considerations it will be certainly preferred with respect to the mode q=π/ρ, this argument fails here since in linearizing around an unstable configuration it turns out to be the slowest mode according to (5.12).

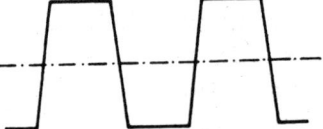

FIG. 7. Schematic representation of the relaxational mode of the array of Fréedericksz walls.

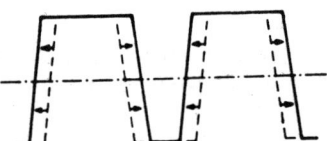

Associated to the most unstable eigenvalue, the corresponding eigenfunction $\chi_0(u) = M^{-1/2}\sum_{j=0}^{n-1} \text{sech}^2(u - j\rho)$ supposes a distorsion of the array of equally spaced interfaces by equal amounts over each wall. The picture we can extract of this relaxation corresponds to the dissolution of one type of domains inside the other one. This is schematically depicted in Fig. 7. Note also the difference with respect to model B where the dominant mode is q=±π/2ρ. As a generic conclusion we could then say that with respect to the initial stages of pattern formation, the model for the Fréedericksz transistion resembles model B in regard to the wavenumber selection q≠0. On the contrary, for late stages it is closer to model A since also in this case the relaxational dynamics would correspond to (5.12) and (5.13), this last term being a purely numerical constant.

From the set of eigenvalues one can get an estimate of the time scale involved in the relaxation in terms of a characteristic time

$$\tau_{ch.}^{-1} \equiv \sum_q |\varepsilon_q| = 64\Gamma^{-1}M \exp(-2\rho) \quad . \tag{5.15}$$

Actually, a domain growth law can be inferred from (5.15) as

$$dM / ds = - \tau^{-1} \quad , \tag{5.16}$$

or in terms of an average domain size $\bar{z} = L_x/M$,

$$d\bar{z}/ds \simeq \Gamma^{-1}\bar{z}\exp(-2\bar{z}) \quad , \tag{5.17}$$

wich indicates that asymptotically $\bar{z}$ grows logarithmically in time with a time scale fixed by the renormalized kinetic coefficient Γ.

A deeper analysis of this dynamics follows the work by Kawasaki et al.[20]. Wall motion is described disregarding collisions and limiting the range of interaction to neighbouring walls. This approach reduces in our case to an equation for the location $u_i(t)$ of the ith wall which reads

$$\sum_j \left(L^{-1}\psi'_{\omega,i}, \psi'_{\omega,j}\right)\dot{u}_j = R(u_{i+1} - u_i) - R(u_i - u_{i-1}) \quad , \tag{5.18}$$

where L is given in (5.3), $\Psi'_{\omega,i} = -(-1)^i \, d\Psi_\omega(u)/du$, and

$$R(u_{i+1} - u_i) = 8 \exp[-2(u_{i+1} - u_i)] \quad . \tag{5.19}$$

Equation (5.18) permits a simple interpretation of the wall dynamics in terms of exponentially decreasing interaction forces between walls given by R, and the kinetic particularities of the problem at hand, here associated to the operator L which, in our case, traces back to the hydrodynamical coupling consideration. For $\rho \gg 1$, (5.18) is simplified to

$$\dot{u}_j = \Gamma^{-1}\left[R(u_{j+1} - u_j) - R(u_j - u_{j-1})\right] \tag{5.20}$$

with Γ given by (5.13). The differences with model B, already pointed out above, are here manifested in the fact that in that case one finds contributions for $i \neq j$ in (5.18) which depend on the distance between walls.

Equation (5.20) is a good basis for studies of late-stage dynamics. Following Kawasaki and Nagai[20] a logarithmic growth law may be obtained as

$$\tilde{z}(s) \simeq (1/2) \ln\left(32\Gamma^{-1}s\right) \quad . \tag{5.21}$$

This result is further modified when including annihilation of pairs of walls upon contact leading to

$$\tilde{z}(s) \simeq \ln s^v \quad , \tag{5.22}$$

with $v \approx 3.5$ as it corresponds to a simple nonconserved dynamics.

ACKNOWLEDGMENT: Financial support from DGICYT (Spain) Projects PB-86-0534 and PB-87-0014 is acknowledged. We thank Raul Toral for producing Fig. 1 for this paper. Also many thanks are due to Agustí Careta for his help during the preparation of the manuscript.

REFERENCES:

1. P. C. Hohenberg and M. C. Cross in *Lecture Notes in Physics, vol. 268*. Ed. L. Garrido, (Springer, Berlin, 1987).
2. H. R. Schöber, E. Allroth, K. Schroeder and H. Müller-Krumbhaar, Phys. Rev. A 33, 567 (1986).
3. J. Swift and P. C. Hohenberg, Phys. Rev. A 15, 319(1977).
4. M. Aguado, R. F. Rodriguez and M. San Miguel, Phys. Rev. A 39, 5686 (1989).
5. J. D. Gunton, M. San Miguel and P. Sahni in *Phase Transitions and Critical Phenomena, vol.8*. Eds. C. Domb and J. L. Lebowitz (Academic, London, 1983).
6. M. San Miguel in *Stochastic Process Applied to Physics*. Eds. L. Pesquera and M. A. Rodríguez. (World Scientific, Singapore, 1985).
   M. San Miguel in *ELAF '87: Connections Among Particle Physics, Nuclear Physics, Statistical Physics and Condensed Matter*. Eds. J. J. Giambiagi et al. (World Scientific, Singapore, 1988).
7. G. F. Mazenko in *Lecture Notes in Physics vol. 319*. Ed. L. Garrido, (Springer, Berlín, 1988).
8. M. San Miguel and F. Sagués, Phys. Rev. A 36, 1883 (1987).
   F. Sagués, F. Arias and M. San Miguel, Phys. Rev. A 37, 3601 (1988).
9. M. C. Torrent, F. Sagués, F. Arias and M. San Miguel, Phys. Rev. A 38, 2641 (1988).
10. F. Sagués and M. San Miguel, Phys. Rev. A 39, 6567 (1989).
11. P. G. de Gennes *The Physics of Liquid Crystals*. (Clarendon, Oxford, 1975).
    S. Chandrasekhar *Liquid Crystal*. (Cambridge University Press. Cambridge, 1977).
12. K. Oki, H. Sagawa and T. Eguchi, J. Physique C7, 414 (1977).
13. H. Tanaka and T. Nishi, Phys. Rev. Lett 59, 692 (1987).
14. D. A. Huse, Phys. Rev. B74, 7845 (1986).
15. M. San Miguel, M. Grant and J. D. Gunton, Phys. Rev. A 31, 1001 (1985).
16. B. D. Gaurin, S. Spooner, and Y. Morii, Phys. Rev. Lett. 59, 668 (1987).
17. S. Katano, M. Iizumi, R. M. Nicklow, and H. R. Child, Phys. Rev. B 38, 2659 (1988).
18. J. G. Amar, F. E. Sullivan, and R. D. Mountain, Phys. Rev. B 37, 9638 (1988).
19. K. Kawasaki and T. Ohta, Progr. Theor. Phys. 67, 147 (1982); 68, 129 (1982).
    K. Kawasaki, Phys. Rev. A 31, 3880 (1985).
    K. Kawasaki, Ann. of Phys. 154, 319 (1984).
    K. Kawasaki in *Progress in Statistical Mechanics*. Ed. C. K. Hu. (World Scientific, Singapore, 1988).
    K. Kawasaki in *Proceedings of Statphys 17* (Rio de Janeiro, 1989).
20. K. Kawasaki and T. Ohta, Physica. 116 A, 573 (1982).
    K. Kawasaki and T. Nagai, Physica 121 A, 175 (1983).

21. J. Viñals, M. Grant, M. San Miguel, J. D. Gunton and E. T. Gawlinskii, Phys. Rev. Lett. 54, 1264 (1985).
    S. Kummar, J. Viñals and J. D. Gunton, Phys. Rev. B 33, 7795 (1986).
22. P. C. Hohenberg and B. I. Halperin, Rev. Mod. Phys. 49, 535 (1977).
23. T. M. Rogers, K. R. Elder and R. Desai, Phys. Rev. B37, 9638 (1988).
    R. Toral, A. Chakrabarti and J. D. Gunton, Phys. Rev. Lett. 60, 2311 (1988).
    J. D. Gunton, R. Toral and A. Chakrabarti, Physica Scripta (1990).
24. G. L. Oppo and R. Kapral, Phys. Rev. A 36, 5820 (1987).
    Y. Oono and S. Puri, Phys. Rev. A 38, 434 (1988).
25. F. Haake, J. W. Haus and R. Glauber, Phys. Rev. A 23, 3255 (1981).
26. M. Grant, M. San Miguel, J. Viñals and J. D. Gunton, Phys. Rev. B 31, 3027 (1985).
27. K. Binder, Physica 140A, 35 (1986).
28. T.Ohta, D. Jasnow and K. Kawasaki, Phys. Rev. Lett 49, 1223 (1982).
29. K. Kaski, M. Grant and J. D. Gunton, Phys. Rev. B 31, 3040 (1985).
30. K. Kawasaki, Phys. Rev. A 31, 3880 (1985).
31. J. S. Langer, Ann. Phys. (NY) 65, 53 (1971).
32. L. Kramer in these Proceedings.
33. E. M. Lifshitz and L. P. Pitaevskii in *Statistical Physics 3rd ed., Part I* Pergamon, Oxford, (1980).
34. F. Lonberg, S. Fraden, A. J. Hurd and R. B. Meyer, Phys. Rev. Lett. 52 , 1903 (1984).
35. E. Guyon, R. Meyer and J. Salán, Mol. Cryst. Liq. Cryst. 54 , 261 (1979).
36 A. J. Hurd, S. Fraden, F. Lonberg and R. B. Meyer, J. Phys. (Paris) 46 , 905 (1985).
37 Y. W. Hui, M. R. Kuzma, M. San Miguel and M. M. Labes, J. Chem. Phys. 83 , 288 (1985).
38 M. R. Kuzma, Phys. Rev. Let. 57 , 349 (1986).
    D. V. Rose and M. R. Kuzma, Mol. Cryst. Liq. Cryst. Lett. 4 , 39 (1986).
39. F. Sagués and M. San Miguel, Phys. Rev. A 33 , 2769 (1986).
    M. San Miguel and F. Sagués in *Recent Developments in Nonequilibrium Thermodynamics: Fluids and Related Topics. Lecture Notes in Physics vol. 253* Ed. J. Casas, D. Jou and M. Rubí , (Springer, Berlin 1986).
40. J. P. Mc. Clymer, M. M. Labes and M. R. Kuzma, Phys. Rev. A 37, 1388 (1988).
41. G. Srajer, S. Fraden, and R. B. Meyer, Phys. Rev. A 39, 4828 (1989).
42. F. Sagués and F. Arias, Phys. Rev. A 38, 5367 (1988).

LOCALIZED STRUCTURES IN REACTION-DIFFUSION SYSTEMS

G. DEWEL* and P. BORCKMANS*
Service de Chimie-Physique
C.P. 231
Université Libre de Bruxelles
Belgium

ABSTRACT. Two mechanisms describing the formation of localized structures are discussed. In the first they result from a symmetry breaking instability in a nonuniform environment. In the second they arise from the existence of front waves and wave blocking phenomena.

1. Introduction

Reaction-diffusion models have been widely used to describe self-organization phenomena in driven systems not only in chemistry and biology but also in physics and in materials science. These models consist of a set of non-linear partial differential equations of the following general form

$$\frac{\partial X}{\partial t} = F(X;B) + D \nabla^2 X \qquad [1.1]$$

where X is the concentration vector, F is a vector function of X representing the local mass-action kinetics and D is the matrix of diffusion coefficients; B stands for a set of parameters describing the external constraints.

In these systems the Turing instability [1] has been presented as a major mechanism for the formation of steady periodic patterns [2]. As a consequence of the absence of any unambiguous observation of such global Turing patterns from chemical reactions in liquid solution, the interest has focused recently on the study of localized structures. In this paper we briefly discuss two types of mechanism that may give rise to such localized patterns.

Most of the theoretical models are based on the pool chemical approximation, i.e. the concentration of some species are supposed to be kept uniform throughout the system and constant in time. Obviously this approximation has to be relaxed if one wants to describe the realistic situations encountered in biology or in the new open reactors [3] where the feeding is restricted to the boundaries. We first study the onset of the localized dissipative structures which may appear in these nonuniform systems.

Front waves of chemical or electrical activity have been observed in various bistable excitable or oscillatory systems [4]. These fronts can sometimes provide the building blocks for the construction of localized patterns even in uniform media. They consist in a region where the system is in the bifurcated state immersed in the bulk of the basic state.

---

* Research Associate with the National Fund for Scientific Research (Belgium)

The stabilization mechanism of these pulse-like solutions is due to a *non-variational* effect, i.e. the non-existence of a Lyapunov functional to minimize. In the second section we illustrate this phenomenon on the formation of stable droplets in the case of a reaction-diffusion model exhibiting bistability. Another example is provided by the numerical simulations of O. Thual and S. Fauve [5] of the complex Ginzburg-Landau equation corresponding to a subcritical Hopf bifurcation.

2. Effects of slow spatial variations on dissipative structures

A large number of time-independent concentration patterns have been observed in thin layers of various chemical mixtures. It has now clearly been shown that these structures do not originate solely from the interplay between chemical reaction and diffusion but that convective or surface effects come into play in all of them [6]. To date, there has been no clear-cut observation of a Turing instability in any chemical experiment. The fundamental reason for this situation was the absence of an adequate tool to maintain and control nonequilibrium conditions in unstirred chemical reactors. Recently experimental tools that overcome these shortcomings have been devised in Austin (TX,USA) and Bordeaux (F) [3]. For instance, the *front reactor* might allow the observation of intrinsic symmetry-breaking instabilities by avoiding spurious convection motions. These gel reactors are fed along opposite boundaries with the reagents. These then diffuse and react when coming into contact with each other. Diffusion and reactions thus act to produce concentration profiles inside the reactor [Fig.1]. These front reactors also mimic some biological processes. We must therefore consider the effects of spatial concentration nonuniformities on the chemical instabilities. The theoretical and numerical analysis of this type of problem has been undertaken some time ago in the case of the Brusselator model [7,8]. A similar problem, the onset of cellular convection in a shallow container heated nonuniformly from below has also received much attention [9].

For the sake of brevity, we limit ourselves here to the case of a spatial symmetry breaking. We suppose that the concentration of the species A, that determines the instability threshold, varies in space. In the corresponding pool reactor this concentration is maintained constant and uniform, and the global Turing instability occurs when $B = B_c(A)$ where B, the other control species, is the bifurcation parameter. We further assume that $\partial B_c/\partial A > 0$.

In order to mimic the profiles occurring in the front reactors, we suppose that A(r) presents a local isolated minimum:

$$A(r) = A(0) [1 + \delta F(\epsilon r)] \quad [2.1]$$

where $F(0) = 0$ and $F(\pm\infty) = 1$ and ϵ^{-1} is the characteristic length scale of the spatial variations of amplitude δ. In the nonuniform system one can define a local threshold $B_c[A(r)]$. It may happen that the control parameter exceeds this local threshold over a certain spatial range, but lies below it over another range. Then provided the local threshold remains close to criticality, a linear analysis is appropriate throughout the whole system. There are two avenues open to us if we want to obtain approximate analytical solutions. First we could make an expansion in powers of the amplitude δ that is assumed to be small. The second possibility, that we discuss here, is to assume that $\epsilon \ll 1$ characterizing a weak spatial variation. The two length scale method is then appropriate to study the stability of the basic profile $X_0(\epsilon r)$.

With our choice of the function F, the instability will start near r = 0 where the threshold is at a minimum. To analyze the onset of the structure we consider a region of width $\varepsilon^{-1/2}$ around r = 0 and we look for solutions in the form

$$X(z) - X_0(z) = \exp i \left[\frac{k_c z}{\sqrt{\varepsilon}}\right] \left[W_0(z) + \sqrt{\varepsilon} W_1(z) + ...\right] \quad [2.2]$$

with $\quad z = \varepsilon^{1/2} r \quad$ and $\quad B_c = B_c^0 + \varepsilon B_\varepsilon \quad$ [2.3]

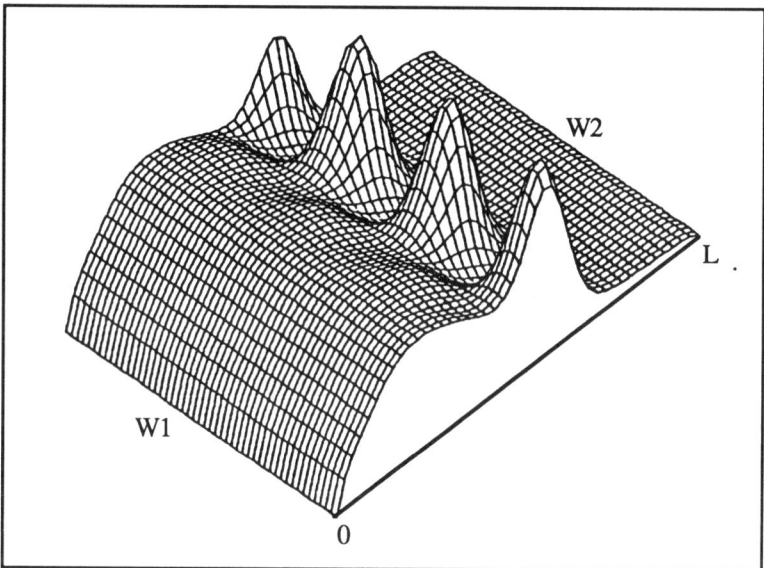

Fig. 1 : Concentration map of species X for the Brusselator model:

$$A \xrightarrow{k_1} X$$
$$B + X \xrightarrow{k_2} Y$$
$$2X + Y \xrightarrow{k_3} 3X$$
$$X \xrightarrow{k_4} P$$

functioning in a rectangular 2d thin film. The concentrations of the major species, A and B, are fixed at the boundaries W1 and W2 [A(0)=3, B(0)=2; A(L)=1, B(L)=6]. The produced intermediate species, X and Y, are maintained at zero concentration at these boundaries. On the orthogonal sides periodic boundary conditions are assumed. Inside the film, transport is achieved only by molecular diffusion [$D_A=D_B=12$, $D_X=4$, $D_Y=20$]. Also $k_1/k_4 = k_2^2/k_3 k_4 = 0.01$. All quantities are expressed in reduced units. {reproduced by courtesy of J. Boissonade, Bordeaux [11]}.

and k_c and B_c^0 are respectively the critical wavenumber and the threshold value of the corresponding homogeneous system where $A(r) \equiv A(0)$ and $B_c^0 = B[A(0)]$.

Eq.[2.2] represents a disturbance of wavelength $2\pi/k_c$ slowly modulated on the scale of the external variation $\delta F(\varepsilon r)$. The slowly varying amplitude W_o is determined by the solvability condition at order ε that yields the following equation [10]:

$$\frac{\partial^2 W_o}{\partial z^2} + \left[\frac{B_c}{4D} - \kappa^2 z^2\right] W_o = 0 \qquad [2.4]$$

where κ can be related to the parameters of the model. The solutions of Eq.[2.4] are well-known. In terms of the original variable r, Eq.[2.2] can be written:

$$X(r) - X_o(r) \propto \exp\left[ik_c r - \frac{\kappa \varepsilon r^2}{2}\right] H_n(r) \qquad [2.5]$$

where $H_n(r)$ is the Hermite polynomial of degree n.

The solution [2.5] corresponds to structures confined in the vicinity of the reaction zone, $r \approx 0$. The wavelength is of the same order of magnitude as in the corresponding uniform system where $A = A(r=0)$. The amplitude decays exponentially away from the center over a distance $(\varepsilon \kappa)^{-1/2}$ which is *shorter* than the scale of the spatial variation. The lowest value of B_ε ($= 4kD$), at which such solutions exist determines the threshold in the presence of gradients. If there exists a local minimum of A, the structures appear at a well defined critical value

$$B_c = B_c^0 + 4\varepsilon\kappa D \qquad [2.6]$$

in contrast with the smooth transitions (imperfect bifurcations) observed when $A(r)$ is a monotonically increasing function of r. This critical value is greater than that in the corresponding uniform system [$A(r)\equiv A(0)$]. This shift plays an important role in the orientation problem in two-dimensional systems as it favors the onset of patterns with a wavevector perpendicular to the concentration gradient [10] [Fig.1]. The wavelength of these localized patterns is uniquely determined contrary to the case of the uniform systems where there is a whole band of possible wavevectors. Similarly it has also been shown that a single wavelength is selected in the presence of a slow monotonic spatial ramp connecting a subcritical to a supercritical region [12].

Localized spatial structures exhibiting most of the properties described above had already been observed by Herschkowitz-Kaufman and Nicolis [7] in their pioneering numerical simulations of the Brusselator model in nonuniform media.

The same technique can be applied to study the effects of slow spatial variations on a Hopf instability [10]. In that case localized chemical waves are generated near r=0. They have also been obtained in the numerical simulations.

On the other hand, in the presence of a very localized heterogeneity $\varepsilon \ll 1$, reaction-diffusion systems can develop local periodic undulations under far less restrictive conditions than for the onset of global patterns [8]. These patterns can play an important role in heterogeneous catalysis where they may mask kinetic data interpreted under the assumption that a uniform state exists on the surface [13].

3. Internal layer solutions of reaction-diffusion models

In this section we describe another mechanism whereby localized patterns can evolve naturally from general initial condition. Contrary to Turing's ideas it does not imply a symmetry breaking instability anymore. The main ingredients behind this mechanism are rather front waves and wave blocking phenomena [14].

3.1. THE PROPAGATOR-CONTROLLER SYSTEMS

Waves of chemical and electrical activity have been observed in a large number of different contexts [4]. The theory of front propagation in active media [15] is usually based on a pair of reaction-diffusion equations which take the following form for a one-dimensional system:

$$\varepsilon \frac{\partial u}{\partial t} = f(u,v) + \delta^2 \frac{\partial^2 u}{\partial r^2}$$

$$\frac{\partial v}{\partial t} = g(u,v) + \frac{\partial^2 v}{\partial r^2}$$

[3.1]

where $\varepsilon = \tau_u/\tau_v$ and $\delta = L_u/L_v$, with the relations $L_i^2 = D_i\tau_i$, are respectively the ratios of the characteristic time and length scales of the two variables. The u-nullcline is S-shaped whereas the v-nullcline is monotonous [Fig.2]. According to their relative position one obtains excitable, oscillatory or bistable kinetics. Pattern formation in such systems is also discussed in the contribution of G.-H. Purwins to this conference.

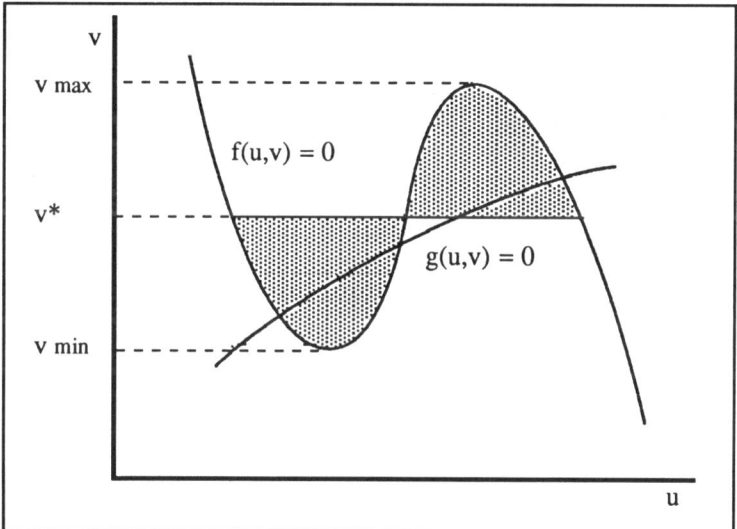

Fig. 2 : Nullclines, where du/dt=0 and dv/dt=0 respectively, in the phase plane for a bistable situation. The line v=v* divides the sigmoidal nullcline into two shaded regions of equal area.

A well known example of a propagator-controller system is the Fitzhugh-Nagumo model for nerve conduction where one takes

$$f(u,v) = u(1-u)(u-a) - v$$
$$g(u,v) = u - \gamma v \qquad [3.2]$$

It is often useful to consider a piecewise linear version of this model:

$$f(u,v) = -u - v + H(u-a) \qquad [3.3]$$

where $H(x)$ is the Heaviside step function.

For many systems, the conditions $\varepsilon \ll 1$ and $\delta \ll 1$ are well satisfied. The variable u is thus fast and short-range, whereas v is slow and long-range. The system can then support fronts in the species u (the *propagator species*), the velocity and the waveform of these fronts being determined by the other species v (the *controller species*).

The dynamical evolution of these fronts does not depend heavily on the local kinetics, i.e. the detailed form of the nonlinearities appearing in $f(u,v)$, but the various behaviors originate from a delicate balance between the diffusivities and the reaction rates of the two species. In this context a crucial role is played by the ratio of the characteristic velocities $\omega = \delta/\varepsilon$ [16].

3.2. TRAVELLING WAVES AND NUCLEATION IN NONRECOVERING MEDIA

We start with a simple case by assuming that the concentration of the controller species is kept constant in space and time. In Fig.2 this would correspond to a horizontal g-nullcline $v = v_0$. For $v_{min} < v_0 < v_{max}$ the system exhibits bistability: it has two stable uniform solutions $u = h_\pm(v_0)$ with $h_-(v_0) < h_+(v_0)$. Because the reaction rate is large, the solution at each point will quickly tend to one of the steady states.

Front waves are self-similar solutions depending on a single propagating coordinate $z = x - ct$, $u(x,t) = U(z)$ that connect the two states. Here we suppose $U(-\infty) = h_+(v_0)$ and $U(+\infty) = h_-(v_0)$.

For $v_{min} < v_0 < v_{max}$, the speed c is uniquely determined by v_0 from the eigenvalue problem

$$\delta^2 U'' + \varepsilon c U' + f(U, v_0) = 0 \qquad [3.4]$$

The direction of propagation may easily be obtained from the sign of the integral

$$J = \int_{h_-(v_0)}^{h_+(v_0)} f(s, v_0) \, ds \qquad [3.5]$$

$$c \begin{cases} > 0 \\ = 0 \\ < 0 \end{cases} \quad \text{according to } J(v) \begin{cases} > 0 \\ = 0 \\ < 0 \end{cases} \qquad [3.6]$$

When the equal-area condition $J(v^*) = 0$ is fulfilled, for $v_0 = v^*$, the system admits a standing front solution corresponding to the coexistence of the two states. On the contrary, when $v_{min} < v_0 < v^*$, thus c>0, the front increases the interval of z values on which u is on the upper branch (*up-jump* wave). We then say that $h_+(v_0)$ is dominant. On the opposite when $v^* < v_0 < v_{max}$, thus c<0, the wave moves transfering the medium from the upper to the lower state (*down-jump* wave). In summary, c is a monotone decreasing function of v_0.

The dynamics of the systems with a single state variable ($v(x,t) = v_0$) can be described from a Lyapunov functional,

$$F = \int dx \left[\frac{\delta^2}{2} \left(\frac{\partial u}{\partial x} \right)^2 + V(u) \right] \qquad [3.7]$$

and the first of the Eq.[3.1] can then be written as

$$\varepsilon \frac{\partial u}{\partial t} = \delta^2 \frac{\partial^2 u}{\partial r^2} - \frac{\delta V}{\delta u} \qquad [3.8]$$

The front moves in such a direction as to increase the territory of the state with the lower potential. Front propagation thus corresponds to a minimization of F.

The potential V can also be used to describe the nucleation phenomena in these one-dimensional potential systems. If the system is initially prepared in the non-dominant state, say $h_-(v_0)$ ($v_{min} < v_0 < v^*$), to nucleate the dominant state $h_+(v_0)$, it has to overcome the short-range attractive interaction between the adjacent interfaces of a nucleus of the state $h_+(v_0)$ embedded into the state $h_-(v_0)$. The width $\sigma(t)$ of such a droplet obeys the following equation, $[V(h_-) > V(h_+)]$,

$$\frac{\partial \sigma}{\partial t} = c[V(h_-) - V(h_+)] - \exp\left[\frac{-2\sigma}{\delta}\right] \qquad (c = \text{constant}) \qquad [3.9]$$

When $v_{min} < v_0 < v^*$, this equation has a steady solution σ_c corresponding to the critical nucleating radius. According to Eq.[3.9] this critical droplet is unstable: a domain smaller than σ_c shrinks and disappears while larger droplets expand and the bifurcating state $h_+(v_0)$ invades the whole system.

Front propagation phenomena are much more complicated when the controller species v is allowed to vary in space and time. The system [3.1] generally does not possess a Lyapunov functional anymore and therefore the quasithermodynamical criterion for the determination of the velocity can, in general, not be applied. Even the notion of dominant state looses its meaning. It has indeed been shown that for the same values of the parameters these systems can exhibit front wave propagation in opposite directions depending on the initial conditions [17].

However when the propagator is the dominant species, i.e. when it imposes on the front its own characteristic velocity ($\omega \gg 1$) it is still possible to apply the method described above for potential systems to study the *inner region* of the front. During the initial time period of order ε the system is again driven to one or the other branch ($h_\pm$) of the slow manifold. In the *boundary layer*, between these two states, the propagator suffers an abrupt change whereas the soft variable v can be considered as constant in the layer of width δ. Therefore at the lowest order the behavior in the inner transition region can still be described by an equation identical to Eq.[3.4] where now v_0 is the concentration of v in the front. Of course one must also add another equation describing the time evolution of v(x,t) in the *outer region* and the *matching conditions* between the regions must be satisfied.

3.3. MOTIONLESS SOLITARY PATTERNS

When v is allowed to vary in space and time, nucleation phenomena are quite different from the potential case we analyzed in paragraph 3.2. When $\omega \ll 1$, i.e. when the velocity of the controller is larger than that of the propagator, spontaneous immobilization of fronts can take place in uniform regions leading to localized patterns in bistable systems.

Let us suppose that after the initial time period a pulse has been formed. Far from this domain the system is on the lower branch. Because there are two characteristic length scales in the problem, small and large pulses can be formed. In the small pulses, of width $\eta_s \approx O(\delta)$, the concentration of the controller is constant, one thus recovers a situation analogous to the gradient case discussed in the preceding paragraph. Here also the small pulses are unstable: pulses of width smaller then critical shrink whereas larger ones grow up. But because the reaction-diffusion model (Eq.[3.1]) does not possess a Lyapunov functional the growing pulse can now be blocked at a larger radius. This makes the essential difference with the classical nucleation and growth problem.

In the larger pulses of width $\eta_s \approx O(1)$ the local state lies on the slow manifold everywhere except in the fronts where the propagator species undergoes an abrupt change whereas the controller remains smooth. The time evolution of such a large pulse can be described by the following set of differential equations

$$\frac{\partial \eta}{\partial t} = \Gamma_c(v_\eta) \qquad [3.10]$$

$$\omega \frac{\partial v}{\partial t} = \frac{\partial^2 v}{\partial x^2} + g(v,\eta) \qquad [3.11]$$

where $\eta(t)$ is the width of the droplet.

The inner equation, [3.10], describes the coupling between the interface motion, characterized by its position coordinate, and the value of the controller concentration in the front v_η; the exact form of Γ_c depends on the model. For large pulses one may neglect the short range attraction appearing in Eq.[3.9]. Eq.[3.11] gives the evolution of the variable v in the outer region far from the two transition layers.

When $\omega \ll 1$, the term on the left hand side of [3.11] may be neglected. This equation can then be solved yielding $v_\eta = V(\eta)$. For instance, in the case of the piecewise linear model, Eq.[3.3], one finds

$$v_\eta = \frac{v_+}{2}\left[1-e^{-2\eta}\right] \qquad [3.12]$$

if v_+ is the value of v on the upper branch of the slow manifold. By substituting this relation into the first equation, one obtains an evolution equation for $\eta(t)$ in a closed form. The stationarity condition $\Gamma_c[V(\eta)] = 0$ provides a transcendental equation, the solution of which gives the stationary width $\eta_L{}^s$ of the large pulse. The stability of this large pulse can be understood by considering the following inequalities: The first

$$\frac{\partial v_\eta}{\partial \eta} > 0 \qquad [3.13]$$

states that if the pulse expands, the concentration of the controller observed at the moving front increases. This in turn acts to slow down the expansion according to the second inequality

$$\frac{\partial\left[\frac{\partial \eta}{\partial t}\right]}{\partial v_\eta} < 0, \qquad [3.14]$$

as we have indeed seen in section 3.2 that the front velocity decreases when the concentration of v in the front, v_η, increases and that $\partial\eta/\partial\tau$ can even reach zero when $v_\eta = v^*$.

In conclusion, one has the following scenario. If the activated domain is initially too small, it will first expand as a result of the triggering effect of the propagator u on the surroundings. But, because the controller has a higher speed ($\omega \ll 1$), its outflow from the activated zone results in an accumulation of the controller ahead of the front and eventually this can block the expansion when v attains the value v^* corresponding to coexistence leading to a stable motionless solitary pattern. The system thus regulates itself until the Maxwell equal area condition is satisfied. The stability of these large pulses can be proved when ω is small enough. However when ω increases, the term on the left-hand side of Eq.[3.11] cannot be neglected anymore. In that case it can be shown that the motionless solitary pulse can be destabilized through a Hopf bifurcation occurring at some $\omega = \omega_c$, leading to a "breathing motion" of the activated droplet [16,18].

A similar analysis can be reproduced for higher dimensional systems [19]. Here one must add a term to [3.9] and [3.11] describing the curvature induced motions.

Such a stabilization, by non-variational effects, of a droplet a the bifurcated state embedded into the basic state has also been reported recently by O. Thual and S. Fauve [5] in the case of the subcritical Hopf bifurcation and further analyzed in [20]. By solving numerically the corresponding amplitude equation, stable pulse-like solutions have been obtained for a large range of the parameters. The phase gradients play in this case the same role as the controller species in the propagator-controller models. They also adjust themselves to maintain coexistence between the two states.

In conclusion we stress that the phenomenon of droplet stabilization by non variational effects provides a general mechanism for the formation of localized structures in nonequilibrium systems.

4. References

01. A. Turing, Phil. Trans. R. Soc. Lond. **237B** 37 (1952)
02. G. Nicolis, I. Prigogine "Self-Organization in Nonequilibrium Systems" (Wiley, New York, 1977)
03. For a review see: J. Boissonade in "Dynamic and Stochastic Processes, Theory and Applications" (R. Lima, L. Streit and Vilela Mendes Eds., Springer, Berlin 1989)
04. R.J. Field, M. Burger (Eds.) "Oscillations and Traveling Waves in Chemical Systems" (Wiley, New York, 1984)
05. O. Thual, S. Fauve, J. Phys. (France) **49** 1829 (1988)
06. P. Borckmans, G. Dewel, D. Walgraef, Y. Katayama, J. Stat. Phys. **48** 1031 (1987)
07. M. Herschkowitz-Kaufman, G. Nicolis, J. Chem. Phys. **56** 1890 (1972)
08. P. Ortoleva, J. Ross, J. Chem. Phys. **56** 4397 (1972)
09. P.M. Eagles, Proc. Roy. Soc. **A371** 359 (1980)
10. G. Dewel, P. Borckmans, Phys. Lett. **138** 189 (1989)
11. J. Boissonade, J. Phys.(France) **49** 571 (1988)
12. L. Kramer, E. Ben-Jacob, H. Brand, M.C. Cross, Phys. Rev. Lett. **49** 1891 (1982)
13. L. Lobban, D. Luss, J. Phys. Chem. **93** 6530 (1989)
14. S. Koga, Y. Kuramoto, Prog. Theor. Phys. **63** 106 (1980)
    P.C. Fife J. Chem. Phys. **64** 554 (1976)
15. J.J. Tyson, J.P. Keener, Physica **D32** 327 (1988)
16. L.M. Pismen, J. Chem.Phys. **71** 462 (1979)
17. P. Ortoleva, J. Ross, J. Chem. Phys. **63** 3398 (1975)
18. Y. Nishiura, M. Mimura, SIAM J. Appl. Math. **48** 481 (1989)
19. T. Otha, M. Mimura, R. Kobayashi, Physica **D34** 115 (1989)
20. V. Hakim, P. Jacobsen, Y. Pomeau, Europhys. Lett. (to appear)
    S. Fauve, O. Thual, Phys. Rev. Lett. (submitted)
    W. Van Saarloos, P. Hohenberg, Phys. Rev. Lett. (submitted)

KINETIC MODELS FOR DEFECT POPULATIONS IN DRIVEN MATERIALS

D. WALGRAEF[†]
Service de Chimie-Physique,
Université Libre de Bruxelles,
Bd du Triomphe, CP 231,
B-1050, Brussels, Belgium.

ABSTRACT. The origin of defect patterns and microstructures in driven materials has recently been studied within the framework of dynamical models for the defect densities. These models take into account the motion and interaction of defects and may present various types of pattern forming instabilities. Some basic properties of these models are reviewed when diffusion or transport compete with annihilation or multiplication processes before any instability is reached. In particular, the possibility of spontaneous segregation of point defects and the origin of effective diffusional behaviors in dislocation dynamics are discussed.

1. Introduction

When crystalline materials are driven away from thermal equilibrium by external mechanical, physical or chemical constraints, they display several types of interesting phenomena which modify their macroscopic properties [1-3]. For example the defect number may increase significantly and homogeneous defect distributions may become unstable. The resulting microstructures are associated with the inhomogeneity of deformation, with strain localization, inhomogeneous swelling, etc. They also act as initiating centers for micro-crack nucleation, influence crack propagation, void lattice formation ,.... and have thus a very practical importance.

[†] Senior Research Associate, National Fund for Scientific Research (Belgium).

Since these phenomena are related to the strong nonequilibrium conditiond imposed to the material, they can usually not be interpreted by classical thermodynamical or mechanical concepts. Furthermore, it appears that defect microstructures result from a dynamical equilibrium between different processes. It is why kinetic models were proposed to describe the collective behavior of defects in driven materials. These models take into account the motion (diffusion, transport, ...) and nonlinear interactions (annihilation, pinning, clustering,...) between defects [3-4].

In this paper I will review some basic properties of such models and their relevance to the description of the macroscopic behavior of defect populations. Some of these properties are generic of reaction-diffusion dynamics : for example when several populations with different mobilities interact via sufficiently high nonlinear processes, pattern forming instabilities may be expected [5]. Such instabilities have been investigated in various fields including hydrodynamics, chemistry, biology and materials science. Besides the determination of the critical wavelength of the microstructures, it is also essential to investigate the geometry, the symmetries, and the stability ranges of the selected structures. As extensively discussed elsewhere, the difficulties of the post-bifurcation analysis lies in the fact that the complexity of the dynamics does not allow, in general, the attainement of analytic solutions for the different variables. However near the instability or bifurcation points, the dynamics may be reduced to much simpler forms by taking advantage of the time and space scale separation between stable and unstable modes and by projecting the dynamics on its unstable manifold [6]. The resulting slow mode dynamics which governs the system evolution on its longest time scale becomes similar to the time dependent Landau-Ginzburg dynamics describing phase transitions in equilibrium systems. This description leads then to amplitude equations for the patterns and allows the derivation of their phase dynamics. Pattern selection and stability may then be discussed in this framework as shown in several contributions of this volume.

In driven materials, the nature of the selection mechanisms and how the selected patterns are influenced by the underlying lattice symmetry, or by other materials and irradiation conditions, remain largely unexplored. For example, as it will be discussed below, macroscopic clustering or segregation may occur when diffusion or transport processes compete with annihilation or multiplication processes before any instability is reached. This effect may influence several aspects of the defect microstructures in the instability regime and should be discussed with special care.

2. Point Defect Dynamics in Irradiated Materials

Irradiated metals and alloys present several types of spatial structures going from dislocation microstructures to void lattices. These structures which originate in the spatial organization of point defect populations have a strong influence on the macroscopic properties of the materials and on their resistance to external constraints. Hence, the understanding of the formation, selection and stability properties of defect patterns in irradiated materials is of primary importance both from fundamental and from technological point of views.

Several attempts have been made to describe these phenomena in the framework of kinetic models for the defect populations [6]. These models are based on the fundamental elements of the collective behavior of each defect population which are :
- their motion (diffusion for point defects such as interstitials and vacancies, glide, climb and cross-slip for dislocations),
- their interactions which correpond for example to recombination of point defects, capture or emission of point defects by microstructures and defect creation mechanisms induced by the external constraint

In irradiated metals and alloys, this leads to at least three coupled nonlinear differential equations for the dynamics of vacancies, interstitials and dislocations or vacancy loops.

A minimal model has been proposed by Murphy to describe the dynamics of defect populations in irradiated metals and alloys [7-8]. It is based on the rate theory of radiation damage originally developed by Bullough, Eyre and Krishan [9], and expanded further by Ghoniem amd Kulcinski [10] to include the dynamics of point defects in the fully dynamic rate theory. The model consists in the following kinetic equations where c_v and c_i are the concentrations of vacancies and interstitials , while ρ_L and ρ_N are the line densities of vacancy loops and network dislocation :

$$\partial_t c_i = K - \alpha c_i c_v + D_i \nabla^2 c_i - D_i c_i (Z_{iN} \rho_N + Z_{iL} \rho_L)$$
$$\partial_t c_v = K(1 - \epsilon) - \alpha c_i c_v + D_v \nabla^2 c_v - D_v (c_v - \bar{c}_v)(Z_{vN} \rho_N + Z_{vL} \rho_L)$$
$$\partial_t \rho_L = \frac{1}{br_L^0}[\epsilon K - \rho_L (D_i c_i Z_{iL} - D_v (c_v - \bar{c}_v) Z_{vL})] \quad (1)$$

The various coefficients which represent defect creation, annihilation, and migration to dislocations and vacancy loops may be obtained from experimental data or theoretical analysis. K is the displacement damage rate and ϵ the cascade collapse efficiency. D_i and D_v are the diffusion coefficients, α the recombination coefficient, b the length of the Burgers vector. $Z_{..}$ are the bias factors and $\bar{c}_{v_i}$ the thermally emitted vacancies from the microstructures.

This system shows interesting features in two limits. In the sink dominated regime, i.e. when the dynamics is dominated by the diffusion of point defects to dislocations and loops, a pattern forming instability occurs when the dislocation bias overcomes the cascade collapse efficiency or, at fixed bias, when the mean loop density is sufficiently high compared to the network dislocation density. The three-dimensional aspects of the post-bifurcation regime have been studied by Walgraef and Ghoniem [11]. In particular, it was shown that, in isotropic media, according to the values of the experimental parameters, the microstructure should correspond to bcc or wall patterns.

The situation is however different for the recombination dominated regime which corresponds to low temperature conditions or to the early stages of the irradiation process. In this case the dynamics is reduced, in the absence of vacancy collapse, to :

$$\partial_t c_i = K - \alpha c_i c_v + D_i \nabla^2 c_i$$
$$\partial_t c_v = K - \alpha c_i c_v + D_v \nabla^2 c_v \qquad (2)$$

It is interesting to note that, according to the properties of the creation term K, this equation can show non classical behavior, namely spontaneous segregation in low-dimensional systems. In the case of a steady deterministic rate of defect creation, the uniform steady state defined by $c_i^0 c_v^0 = K/\alpha$ is marginally stable. Since the dynamics of $c_i - c_v$ is purely diffusive, density fluctuations are expected to lead to long-time anomalous kinetics in $d \leq 2$ systems (this variable plays here the same role as the Goldstone modes of critical phenomena) and should be incorporated in the dynamics. If one represents these fluctuations by a gaussian white noise, the system (2) becomes, on writing the sum of the defect densities c_i and c_v as ρ and their difference as δ :

$$\partial_t \rho = 2K - \frac{\alpha}{2}(\rho^2 - \delta^2) + D_+ \nabla^2 \rho + D_- \nabla^2 \gamma + \eta_\rho$$
$$\partial_t \gamma = D_+ \nabla^2 \gamma + D_- \nabla^2 \rho + \eta_\gamma \qquad (3)$$

where $< \eta_i > = 0$ and $< \eta_i(\vec{r}_1, t_1)\eta_i(\vec{r}_2, t_2) > = \Gamma_i \delta(\vec{r}_1 - \vec{r}_2)\delta(t_1 - t_2)$. It is now possible to transpose the argument developed by Lindenberg et al. in their study of steady-state segregation in diffusion limited chemical reactions [12]. Effectively, the γ correlation function behaves, in a system of linear spatial extension L, as :

$$< \gamma(\vec{r}_1, t_1)\gamma(\vec{r}_2, t_2) > = \frac{\Gamma_\gamma}{D_+} \Sigma_{\vec{k} \neq 0} \frac{e^{i\vec{k}(\vec{r}_1 - \vec{r}_2)}(1 - e^{-2D_+ k^2 t})}{k^2} \qquad (4)$$

where $\vec{k} = \frac{2\pi \vec{m}}{L}$, $\vec{m}$ being a d-component vector with integer components. In one-dimensional systems, $(d = 1)$, the asymptotic limit of (4) may evaluated exactly and is equal to

$$\frac{\pi L}{12 D_+}(1 - \frac{6}{L}(x_1 - x_2) + \frac{6}{L^2}(x_1 - x_2)^2) \qquad (5)$$

Hence the species present at $x = x_1$ is dominant over a length equal to $(1 - \frac{1}{\sqrt{3}})L$ and the other species spreads over a consecutive length $\frac{L}{\sqrt{3}}$. This type of pattern persist in higher dimensions but the difference resides in the amplitude of the correlation function which goes as L^{2-d}. However these results are not sufficient to determine whether macroscopic segregation should be observable or not. In fact, one has to evaluate the steady state segregation index

$$S = lim_{t \to \infty} S(t) = lim_{t \to \infty} \frac{< \gamma^2(\vec{r}, t) >}{< \rho^2(\vec{r}, t) >} \qquad (6)$$

The long time limit for $<\rho^2>$ is evaluated through the steady state condition

$$0 = \nabla^2 <\rho> + K - \frac{\alpha}{2}[<\rho^2> - <\gamma^2>] \qquad (7)$$

Since $<\gamma^2>$ provides an extra creation rate of order L^{2-d}, it is easy to see that, in order to balance this steady state condition, $<\rho^2> \simeq L$ for $d = 1$, while $<\rho^2> \simeq L^0$ for $d > 1$. Hence, the segregation index vanishes at large L for $d > 2$ and spontaneous segragation should occur for $d \leq 2$. Of course, the resulting pattern has no regularity, nor well defined wavelengthes. Its influence on the further evolution of the system should nevertheless be worth studied.

3. Dislocation Dynamics during Plastic Deformation

The phenomenological laws of plasticity tell us that the plastic deformation of metals is related to the balance between dislocation motion and interactions. For example, Orowan's law expresses the fact that the plastic strain rate $\dot{\epsilon}$ is proportional to the product of the density of mobile dislocations ρ_m and of their mean velocity v: $\dot{\epsilon} = b\rho_m v$. The velocity itself depends on the density of the nearly immobile dislocations of the forest ρ_i ($v = F(\sigma/\sigma_f)$, with $\sigma_f - \sigma_0 \propto \sqrt{\rho_i}$, where σ_0 is the yield stress and F is a phenomenological function which may be of the Arrhenius type).

Hence the knowledge of dislocation dynamics is required to understand and correctly describe the various aspects of plasticity, including the formation of dislocation microstructures. It is why dynamical models of the reaction-transport type have recently been proposed to describe the behavior of dislocation densities in crystalline materials under cyclic or monotonous loadings [13-15]. Despite the fact that they correctly describe various aspects of dislocation patterns, some ingredients of these models are still controversial and require further justification. For example, one should know the conditions required to have an effective diffusion for the mobile dislocations and what is the dependence of the effective diffusion coefficient on the dislocation mean velocity. Another point is to clarify the role of the few nonlinearities (annihilation, clustering, pinning) which are thought to be essential in the destabilization of uniform dislocation distributions.

In the first stages of plastic deformation, very elementary kinetic models may already give some insight into the local behavior of dislocation densities. Let us consider the simple case of a family of parallel, straight, infinite dislocations of the same character and of Burgers vector $\pm b\vec{1}_x$. These dislocations glide along the x-direction and may eventually cross-slip in the y-direction. The balance equations which describes the local evolution of the corresponding densities ρ^+ and ρ^- may then be written as :

$$\partial_t \rho^+ = -\partial_x v\rho^+ + D\partial_y^2 \rho^+ + g^+$$
$$\partial_t \rho^- = +\partial_x v\rho^- + D\partial_y^2 \rho^- + g^- \qquad (8)$$

If we consider the sinks and source terms associated with the multiplication (e.g. Frank-Read mechanisms) and annihilation of mobile dislocations, we may write :

$$g^+ = g^- = \kappa(\rho^+ + \rho^-) - \alpha\rho^+\rho^- \tag{9}$$

and the system (8) may be transformed into:

$$\partial_t \rho_m = -\partial_x v \delta_m + D\partial_y^2 \rho_m + \kappa \rho_m - \frac{\alpha}{2}(\rho_m^2 - \delta_m^2)$$
$$\partial_t \delta_m = -\partial_x v \rho_m + D\partial_y^2 \delta_m \tag{10}$$

where $\rho_m = \rho^+ + \rho^-$ and $\delta_m = \rho^+ - \rho^-$. If one imposes $\delta_m = 0$ at any time and if one considers uniform dislocation distributions only, this system is reduced to the Verhulst equation which was studied in the context of plasticity by several authors [16-18].

However, since nonuniformities and fluctuations may be important in this type of dynamics, we need to consider the total system and investigate the stability of the uniform steady state $\rho_m^0 = 2\kappa/\alpha$, $\delta_m^0 = 0$. Its linear evolution matrix is given, in Fourier space, by:

$$\begin{pmatrix} -\kappa - q_y^2 D & -iq_x v \\ -iq_x v & -q_y^2 D \end{pmatrix} \tag{11}$$

Since its eigenvalues are given by

$$\omega = \frac{1}{2}[-\kappa - 2q_y^2 D \pm \sqrt{\kappa^2 - (q_x v)^2}] \quad , \tag{12}$$

density fluctuations have a diffusive decay in the y direction while in the x direction their decay is diffusive for $|q_x v| < \kappa$ and propagative for $|q_x v| > \kappa$ [19]. According to the fact that glide is much faster than cross-slip, the effective diffusion coefficient in the x direction (which is proportional to v^2/κ) is orders of magnitude larger than in the y direction, and the homogeneisation is much faster in the slip direction. Furthermore, due to the nonlinear coupling between ρ_m and δ_m, local fluctuations in δ_m increase locally and temporally the effective dislocation production.

Besides pair annihilation, dislocations of opposite Burgers vectors may also form dipoles which are far less mobile, and in order to take this process into account, we may modify the dynamics (10) in the following way:

$$\partial_t \rho_m = -\partial_x v \delta_m + D\partial_y^2 \rho_m + \kappa \rho_m - \frac{\alpha}{2}(\rho_m^2 - \delta_m^2)$$
$$\partial_t \delta_m = -\partial_x v \rho_m + D\partial_y^2 \delta_m$$
$$\partial_t \rho_d = \frac{\bar{\alpha}}{4}(\rho_m^2 - \delta_m^2) \tag{13}$$

where $\bar{\alpha} < \alpha$ since $\bar{\alpha}$ is the kinetic rate of dipole formation and $\alpha - \bar{\alpha}$ the kinetic rate associated with pure annihilation. Within this framework, the evolution of the dipole density is triggered by the dynamics of mobile dislocation density. As a result local fluctuations in the mobile dislocation density induce nonuniformities in the dipole density which are rapidly elongated in the slip direction x and which slowly expand in the y direction, just as the formation of coarse slip proceeds.

If the stress level is sufficiently high, the dislocation density increases and many dipoles are formed. They may form clusters which contribute to the forest of immobile dislocations and act as pinning centers for mobile dislocations. One may take this effect into account in the dynamics of mobile dislocations on writing the corresponding equations as :

$$\partial_t \rho_m = -\partial_x v \delta_m + D \partial_y^2 \rho_m + \kappa \rho_m - \frac{\alpha}{2}(\rho_m^2 - \delta_m^2) - \gamma(\rho_i)\rho_m$$
$$\partial_t \delta_m = -\partial_x v \rho_m + D \partial_y^2 \delta_m - \gamma(\rho_i)\delta_m \qquad (14)$$

where ρ_i is the density of immobile dislocations. For monotonous loadings and constant glide velocity, the linear evolution of the perturbations of the uniform steady state is given in Fourier space, by :

$$\partial_t \rho_{m,\vec{q}} = -iq_x v \delta_{m,\vec{q}} - q_y^2 D \rho_{m,\vec{q}} - (\kappa + \gamma(\rho_i^0))\rho_{m,\vec{q}}$$
$$\partial_t \delta_{m,\vec{q}} = -iq_x v \rho_{m,\vec{q}} - q_y^2 D \delta_{m,\vec{q}} - \gamma(\rho_i^0)\delta_{m,\vec{q}} \qquad (15)$$

The adiabatic elimination of the rapidly evolving local fluctuations in the difference between positive and negative dislocations gives :

$$\partial_t \rho_{m,\vec{q}} = -\frac{(q_x v)^2}{q_y^2 D + \gamma(\rho_i^0)} - q_y^2 D \rho_{m,\vec{q}} - (\kappa + \gamma(\rho_i^0))\rho_{m,\vec{q}}$$
$$\delta_{m,\vec{q}} = -\frac{iq_x v \rho_{m,\vec{q}}}{q_y^2 D + \gamma(\rho_i^0)} \qquad (16)$$

Hence, on large space scales, the mobility of dislocations gliding in a forest of pinning centers is effectively diffusive. The effective diffusivity of mobile dislocations may also be justified in the dynamical description of fatigue processes [4]. As in other systems presenting this effect (Taylor diffusion in hydrodynamic flows, chemically reacting electrolytes in the presence of electric fields,...[20-21]), the effective diffusion coefficient is proportional to the square of the mean velocity. In the problem discussed here, this allows one to relate the effective diffusion coefficient to the flow stress via the phenomenological laws mentioned above. Since this stress depends on the density of immobile dislocations, we see that the approximation made up to now and which consists in taking a constant uniform glide velocity may fail in several circumstances.

For example, let us first consider an explicit dependence of the glide velocity on the flow stress and consequently on the density of immobile dislocations. In this case, for a uniform steady state, the adiabatic elimination of $\delta_{m,\vec{q}}$ in eq.15 leads to the following contribution to the dynamics of $\rho_{m,\vec{q}}$ in real space :

$$\nabla_x v \frac{1}{\gamma(\rho_i^0)} \nabla_x v \rho_{m,\vec{q}} =$$

$$= \frac{1}{\gamma(\rho_i^0)} [v_0^2 \nabla_x^2 \rho_{m,\vec{q}} + v_0 v_0^{(1)} \rho_m^0 \nabla_x^2 \rho_{i,\vec{q}}$$

$$+ 3 v_0 v_0^{(1)} \nabla_x \rho_{m,\vec{q}} \nabla_x \rho_{i,\vec{q}} + v_0^{(1)2} \rho_m^0 (\nabla_x \rho_{i,\vec{q}})^2] \quad (17)$$

where $v_0 = v(\rho_i^0)$ and $v_0^{(1)} = dv/d\rho_i|_{\rho_i^0}$. We see that in this case, we obtain nonlinear couplings between the spatial gradients of the dislocation densities and cross-diffusion terms between the different densities. Since $v_0^{(1)}$ is negative and since an increase in ρ_m increases the growth of ρ_i via the pinning effects, the cross-diffusion has a destabilizing effect on uniform densities. The importance of this phenomenon has been recognized by Franek et al. [22] and can easily be incorporated in the approach introduced by Walgraef and Aifantis [13], but does not affect qualitatively the results. Effectively, their reaction-diffusion model becomes, in this case :

$$\partial_t \rho_{m,\vec{q}} = [D_{x,m} \nabla_x^2 + D_{y,m} \nabla_y^2] \rho_{m,\vec{q}} - E_m \nabla_x^2 \rho_{i,\vec{q}} + (\beta - \gamma \rho_i^{02}) \rho_{m,\vec{q}}$$

$$\partial_t \rho_{i,\vec{q}} = D_s [\nabla_x^2 + \nabla_y^2] \rho_{i,\vec{q}} + f(\rho_{i,\vec{q}}) - (\beta - \gamma \rho_i^{02}) \rho_{m,\vec{q}} \quad (18)$$

and the linear stability analysis of the uniform steady state is given by the eigenvalues of the matrix :

$$\begin{pmatrix} -\gamma \rho_i^{02} - D_{x,m} q_x^2 - D_{y,m} q_y^2 & -\beta + q_x^2 E \\ +\gamma \rho_i^{02} & -\alpha + \beta - q^2 D_s \end{pmatrix} \quad (19)$$

Hence one finds a pattern forming instability at $\beta = \beta_c = \beta_c^0 - \gamma \rho_i^{02} \frac{E}{D_{x,m}}$, where β_c^0 is the instability threshold in the absence of cross diffusion, and for $\vec{q} = q_c \vec{1}_x$ with $q_c^4 = \frac{\alpha \gamma \rho_i^{02}}{D_{x,m} D_s}$. Thus, we see that the critical wavelength is not affected by the cross-diffusion effect while the instability occurs at lower freeing rates of immobile dislocations.

When the distribution of pinning centers is nonuniform (e.g. when immobile dislocations form cellular structures) the effective diffusion of mobile dislocations is expected to be *anomalous*. Effectively, this problem is similar to the hydrodynamical case of particle transport in a nonuniform flow field. For example, when the flow is non uniform as in Rayleigh-Benard or Taylor-Couette systems , it may be shown that , on length scales much larger than the typical wavelength of the rolls, the effective diffusion coefficient is proportional to v^α (where v is the amplitude of the flow velocity variations and $\alpha < 2$) instead of v^2 [23]. The same behavior should be expected here and the dependence of the effective diffusion coefficient of mobile dislocations on the flow stress should be affected by the structure of the forest. A detailed study of this phenomenon should thus have a very practical interest for a quantitative study of the dependence of the instability threshold and critical wavelength on the density of immobile dislocations.

4. Conclusion

It has been shown that, in the collective behavior of defects in driven materials, the coupling between reaction and transport may lead to specific effects. For example, when the dynamics of point defects is dominated by recombination processes, the coupling with diffusion may lead, in irradiated materials, to the spontaneous segregation of interstitials and vacancies. On the other hand, in the case of deformed materials, gliding dislocations may acquire a diffusive behavior when annihilation or pinning effects are sufficiently important. According to the characteristics of the distribution of pinning centers, the diffusion may be normal or anomalous. Furthermore, when the pinning centers are the immobile dislocations of the forest, cross-diffusion terms appear which facilitate the occurence of the pattern forming instabilities.

References

1. F.Baras and G.Nicolis, "Chemical Instabilities," Reidel, Dordrecht, 1983.
2. D.Walgraef, "Patterns, Defects and Microstructures in Nonequilibrium Systems," Martinus Nijhoff, Dordrecht, 1987.
3. G.Martin and L.P.Kubin, "Nonlinear Phenomena in Materials Science," Trans tech, Aedermannsdorf (Switzerland), 1988.
4. D.Walgraef and E.C.Aifantis, Res Mechanica **23** (1988), p. 161.
5. D.Walgraef, in "Nonlinear Phenomena in Materials Science," G.Martin and L.P.Kubin eds., Transtech, Aedermannsdorf (Switzerland), 1988, p. 77.
6. a)G.Martin, Phys.Rev. **B30** (1984), p. 1424.
6. b)K.Krishan, Radiat.Eff. **66** (1982), p. 121.
7. S.M.Murphy, Europhys.Lett **3** (1987), p. 1267.
8. S.Murphy, in "NonlinearPhenomena in Materials Science," G.Martin and L.P.Kubin eds., Transtech, Aedermannsdorf (Switzerland), 1988, p. 295.
9. R.Bullough, B.L.Eyre and K.Krishan, J.Nucl.Mat. **44** (1975), p. 121.
10. N.M.Ghoniem and G.L.Kulcinski, Radiation Effects **39** (1978), p. 47.
11. D.Walgraef and N.M.Ghoniem, Phys.Rev. **B39** (1989), p. 8867.
12. K.Lindenberg, B.J.West and R.Kopelman, Phys.Rev.Lett. **60** (1988), p. 1777.
13. D.Walgraef and E.C.Aifantis, J.Appl.Phys. **58** (1985), p. 688.
14. G.Ananthakrishna and D.Sahoo, J.Phys. **D14** (1981), p. 2081.
15. Y.Estrin and L.Kubin, Res Mechanica **23** (1988), p. 197.
16. W.G.Johnston and J.J.Gilman, J.Appl.Phys. **30** (1959), p. 129.
17. U.Essmann and H.Mughrabi, Philos.Mag. **A40** (1979), p. 731.
18. M.Bocek, in "Nonlinear Phenomena in Materials Science," G.Martin and L.P.Kubin eds., Transtech, Aedermannsdorf (Switzerland), 1988, p. 369.
19. D.Walgraef and E.C.Aifantis, *Dislocation Inhomogeneity in Cyclic Deformation*, preprint, 1989.

20. G.I.Taylor, Proc.Roy.Soc.London **A225** (1954), p. 473.
21. P.J.Ortoleva and S.L.Schmidt, in "Oscillations and Traveling Waves in Chemical Systems," R.J.Field and M.Burger, eds., Wiley (New York), 1985, p. 333.
22. A.Franek, R.Kalus and J.Ktatochvil, *On the Stability of Dislocation Structures in cyclically deformed Metal Crystals and the Formation of Persistent Slip Bands*, preprint, 1989.
23. Ya B.Zeldovich, Sov.Phys.Dokl. **27** (1982), p. 797.

EXTERNAL NOISE AND PATTERN SELECTION IN CONVECTIVELY UNSTABLE SYSTEMS

ROBERT J. DEISSLER
Center for Nonlinear Studies, MS-B258
Los Alamos National Laboratory
Los Alamos, NM 87545, USA

In this note I briefly review work on external noise and pattern selection in the complex Ginzburg-Landau equation under conditions when the system is convectively unstable; and on slugs and confined states in this equation (with a destabilizing cubic term and a stabilizing quintic term) under conditions when the system is subcritical.

1. Convective Instability, External Noise, and Pattern Selection

Consider the stationary state of some system and a small spatially localized perturbation about that state. If the perturbation grows with time at a given stationary point, the state is *absolutely unstable*. However if the perturbation grows only in a moving frame of reference, eventually damping at any given stationary point, the state is *convectively unstable*.[1-7] For example, in open-flow systems such as fluid flow over a flat plate and plane Poiseuille flow,[7] a perturbation is convected with the flow so that it grows only in a moving frame of reference. Although the distinction between convective and absolute instability is not often explicitly made, it is an important distinction since, in convectively unstable systems: 1) external noise (or other external perturbation) is necessary for an asymptotic state different from the stationary state – giving rise to a *noise-sustained structure*;[1-3] 2) the external noise is selectively and spatially amplified, giving rise to spatially growing waves – this being a mechanism for *pattern selection*;[1-3] and 3) the external noise can play an important role in the macroscopic dynamics such as *spatiotemporal intermittency*.[1-3,8] Therefore, in convectively unstable systems, external noise can play a crucial role in both the origin and dynamics of structures (or patterns) which form.

For example, in fluid flow over a flat plate,[9] external noise near the leading edge of the plate can be selectively and spatially amplified giving rise to spatially growing waves. Further downstream, spatiotemporal intermittency occurs which is most probably correlated with the noise at the leading edge of the plate. Another example is the growth of sidebranches on a dendrite.[10-12] Noise near the tip of the dendrite could be selectively and spatially amplified as it is convected along the sides of the dendrite, giving rise to the sidebranches. The selective amplification would be responsible for the wavelength of the pattern which is selected and the random nature of the noise would be responsible for the irregularities in the sidebranches. As pointed out in ref. 13, in any system where noise is suspect, a good test to study the effect of noise is to add noise and see how the system responds.

An excellent model system to study these phenomenon is the complex Ginzburg-Landau equation:

$$\frac{\partial A}{\partial t} = aA - v_g \frac{\partial A}{\partial x} + b\frac{\partial^2 A}{\partial x^2} - c|A|^2 \psi \qquad (1)$$

The state $A = 0$ can be shown to be convectively unstable if $0 < a_r < v_g^2 b_r/(4|b|^2)$.[1,2] Under these conditions, external noise near the left boundary of the system is selectively and spatially amplified as it is convected to the right ($v_g > 0$) giving rise to spatially growing waves. The waves eventually saturate as a result of the nonlinearity forming some pattern. The wavelength of the pattern selected is related to the frequency of the spatially growing waves. In fact the selected wavelength can be calculated analytically for this equation.[1,2] For parameter values for which the pattern is convectively unstable (a secondary convective instability), the pattern breaks up at some spatial point (which changes with time) giving rise to spatiotemporal intermittency[1-3] and *convective chaos*.[2,3,14] These phenomena are most clearly seen in a movie which was shown at the conference.

2. Slugs and Confined States

Consider the stationary state of some system and a finite spatially localized perturbation about this state. Assume that the stationary state of this system is subcritical, meaning that only a perturbation of sufficient amplitude will grow whereas a perturbation of smaller amplitude will damp. Therefore, if the perturbation is sufficiently large it will grow and eventually saturate as a result of nonlinearities, producing a localized structure which will be surrounded by the stable stationary state. I will call such a structure a slug, after the term used for the subcritical structures occurring in pipe flow.

By taking, in eq. (1), $a_r < 0$ (so that sufficiently small perturbations damp), $c_r < 0$ (so that larger perturbations grow), and adding a quintic term to cause saturation, the complex Ginzburg-Landau equation can produce slugs – both chaotic and regular depending on parameter values.[3,8] By letting a_r be a function of space to simulate a convectively unstable inlet region and a subcritical region further downstream, this equation has been used to model the formation of turbulent slugs in pipe flow.[8] A movie of this system was shown at the conference. Also, for a certain set of parameter values, slugs (or confined states) form which have a unique size and shape for a wide range of initial conditions.[15] This feature of uniqueness makes these states very particle-like.

One of the least understood features of the confined states occurring in binary fluid convection is the fact that the velocity of the envelope of the state is orders of magnitude smaller than the phase velocity of the rolls within the state.[16,17] Brand and I[18] studied this problem by incorporating nonlinear gradient terms (which occur to the same order as the quintic term) into the Ginzburg-Landau equation. We found that indeed these terms can greatly affect the group velocity of the confined states.

3. Coupled Ginzburg-Landau Equations

Some systems such as binary fluid convection (convection in two miscible fluids) can be reduced near criticality to a set of two coupled complex Ginzburg-Landau equations.[19] The new features, as compared to a single Ginzburg-Landau equation, are equal and opposite group velocities which allow for counterpropagating waves; and a nonlinear cross-coupling term. This crosscoupling term was found to produce a number of qualitatively new phenomena[20] (depending on whether it is stabilizing or destabilizing) such as: 1) transitions between different stability regimes (e.g. convective to absolute and vice versa, and subcritical to supercritical and vice versa); and 2) annihilation of colliding slugs. A movie was shown which demonstrates these phenomena.

Brand and I[21] have also studied collisions between the particle-like states[15] mentioned above. Depending on the cross-coupling, these states can annihilate upon interaction or interpenetrate and emerge unchanged from their state before interaction. This later behavior is very reminiscent of the interaction between solitons (which interpenetrate and emerge unchanged from their state before the interaction) which occur in integrable systems. However, I stress that the particle-like states which occur in the Ginzburg- Landau equation are not solitons since the system is highly dissipative. So this is the first example of this soliton-like behavior occurring in both a highly dissipative and dispersive system. The collisions of particle states was shown in a movie.

Also, Sullivan and I[22] have studied coupled Ginzburg-Landau equations with a destabilizing cubic term and with reflections.[23] We find that as a result of the convective nature of the instability and a stabilizing cross-coupling, hysteresis (which one would expect for a destabilizing cubic term) can be eliminated. These results may explain the lack of hysteresis in some binary fluid convection experiments.[24−26]

4. Other Model Equations

Other model equations in which some of the behavior discussed above can occur are phase equations. In contrast to amplitude equations which describe the slowly varying amplitude of a plane wave (i.e. physical varible $\propto A(\chi,\tau)e^{ik_c x - i\omega_c t}$), phase equations describe the slowly varying phase of a plane wave (i.e. physical varible $\propto e^{ik_o x - i\omega_o t + \phi(\chi,\tau)}$). Brand and I[27] have shown under what conditions the Kuramoto-Sivashinsky equation is convectively unstable and have found noise-sustained structures in numerical simulations. We have also studied another phase equation which can produce the analogue of slugs (or confined states) which occur in amplitude equations.[28-30] [See the contribution by H. Brand in these proceedings.] The states we find correspond to rolls of a particular wavelength coexisting with rolls of a different wavelength.

References

1) R. J. Deissler, *J. Stat. Phys.* **40**:371 (1985); *Physica* **18D**:467 (1986).
2) R. J. Deissler, *Physica* **25D**:233 (1987).
3) R. J. Deissler, *J.Stat. Phys.* **54**:1459 (1989).
4) L. D. Landau and E. M. Lifshitz, *Fluid Mechanics* (Pergamon Press, London, 1959) p. 111.
5) R. J. Briggs, *Electron-Stream Interaction with Plasmas* (M. I. T. Press, Cambridge, Massachusetts, 1964).
6) P. Huerre, in *Instabilities and Nonequilibrium Structures*, E. Tirategui and D. Villaroel, ed. (Reidel, Dordrecht, Holland, 1987).
7) R. J. Deissler, *Phys. Fluids* **30**:2303 (1987).
8) R. J. Deissler, *Phys. Lett.* **120A**:334 (1987).
9) F. M. White, *Viscous Fluid Flow* (McGraw-Hill, New York, 1974).
10) A. Dougherty, P. D. Kaplan, and J. P. Gollub, *Phys. Rev. Lett.* **58**:1652 (1987).
11) R. Pieters and J. S. Langer, *Phys. Rev. Lett.* **56**:1948 (1986).
12) D. A. Kessler and H. Levine, *Europhys. Lett.* **4**:215 (1987).
13) J. D. Farmer, Los Alamos Preprint LA-UR-83-1450 (1982); R. J. Deissler and J. D. Farmer, submitted to Physica D (1989).
14) R. J. Deissler and K. Kaneko, *Phys. Lett.* **119A**:397 (1987).
15) O. Thual and S. Fauve, *J. Phys. (Paris)* **49**:1829 (1988).
16) P. Kolodner, D. Bensimon, and C. M. Surko, *Phys. Rev. Lett.* **60**: 1723 (1988).
17) J. Niemela, G. Ahlers, and D. S. Cannell, *Bull. Am. Soc.* **33**:2261, and to be published.

18) R. J. Deissler and H. R. Brand, submitted for publication (1989).
19) H. R. Brand, P. S. Lomdahl, and A. C. Newell, *Phys. Lett.* **118A**:67 (1986); *Physica* **23D**:345 (1986).
20) R. J. Deissler and H. R. Brand, *Phys. Lett.* **130A**:293 (1988).
21) H. R. Brand and R. J. Deissler, *Physical Review Letters* (accepted for publication, 1989).
22) T. S. Sullivan and R. J. Deissler, *Phys. Rev. A* (to appear, 1989).
23) M. C. Cross, *Phys. Rev. Lett.* **57**:2935 (1986); *Phys. Rev. A* **38**:3593 (1988).
24) J. Fineberg, E. Moses, and V. Steinberg, *Phys. Rev. Lett.* **61**: 838 (1988).
25) P. Kolodner and C. M. Surko, *Phys. Rev. Lett.* **61**: 842 (1988).
25) T. S. Sullivan and G. Ahlers, *Phys. Rev. A* **38**: 3143 (1988).
27) H. R. Brand and R. J. Deissler, *Phys. Rev. A* **39**:462 (1989).
28) H. R. Brand and R. J. Deissler, *Phys. Rev. Lett.* **63**: 508(1989).
29) H. R. Brand and R. J. Deissler, submitted for publication.
30) R. J. Deissler, Y. C. Lee, and H. R. Brand, to be published.

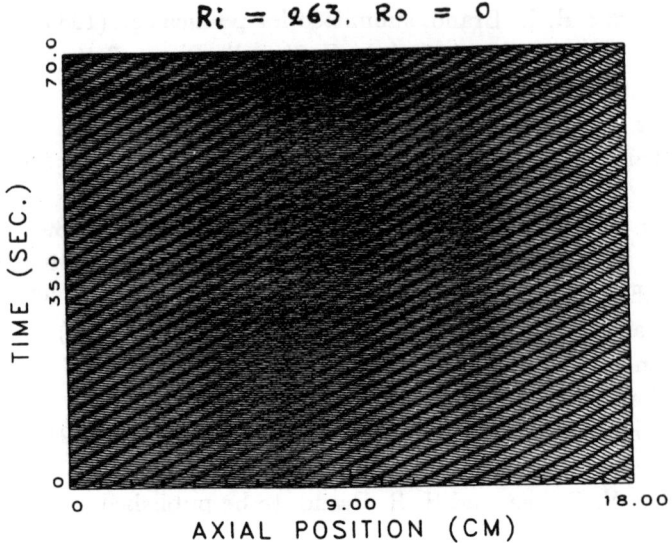
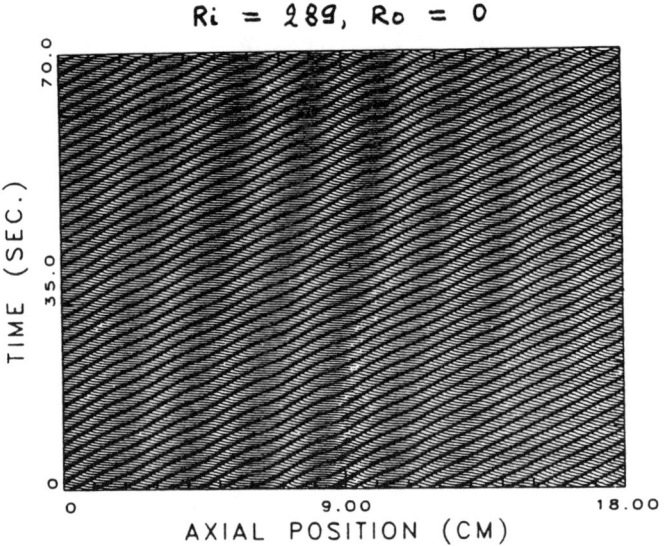

SECONDARY INSTABILITY OF TRAVELING INCLINED ROLLS IN TAYLOR-DEAN SYSTEM

I. MUTABAZI
Laboratoire d'Hydrodynamique et Mécanique Physique
URA 857 CNRS
10, rue Vauquelin
F-75231 Paris Cedex 05, France

The Taylor-Dean system consists of two rotating coaxial horizontal cylinders with a partially filled gap. It has been shown recently [1,2] that flow in this system exhibits stationary or Hopf bifurcation from laminar state depending on the velocities ratio of the rotating cylinders. The stationary bifurcation gives rise to axisymmetric stationary Dean and Taylor rolls while the Hopf bifurcation gives rise to a pattern of traveling inclined rolls. We study the instability of those traveling rolls.

The experimental system has the following characteristics : the cylinder radii ration $\eta = 0.883$, the gap size $d = 0.594$ cm and the aspect ratio $\Gamma = 90$. The gap between the two horizontal cylinders is filled partly with water containing 1% of Kalliroscope polymeric flakes for visualization of the flow. We define the inner and outer cylinders Reynolds numbers in following way: $R_i = \Omega_i r_i d/\nu$ and $R_o = \Omega_o r_o d/\nu$ with r_i, r_o the inner and outer cylinders radius respectively. Experimentally, one fixes the outer cylinder Reynolds number and varies the inner cylinder Reynolds number.

We investigate the traveling inclined rolls observed at threshold for the values or $R_o \in [-45, 45]$. The frequency and the velocity of the rolls varies along the cylinder axis (this is the manifestation of the phase instability) giving rise to nonperiodic collision events for ε 0.07 for $R_o = 0$. The second instability appears in the form of wavy modes in the azimuthal direction, and as the rolls travel along the axis of the system, the wavy instability induces a second wavelength in the axial direction, giving rise to a short wavelength modulation of 3 rolls. This wavy instability is forward and its threshold depends on the outer Reynolds number R_o. It is however difficult to measure the period of the azimuthal perturbations because of the partial filling but one expects that period to be a noninteger number because of the externally broken rotational symmetry. The figure shows the space-time plots for $R_o = 0$ at the threshold of the primary and secondary instability. The wavy instability is associated with a low frequency 20 times smaller than the fundamental one. For $\varepsilon = 0.13$, the roll pattern looses coherence between individual rolls but it conserves a coherent spatial structure with a

wavelength of 3 rolls even for higher ε before the turbulent states enter the system [3].

References

1. I. Mutabazi, J.J. Hegseth, C.D. Anderteck and J.E. Wesfreid, Phys. Rev. A, **38**, 4562 (1988).
2. I. Mutabazi, C. Normand, H. Peerhossaini and J.E. Wesfreid, Phys. Rev. A 39, **763** (1989).
3. I. Mutabazi, J.J. Hegseth, C.D. Andereck and J.E. Wesfreid, preprint (1989).

* The author wishes to thank the Nato financial support for travel expenses.

EXPERIMENTS ON THE FORMATION OF STATIONARY SPATIAL STRUCTURES ON A NETWORK OF COUPLED OSCILLATORS

T. Dirksmeyer, R. Schmeling, J. Berkemeier, H.-G. Purwins
Institut für Angewandte Physik, 4400 Münster, West-Germany

ABSTRACT. We describe an electrical network of coupled nonlinear oscillators exhibiting various spatial structures. Different routes into such structures have been observed experimentally. The route which is understood best is that via diffusive destabilization: by changing the diffusive coupling term a stable homogeneous system bifurcates into a stationary spatially periodic pattern. Wavelengths of the resulting patterns can be analytically predicted and agree well with the experiment. The discrete electrical circuit can be used to model pattern formation in continuous physical systems.

1. Introduction

The study of the dynamics of nonlinear electrical oscillators goes back to the beginning of this century and is closely related to the work of van der Pol /1/. A deeper understanding, however, has been achieved only in recent years. Typical nonlinear behaviour as period doubling, frequency locking and deterministic chaos including intermittency have been observed and studied by various authors and on various kinds of electrical oscillators. For a review of the experimental work on real physical systems see e. g. the recent book by F. C. Moon /2/.

Large arrays of coupled oscillators have been used as discrete equivalent circuits for continuous systems in the classical work of Nagumo et. al. for the study of nerve pulse transmission /3/, by Jäger /4/ and Yoshinaga /5/ for soliton experiments and by Geist and Lauterborn /6/ for numerical studies on soliton propagation. In these systems travelling waves and resonance effects have been of interest. Kuramoto et. al. /7/ treated weakly coupled oscillators using an array of coupled phase equations leading to self-organization and spatial structures. The reduction of the dynamics of the oscillators to phase equations, however, relies on the assumption of weak coupling and can probably not be applied to most continuous systems.

In general considered as another field of nonlinear dynamics, however strongly related to an array of identical coupled oscillators are systems described by reaction-diffusion equations (r-d-equations), which are of the form

$$\frac{\partial u}{\partial t} = D \Delta u + F(u) \quad ,$$

where u is a vector-valued function of space and time, D a diagonal matrix of diffusion coefficients, Δ the Laplace operator for the components of u, and F(u) a nonlinear function of u. R-d systems are intensively studied in various fields of physics, chemistry and other sciences. Of particular interest for the present paper is the work of Rashevski /8/ and Turing /9/, who applied this type of partial differential equation to the problem of biological morphogenesis. Stimulated by this work, r-d equations enjoy widespread interest today for the study e.g. of moving fronts in reactive chemical systems, flame propagation, morphogenesis, population dynamics and various other problems /10/. As these are systems with an infinite number of degrees of freedom, a thorough mathematical understanding is much more difficult than for systems described by ordinary differential equations where in general we deal with few degrees of freedom. Fundamental results on the pure mathematics are given by Smoller /11/. We also mention the work by Mimura, Nishiura, Fujii and coworkers /12/, Maginu /13/ and Rothe /14/, who treated a system equivalent to the one dealt with in this paper.

Though many qualitative features of real systems compare well with theoretical results, the quantitative description is still very poor except for certain soliton systems. Therefore we constructed an electronic network that obeys relatively well-defined dynamical equations. We demonstrate that the experimental results show quantitative agreement with numerical and analytical predictions. The network is a chain of van der Pol like oscillators with a strong diffusive coupling, so that the system can be viewed upon as a discrete version of a spatially continuous r-d-equation /15/. Experiments on weakly coupled oscillators have also been carried out. These will be communicated in a different paper.

2. The experimental setup and the network equations

The experimental setup used for this work consists of 128 coupled oscillators arranged on a line. The single oscillator -we call it a "cell" - is an LC-circuit with an s-shaped negative differential resistance (SNDC) (see fig. 1a). A single cell can be mathematically described by two first-order differential equations for the current I_i and the voltage U_i of fig. 1a. From Kirchhoff's rules we obtain:

$$L\frac{dI_i}{dt} = U_i - S(I_i)$$

$$C\frac{dU_i}{dt} = \frac{1}{R_V}(U_V - U_i) - I_i$$

2.1

$S(I_i)$ is the voltage-current characteristic of the SNDC shown in fig.1c and is realized by the two-transistor circuit fig. 1b that was designed to have a characteristic which can be approximated by a cubic polynomial in the current range where we do our experiments. We note that by eliminating the variable U_i from equations (2.1), we obtain a second-order differential equation similar to a van der Pol equation.

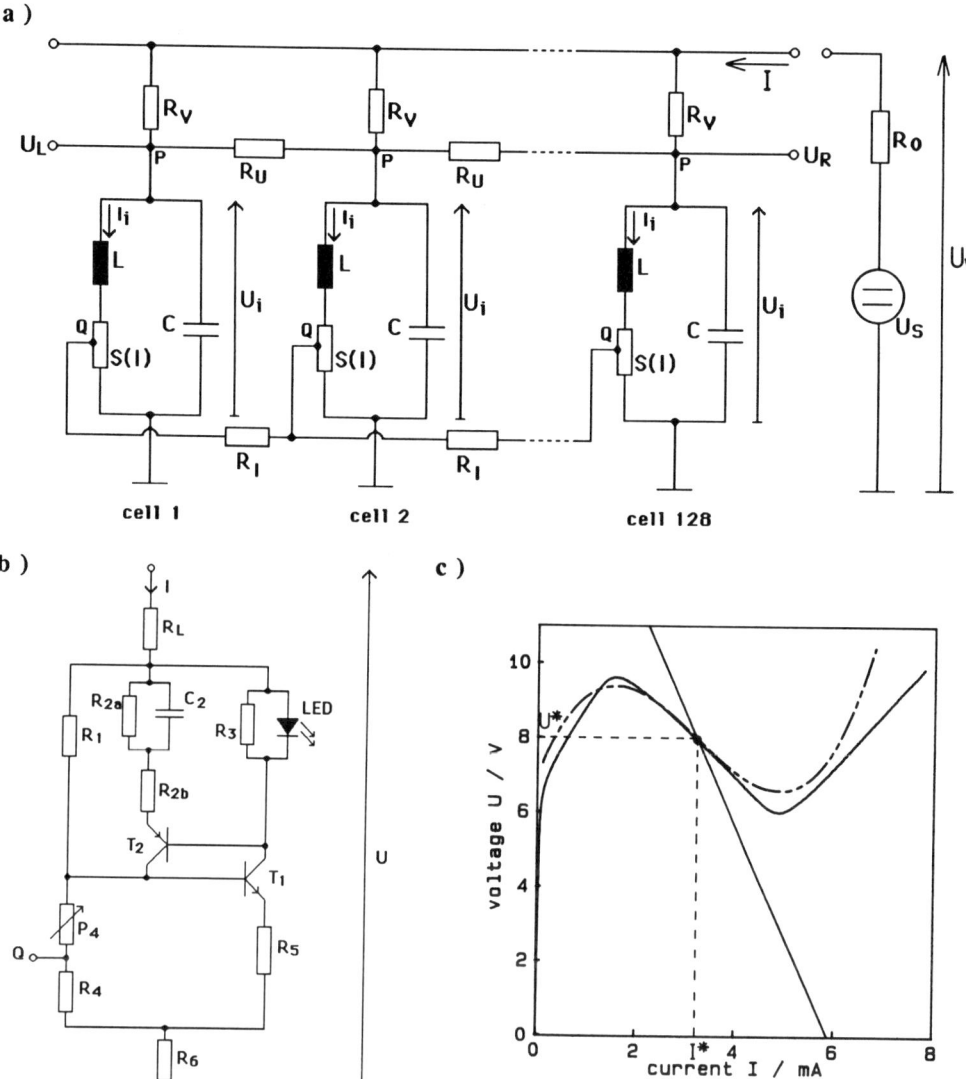

Figure 1. The experimental setup:
a) electrical network with supply voltage U_S and left and right hand boundary voltages U_L and U_R, L = 33mH,
b) circuit of the nonlinearity S(I) and cubic approximation(for details see /19/),
c) voltage-curent characteristic S(I) of the nonlinear resistance of a) (———), its cubic approximation (— — —) and a typical load line for R_V intersecting S(I) at the inflection point (I*, U*).

The single oscillators are voltage and current coupled by connecting points P of neighbouring oscillators via resistors R_U and points Q via R_I. For a detailed discussion of the current coupling we refer to the appendix. The R_U are realized by semiconductor photoresistors in an illuminated box and can be varied simultaneously in a range of two decades. Typical variances of R_U in the network are of the order of 5%. As we show in the appendix, the coupling resistances R_I result in a coupling term of the form $\gamma/R_I(I_{i+1}-2I_i+I_{i-1})$ between the currents of adjacent cells. The constant γ is $11.5 \cdot 10^6 \Omega^2 \pm 5\%$.

The whole system of fig. 1a is finally described by the following equations:

$$L \frac{dI_i}{dt} = U_i - S(I_i) + \gamma \frac{I_{i+1} - 2I_i + I_{i-1}}{R_I} , \qquad 2.2$$

$$C \frac{dU_i}{dt} = \frac{1}{R_V} \left(\frac{R_V}{R_V + NR_0} U_S - U_i + \frac{R_0}{R_V + NR_0} \sum U_i \right) - I_i + \frac{U_{i+1} - 2U_i + U_{i-1}}{R_U} .$$

Equations 2.1 and 2.2 may be scaled into dimensionless variables v and w (see table 1), where (I^*, U^*) is the inflection point of the characteristic in fig. 1c and the characteristic may be approximated by a symmetric cubic polynomial (see fig. 1c). Finally equ. 2.2 has the form

$$\dot{v}_i = D_v (v_{i+1} - 2v_i + v_{i+1}) + f(v_i) - w_i ,$$
$$\delta \dot{w}_i = D_w (w_{i+1} - 2w_i + w_{i+1}) + v_i - w_i - \kappa_1 + \kappa_2 \sum w_i \qquad 2.3$$

The parameter κ_1 is the rescaled dimensionless supply voltage such that at $\kappa_1 = 0$ the load line intersects the characteristic at the inflection point U^*, I^*. The set of coupled differential equations 2.3 can be interpreted as the second-order discrete approximation of a reaction-diffusion equation in the variables $v(\xi,t)$ and $w(\xi,t)$, if we identify $v_i(t)$ with $v((i-.5)\Delta\xi, t)$ and $w_i(t)$ with $w((i-.5)\Delta\xi, t)$, $\Delta\xi$ being the spatial resolution. With $\sigma = D_v/D_w$, $D_w = 1/(\Delta\xi)^2$ we obtain

$$\frac{\partial}{\partial \tau} v = \sigma \Delta v + f(v) - w \quad ; \qquad f(v) = \lambda v - v^3 ,$$
$$\delta \frac{\partial}{\partial \tau} w = \Delta w + v - w - \kappa_1 + \kappa_2 \int w \, d\xi \qquad 2.4$$

Without the integral term this equation has been treated in another context for instance by Maginu and Rothe /13/, /14/. For the relations between physical systems and these equations see Purwins and Radehaus /16/. We note that the dimensionless variable σ is proportional to the ratio of the voltage coupling resistance R_U and the current coupling resistance R_I.

The dynamics of one single cell can be visualized in the I-U- or v-w-phase space, see fig. 2. In the v-w-phase space the equilibrium points of one cell are the intersection points of the characteristic f(v) with the "load line" of the voltage source $v = w + \kappa_1$. The linear stability of these equilibrium states can be calculated from linear stability analysis. The intersection points (v_0, w_0) are in stable equilibrium if $f'(v_0) \delta < 1$ <u>and</u> $f'(v_0) < 1$. Points with negative $f'(v_0)$ (i.e. positive differential resistance) are necessarily stable, points with negative differential resistance may be stable or unstable. If no stable equilibrium exists, the system has a stable limit cycle.

TABLE 1. Scaling equations used to scale equation 2.1 and 2.2 into dimensionless variables v and w.

$$S(I) \approx U^* - \chi(I-I^*) + \varphi(I-I^*)^3 \quad \chi = 1270 \ \Omega \quad \varphi = 165 \cdot 10^6 \ V/A^3$$

$$v_1 = \alpha(I-I^*) \qquad\qquad w_1 = \beta(U^* - U)$$

$$\alpha = \sqrt{(\varphi/R_v)} \qquad\qquad \beta = \sqrt{(\varphi/R_v^3)}$$

$$\tau = (R_v/L)t \qquad\qquad \delta = R_v^2 \, C/L$$

$$\lambda = \chi/R_v \qquad\qquad \sigma = D_v/D_w$$

$$D_v = \frac{\gamma}{R_I R_v} \qquad\qquad D_w = \frac{R_v}{R_U}$$

$$\kappa_1 = \alpha\left(\frac{U_s - U^*}{R_v - N R_0} - I^*\right) \qquad \kappa_2 = \frac{R_0}{R_v + N R_0}$$

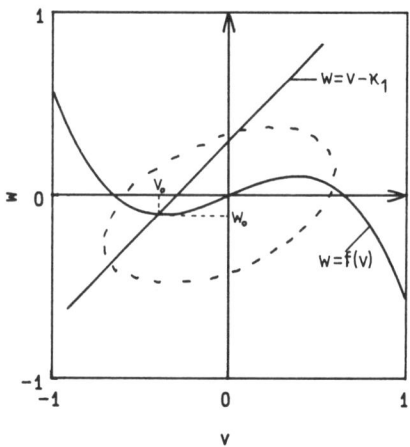

Figure 2. Phase diagramm of a single oscillator with $f'(v_0) < 1$. The intersection point (v_0, w_0) is in stable eqilibrium if $f'(v_0) \delta < 1$, otherwise the system has a stable limit cycle (-----).

3. Formation of stationary periodic patterns from a homogeneous oscillating state

When doing experiments the interesting components I_i and U_i or v_i and w_i can be measured easily. The dynamical behaviour is measured and stored with a digitizing oscilloscope, and stationary time-independent patterns are read with an integrated multiplexer and a minicomputer with analog-digital converter.

For the experiments described in this paragraph we have always $R_0 = 0\ \Omega$ and choose $\lambda < 1$, which means that there is only one stationary state for one isolated oscillator. Also we take $\lambda\delta > 1$ and $\kappa_1 = \kappa_2 = 0$. This means that the stationary solution $(v,w) = (0,0)$ of the isolated oscillator is unstable and a limit cycle is a stable motion. Note that although the cells are designed to be very similar, there is a variation of the frequency of about 2 % due to the scatter of properties of the electronic components. Doing experiments in the mentioned parameter range and choosing free boundaries which represent Neumann boundary conditions we observe oscillations if the voltage is switched on rapidly compared to the time scale of the oscillation, which is of the order of $(LC)^{-1/2} \approx 10^{-4}$ sec.

With sufficiently strong coupling between the cells, the oscillations of the single oscillators synchronize into an almost homogeneous phase locked state at one single frequency for all 128 cells. The formation of a spatially periodic stationary state is induced by applying a constant external voltage to one of the oscillators, preferably at the boundary. This results in an additional damping of the neighbouring oscillator, so that oscillations die out in a few periods. The resulting stationary state acts as a new boundary condition for the next oscillator cell, which in turn stops oscillating. Thus a front propagates through the lattice leaving behind a stationary state of cells. Finally all oscillators are in stable stationary states, and a spatially periodic pattern is observed. Returning to free boundary conditions has only minor effects very near to the perturbed boundary sites. Figure 3 shows the time evolution into equilibrium for two representative cells. Typical stationary voltage and current structures resulting from these experiments are shown in fig. 4. The parameters for these experiments are given in the figure captions.

Detailed measurements of the dynamics of these patterns show that the structurizing front moves through the system at constant speed. The delay between two adjacent oscillators is proportional to the square root of R_U (fig. 5). This is plausible because the square root of R_U scales with the length scale $\Delta\xi$ of the continuous system. Thus the observed behaviour is to be expected if the discrete experimental system approximates the properties of the continuous equations. For the same reason the period length of the resulting spatial structures scales with $1/\sqrt{R_U}$ (fig. 6). The inhomogeneous structures induced in this way show a remarkable stability with respect to disturbances of any kind. Even relatively large variations of the coupling and changes from Neumann to Dirichlet boundary conditions have only minor effects on the pattern. The numerical calculations are in good agreement with the experiments.

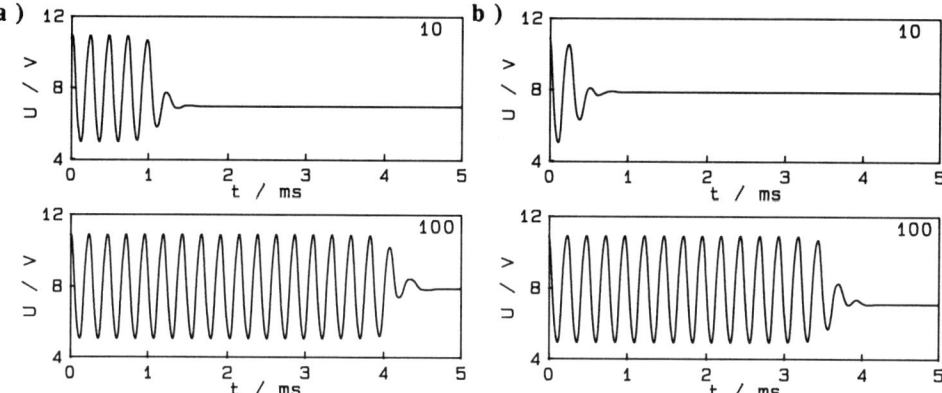

Figure 3. Time evolution of U_i into stationary states at cell No. 10 and 100 after applying a boundary voltage U_L to cell No.1. $R_U = 135\Omega$, $R_I = 47k\Omega$, $R_V = 3k\Omega$, $C = 33nF$, $U_S = 18V$; a) experiment, b) calculation.

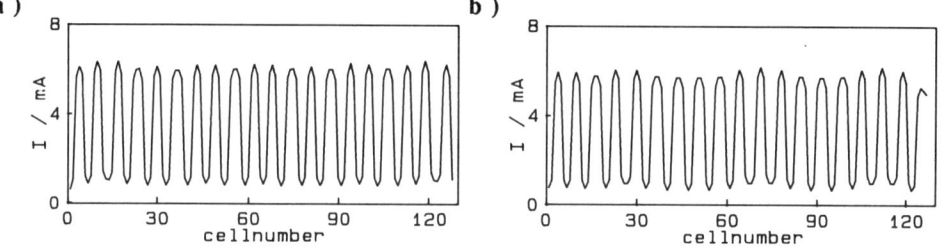

Figure 4. Stationary current structures found after front propagation as indicated in figure 3. Parameters as in figure 3.
a) experiment,   b) calculation.

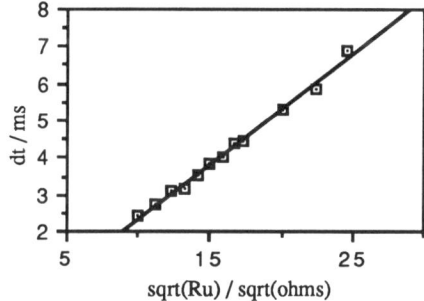 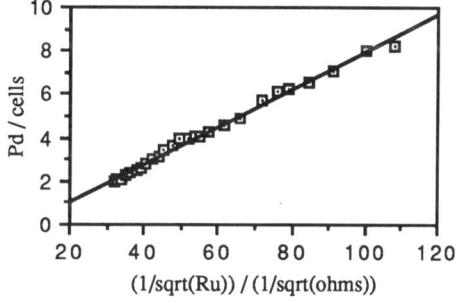

Figure 5. Dependence of delay time dt between two oscillators on the voltage coupling R_U. Other Parameters as in figure 3.

Figure 6. Dependence of the stationary period length P_d on the voltage coupling R_U. Other Parameters as in figure 3.

4. Stationary periodic patterns through diffusive instability

Pattern formation through diffusive destabilization of the homogeneous state is possibly the best known mechanism for the formation of spatial inhomogeneous structures. The experiments are carried out with the same parameters as in chapter 3, except that δ is chosen smaller by using a different capacitor so that $\lambda\delta < 1$ and the single oscillator has one stable equilibrium at the point (U^*, I^*), i.e. $\kappa_1 = 0$ [1]. This state is certainly also an equilibrium point for the coupled system. It is, however, not necessarily stable, as Turing showed in his pioneering paper /6/.

Linear stability analysis of the system at the point (U^*, I^*) results in an analysis of the stability of all eigenfunctions of the Laplacian. We treat first the continuous system with infinite extension, where the eigenfunctions are simply all periodic exponentials and where we can perform a Fourier transform instead. We then get an ordinary differential equation and a condition for the stability of each spatial mode q:

$$\lambda < \sigma q^2 + \frac{1}{1+q^2} \ . \qquad 4.1$$

A typical curve of this relation is shown in fig. 7a for $\sigma = 0.065$. Values (λ, q) lying below this curve correspond to spatial modes decaying to 0 in time, whereas values (λ, q) lying above correspond to modes which increase in time until saturation has occured. Therefore for an infinite system the homogeneous state is stable if λ is less than the minimum of that curve.

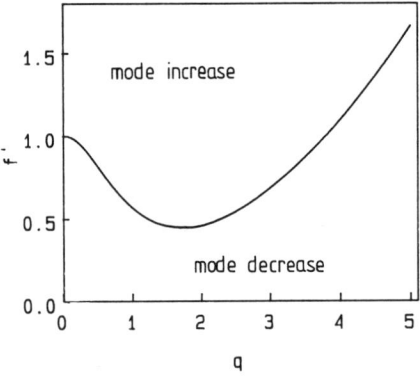 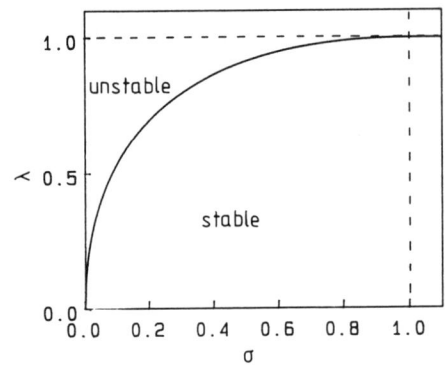

Figure 7a. Stability of the homogeneous state with respect to perturbations with mode q for $\sigma = 0.065$.

Figure 7b. Stability courve $\lambda_c = 2\sqrt{\sigma_c} - \sigma_c$ of the homogeneous state in the $\lambda - \sigma$ parameter space.

---

[1] For $\kappa_1 \neq 0$ substitute the value of λ by the slope of the characteristic f(v) at the intersection point. Though the linear stability analysis is identical, the evolving patterns may be different.

For a given value of λ ($\lambda < 1$), there will be a value σ_c where the homogeneous state is unstable with respect to in general exactly one spatial mode q_c. It is easily shown that this is the case for $\lambda_c = 2\sqrt{\sigma_c} - \sigma_c$ (see figure 7b). For the mode q_c we then have :

$$q_c^2 = \frac{1}{\sqrt{\sigma_c}} - 1 = \frac{1}{\lambda_c}\left(1 - \lambda_c + \sqrt{1-\lambda_c}\right) . \qquad 4.2$$

At the point λ_c, σ_c bifurcation will occur and the spatial mode q_c will grow exponentially in time until the growth is limited by nonlinearities. The bifurcation point can be reached by varying either σ or λ or a combination thereof. For finite systems we must take into account the fact that there are only discrete modes q_i. The first mode to be destabilized is still the one for which the right hand side of Eq. 4.1 is minimal. In small systems, however, the modes q_i may be so far apart from each other that the homogeneous mode $q_0 = 0$ is the first mode to be destabilized and so a spatial pattern formation will not occur. This is the mathematical aspect of the experience that a system may be "too small" to exhibit spatial structures.

What does this mean for the experimental system? We saw from the theoretical considerations above that by variation of σ a homogenous stable stationary state could be destabilized at a critical spatial wavenumber q_c. In the experiment the variation of σ is realized by the variation of the coupling resistance R_U. This can be done continuously by changing the photoresistors. We start with a relatively large R_U and establish a homogeneous equilibrium at the inflection point of the characteristic $S(I)$. When we reduce R_U below the threshold value, we observe the evolution of a periodic pattern in the current distribution of the system. The amplitude of these spatial oscillations increases with the distance from the bifurcation value. It turns out that the intensity - i.e. the square of the amplitude - of these oscillations grows almost linearly with the distance from the bifurcation value. This is shown in figure 8 together with a series of current distributions with decreasing resistance R_U. The critical value for R_U can be calculated from equ. 4.1, 4.2 and table 1:

$$R_U = \frac{R_V^2 R_I}{\gamma}\left(2 - \lambda - 2\sqrt{1-\lambda}\right) \quad , \quad \lambda = \frac{\chi}{R_V} . \qquad 4.3$$

Several experiments have been performed with different values of R_I. The observed critical values of R_U agree well with Eq. 4.3 and also with numerical simulations of the network equations, as is shown in the left part of table 2.

Is there a quantitative correspondence between the first unstable wavenumber and the observed patterns on the network? The natural equivalent of the wavenumber in the discrete system is the phase angle q between two adjacent cells, which is simply 2π/wavelength. Thus a continuous wavenumber q translates with help of table 1 into a phase angle $\theta = q \, \Delta x = q \, D_w^{-1/2} = q \, (R_U/R_V)$. In the right part of table 2 we compare the experimentally observed phase angles with those calculated in a numerical simulation on a digital computer and the phase angles obtained analytically from the first unstable mode of the continuous system. There is an excellent correspondence for small values of Δx, and deviations occur only in the more "coarse-grained" cases.

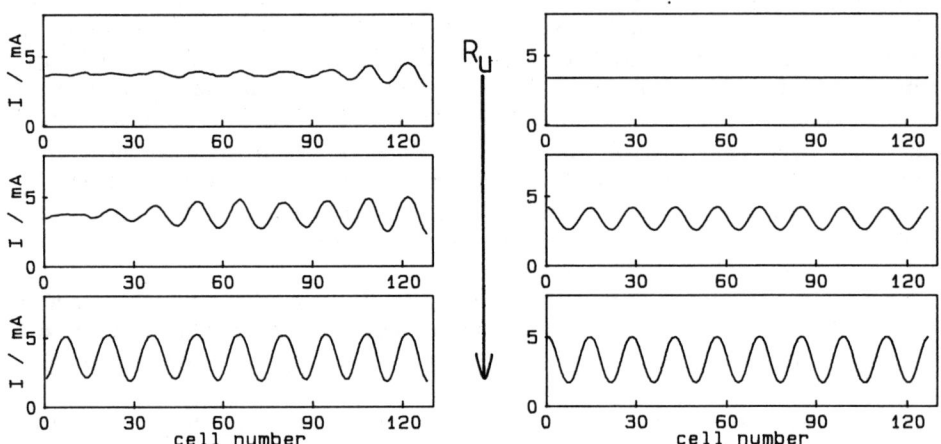

Figure 8a. Current in the network for different R_U. (Left : experiment, right : calculation) Parameters are: $R_U = 140\Omega$, 190Ω, 210Ω, $R_V = 3k\Omega$, $R_I = 4k\Omega$, $C = 6.8$ nF, $U_S = 18V$, $\lambda = 0.423$, $\kappa_1 = \kappa_2 = 0$, $\delta = 1.85$ and so $\lambda\delta < 1$.

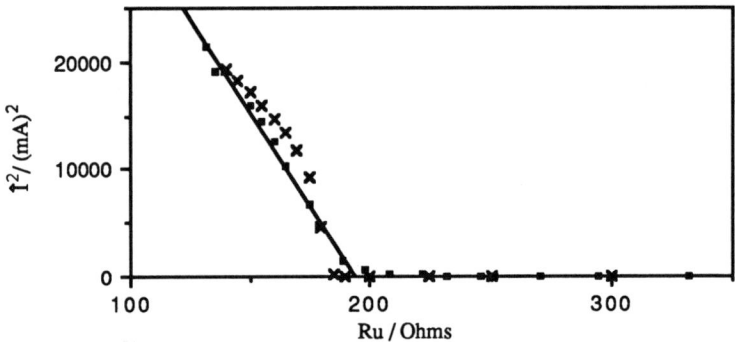

Figure 8b. Square of the amplitude of the periodic structure as a function of the bifurcation parameter R_U. (Squares : experiment, crosses : calculation) Parameters as in fig. 8a.

What corrections must be taken into account, if we want to be more accurate in the description of the discrete network? The linear stability analysis should have been done for a system with a finite difference operator instead of the Laplacian. This yields eigenvalues 2 (cosθ - 1) instead of the eigenvalues $-q^2$ of the Laplacian. The rest of the calculation is completely identical. For small values of θ the two expressions are identical up to second order, this is when the length of the patterns is large compared with the discretisation length Δx. For large θ we see increasing deviations, 2π being the upper limit for wave-numbers in the discrete system because of the periodicity of the cosine function. The finite size of the system and Neumann boundary conditions also set a lower limit for possible wave-numbers in our experiment. These corrections for discreteness are taken into account in the last column in table 2, resulting in a further improvement of the predictions.

TABLE 2. Experimentally observed, numerically calculated, and theoretically predicted (Eq. 4.2) critical coupling resistances R_U and spatial phase angles θ. The parameter λ was 0.423 thus σ = 0.058. The physical parameters were R_V = 3kΩ, C = 6.8 nF, U_S = 18V.

R_I / kΩ	R_U / kΩ exp.	R_U / kΩ num.	R_U / kΩ theor.	θ exp.	θ num.	θ cont.	θ discr.
4.02	0.19	0.19	0.18	0.44	0.44	0.44	0.44
8.20	0.40	0.37	0.37	0.62	0.62	0.63	0.64
15.00	0.45	0.68	0.68	0.84	0.89	0.85	0.87
22.00	1.07	1.00	1.00	1.05	1.08	1.02	1.08
30.00	1.50	1.37	1.36	1.23	1.28	1.20	1.28
39.00	1.94	1.80	1.76	1.46	1.46	1.36	1.51

How can we predict the amplitudes of the spatially inhomogeneous structures after bifurcation - but near the bifurcation point? The wavelength of the structure does not change much during the evolution of a steady state, as we saw in fig. 8. The exponential growth of the amplitude will be limited by the cubic nonlinearity. With the ansatz [2]

$$v(x,t) = \tilde{v}(t) \cos q_c x , \quad w(x,t) = \tilde{w}(t) \cos q_c x$$

we can approximate the new steady state solution by harmonic balance; the coefficients $\tilde{v}, \tilde{w}$ then obey the equations

$$\dot{\tilde{v}}(t) = -\sigma q_c^2 \tilde{v} + \lambda \tilde{v} - \frac{3}{4}\tilde{v}^3 - \tilde{w}$$

$$\dot{\tilde{w}}(t) = -q_c^2 \tilde{w} + \tilde{v} - \tilde{w}$$

---

[2] The phase difference between the two periodic functions can be shown to be zero.

These have stationary solutions:

$$\tilde{v}^2 = \frac{4}{3}\left[\lambda - \sigma q_c^2 \cdot \left(1 + q_c^2\right)^{-1}\right]$$

$$\tilde{w} = \left(1 + q_c^2\right)^{-1} \tilde{v}$$

If we vary only λ, keeping σ fixed, the above equation becomes

$$\tilde{v}^2 = \frac{4}{3}\left[\lambda - \lambda_c\right].$$

If we vary σ with fixed λ we have

$$\tilde{v}^2 = \frac{4}{3}\left[\sigma_c - \sigma\right]q_c^2.$$

In both cases the result is that the square of the resulting amplitudes is proportional to the distance from the bifurcation point, a result that is typical for pitchfork bifurcations. The stability of this solution is easily checked. As we showed already in figure 8, this behaviour is indeed observed in experiment as well as in numerical simulations.

5. Stationary aperiodic patterns (filaments)

Filamentary nonperiodic patterns in reaction-diffusion systems have found wide interest, e. g. in the study of semiconductors and gas discharge systems /17/. With a few modifications the system described above exhibits structure formation of this kind as well.

As experimental setup we use again a circuit as shown in figure 1a but this time with $R_0 \neq 0\Omega$ (typical values are 30 Ω). This resistor introduces a voltage drop proportional to the total current in the system. Thus an increase of the current in one single cell will induce a decrease of the supply voltages U_V at all other cells in the system. This is a nonlocal coupling of the cells that may act as a global inhibition. Thus equations 2.2, 2.3 and 2.4 apply. In the experiment we use the supply voltage U_S as a bifurcation parameter. When increasing U_S from 0 V, voltages and currents through the cells will increase monotonocally until the maximum of S(I) is reached. If $S'(I) > S'(I)_{crit}$ for the system (see preceding chapter), diffusive instability must occur at some point between the maximum of S(I) and the inflection point (U*,I*), - unless inhomogeneities destroy the homogeneous state before reaching this point. The bifurcation at the critical point may be supercritical or subcritical. In the supercritical case a periodic solution grows continuously out of the homogeneous state, as described in the preceding chapter. In the subcritical case the new solution will evolve in a discontinuous jump. For our system this will be the case if $\sigma < 0.035$, as can be derived from a calculation by Elmer /18/.

In the experiment we observe the spontaneous evolution of a high-current region out of random fluctuations at the bifurcation point as can be seen from figure 9. As could be expected from a subcritical bifurcation, the new stationary solution is formed in a sudden discontinuous jump. Thus a "filament" is formed, namely a localized structure in the system with high current density in a surrounding with low current density. The filament remains spatially extended because of the selfactivating of the current in the neighbourhood of high current cells, whereas the voltage coupling, - stronger than the current coupling - , has a long-range lateral inhibiting effect, that keeps the structure localized. This results in a relatively smooth space dependence of the voltages in combination with sharper gradients in the current distribution. The common resistor R_o is essential in so far as it suppresses the formation of concurring filaments by the effective supply voltage drops as a filament is formed. Only after some "recovery" a new filament may form.

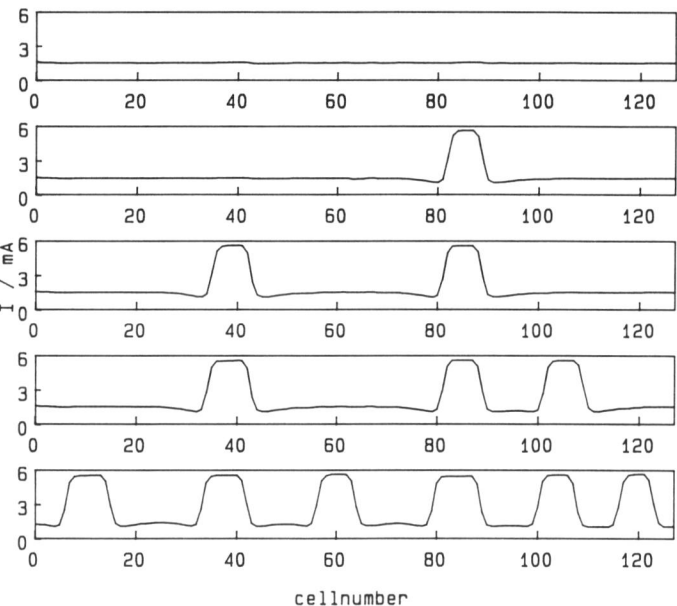

Figure 9. Experimentally observed formation of current filaments in the network at increasing the value of the supply voltage U_S from 17 V (top) to 22 V (bottom). Parameters are: $R_o = 30\Omega$, $R_V = 1100\Omega$, $R_U = 100\Omega$, $R_I = 15k\Omega$, $C = 6.8nF$, and $\lambda = 1.15$, $\delta = 0.037$, $\sigma = 0.063$, $\kappa_2 = 0.0061$.

When the supply voltage is increased further, finally existing filaments merge until the whole system is in a homogeneous high current state. When the voltage is reduced, the inverse process occurs. "Inverse filaments" of switched off cells occur, growing in number until they merge to form a homogeneous low current state. The current-voltage characteristic of the whole system is shown in fig. 10. It compares well with observed characteristics in gas discharge systems and in pin-diodes that can be described by the same reaction diffusion equation and where the same process of filamentation can be observed /17/. The point symmetry of the characteristic is a natural result of the symmetry of our local characteristic S(I) and since of equations 2.2.

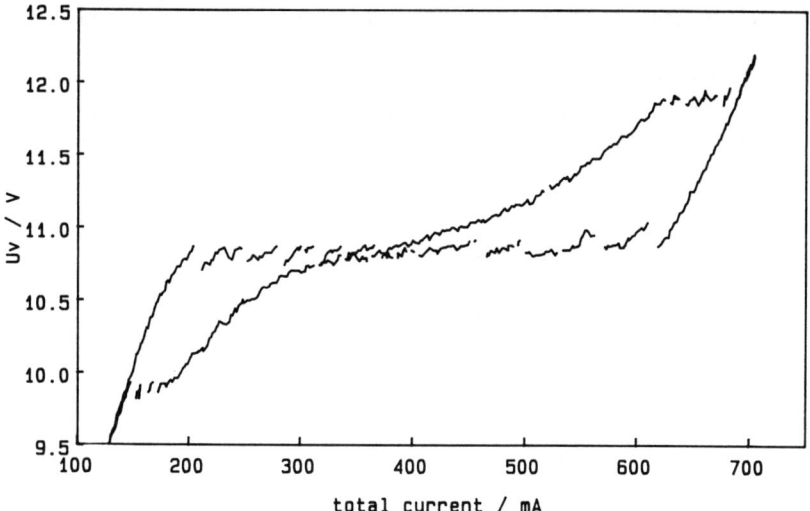

Figure 10. Global voltage-current characteristic of the network from experimental results. Parameters as in figure 9.

A detailed phase diagram of the qualitative behaviour of our systems has not yet been obtained. However, the influence of the parameters can be characterized as follows: a most essential feature is the combination of a strong inhibitor diffusion and a comparatively weak but finite activator diffusion. Activator diffusion must be small compared to inhibitor diffusion in order to allow stable stationary structures. It must have a finite value, however, so that initial fluctuations can grow into full-size filaments. In earlier experiments without activator diffusion point-like patterns appeared instead. The common resistor R_0 is essential for the selection of one filament from an initial state and for the suppression of concuring filaments but not for the evolution of an inhomogeneous state in general. The value of $\lambda \delta$ should be less than 1 to avoid autonomous homogeneous oscillations that suppress the evolution of spatial patterns. The relative slope of the characteristic λ may be smaller or larger than 1. If it is larger than 1, the formation of a periodic structure by diffusive instability may occur in a very small range of supply voltages and can be completely suppressed if inhomogeneities in the system allow one single cell to switch into the high-current state before. Then the evolution of a filament starts from this activated cell and the bifurcation point is not reached.

Is such a filament also a stable solution of the continuous system equ. 2.4 ? This question is not easily answered. In the case of diffusive destabilization as discussed in the preceding chapter we could derive the existence of periodic spatial solutions for the continuous system analytically. This could not be achieved for the formation of filaments. It is not sure if the typical finite maximum size of a filament is indeed a property of the continuous system. However, it seems that the maximum size of a filament is associated with the wavelength of the stationary patterns in diffusive destabilization, a phenomenon that is known to exist in the continuous system.

6. Conclusion

We could show three different typical forms of pattern formation in an experimental electrical network. The results are typical for the behaviour of reaction-diffusion systems and compare well with analytical predictions and numerical simulations. The experimental system allows quick answers to variety of problems in reaction-diffusion equations and is particularly useful in combination with numerical work. It is plausible in its design and easily adapted for different problems.

The network was also used as a first approximation for the understanding of such complex phenomena as the filamentation in gas-discharge systems and semiconductors. Currently work is also done on front-propagation as it is found in flame propagation, chemical systems and bistable electrical media.

Appendix

THE REALIZATION OF THE CURRENT COUPLING IN THE SYSTEM

In the experimental setup described above it is not possible to realize a coupling between the currents through adjacent cells as simply as it is for the respective voltages. However, we can imagine the current coupling as an influence of adjacent currents on the nonlinearity S(I) in equ.2.2:

$$L\ I_i = U_i - S(I_i) + \gamma / R_I\ (I_{i+1} - 2\ I_i + I_{i-1}) = U_i - S'(I_i, I_{i+1}, I_{i-1}) \qquad A.1$$

This means, the voltage over the nonlinearity must be shifted proportional to the current gradient. How can we achieve this? First we note that at point Q_i the voltage $U_i{}^Q$ depends linearly on the total current I_i through the device:

$$U_i^Q \approx U_c + \kappa I_i \qquad U_c = \text{const.} \qquad \text{A.2}$$

Second an additional small current j impressed at point Q shifts the voltage S(I) over the device:

$$S'(I, j) \approx S(I) - \varphi j \qquad \varphi > 0. \qquad \text{A.3}$$

If we now introduce a resistive coupling R_I between respective points Q_i, we expect a current j at point Q_i:

$$j = 1/R_I (U_{i+1}^Q - 2 U_i^Q + U_{i-1}^Q) = \kappa / R_I (I_{i+1} - 2 I_i + I_{i-1}). \qquad \text{A.4}$$

and finally

$$S(I_i, I_{i+1}, I_{i-1}) = S(I_i) - \varphi \kappa / R_I (I_{i+1} - 2 I_i + I_{i-1}),$$

just the term we needed for diffusive coupling of the currents.

The validity of the approximations depends critically on the parameters of the device. The circuit in use could be optimized, however, such that the coupling currents are very small and the deviation from equ. A.3 is less than 1.5 % over the whole range of currents in our experiments. Equation A.2 is correct with a deviation of less than than 1 % except for points in the characteristic where the transistor T_1 is blocked. This is the case, however, only for currents less than 100 μA which are not of interest in our experiments. So a simple but effective method could be established to simulate diffusive coupling also in the current variable.

References:

1. B. v. d. Pol; Proc. IRE, Vol. 22, No. 9, 1051-1086, 1934
2. F.C. Moon, Chaotic Vibrations, An Introduction for Applied Scientists and Engineers; J. Wiley, New York 1987
3. J. Nagumo, S. Arimoto, S. Yoshizawa; Proc. IRE, 50, 2061-2070, 1962
4. D. Jäger; J.Phys.Soc.Japan 51, 1686, 1982 and references therein
5. T. Yoshinaga, T. Kakutani; J.Phys.Soc.Japan 53, 85, 1984
   56, 3447, 1987
6. K. Geist, W. Lauterborn; Physica D31, 103-116, 1988
7. Y. Kuramoto; Chemical oscillations, waves, and turbulence; Springer 1984
8. N. Rashevski; Bull. Math. Biophys., Vol. 2, 15-25, 1940
9. A.M. Turing; Phil.Trans.Roy.Soc. 237, 37-72, 1952
10. P.C. Fife; Mathematical aspects of reacting and diffusing systems; Springer 1979
    H. Meinhardt, A. Gierer; J.Cell Sci. 15, 321-346, 1974
    D.G. Aronson, H.F. Weinberger; Nonlinear diffusion in population genetics, combustion and nerve pulse propagation. in: Proceedings of the Tulane Program in Partial Differential Equations and Related Topics, Springer 1975
11. J. Smoller; Shock waves and reaction diffusion equations; Springer, Berlin 1983
12.. H. Fujii, M. Mimura, Y. Nishiura; Physica 5D, 1-42, 1982
    T. Ohta, M. Mimura, R. Kobayashi; to appear 1989
13. K. Maginu; Math. Biosc. 27, 17-98, 1975
14. F. Rothe, P de Mottoni; Annali Mat. pura appl. 122, 141-157, 1979
15. J. Berkemeier,T. Dirksmeyer, G. Klempt, H.-G. Purwins; Z.Phys.B, 65, 255-258, 1986
    H.-G. Purwins, G. Klempt, J. Berkemeier; in: Festkörperprobleme - Advances in Solid State Physics - 27, 27-61, Vieweg, Braunschweig 1987
16. H.-G. Purwins, C. Radehaus; in: Neural and synergetic computers, international workshop on synergetics; Springer 1988
17. H. Baumann, T.Pioch, H.Dahmen, D.Jäger; Scann. Electr. Microsc. 1986 II, 441-446, 1986
    C.Radehaus, K.Kardell, H.Baumann, D.Jäger, H.-G.Purwins; Z.Phys.B, 65 515-525, 1987
    C.Radehaus, T.Dirksmeyer, H.Willebrand, H.-G.Purwins; Phys.Lett. A125, No. 2, 92-94, 1987
    J.Parisi, J.Peinke, B.Röhricht, K.M.Mayer; Z.Naturf. 42a, 329-332, 1987
18. F.J.Elmer; Physica D 3, 321-342, 1988
19. R. Schmeling, Diplomarbeit , Münster 1988

STUDIES ON INSTABILITIES AND PATTERNS IN EVAPORATING LIQUIDS AT REDUCED PRESSURE AND/OR MICROWAVE IRRADIATION

G. BERTRAND, M. LALLEMANT, A. STEINCHEN, P. GILLON, P. COURVILLE, D. STUERGA.
Laboratoire de Recherches sur la Réactivité des Solides
Université de Bourgogne (Unité Associée au CNRS U.A. 23) BP 138
F. 21004 DIJON Cedex

ABSTRACT. This paper summarizes our recent experimental and theoretical work on the instabilities in liquids and at interfaces which form during evaporation at reduced pressure and/or microwave irradiation. We have observed a variety of patterns (Benard rolls, Marangoni waves, Hickman interface deformations) which depend on the value of the reduced pressure and the power of the incident beam.

1. INTRODUCTION.

The evaporation of liquids under reduced pressure of their vapour has been very widely studied. The non equilibrium state in the liquid phase and at the interface, is the driving force of the evaporation. It is also a source of well-known hydrodynamical instabilities.

One finds again the famous volumic phenomenon, the so-called Rayleigh-Benard convection [1, 2]. One finds also the surface Marangoni effect [3]. But, as the liquid evaporates endothermically, exchanges of matter and heat occur at the interface and the boundary moves. So, these classical topics cannot be treated with the standard methods. They can be rather studied in comparison with what is made in crystal growth.

2. ETHANOL EVAPORATION UNDER REDUCED PRESSURE.

Figure 1 schematizes the apparatus [4, 5]. An evaporating tube, with a circular or square section contains only the pure liquid and its pure vapour, at the beginning in the equilibrium state at an outside controlled temperature. Some observations and measurements can be made : the motion of fluid particles by a video camera, the surface temperature by an infra-red radiometer or an infra-red camera, the normal temperature gradient by a set of mobile, fine thermocouples, the evaporation rate by using an electronic device maintaining the interface at a constant level. The reduced pressure is imposed and controlled by an isothermal cold point.

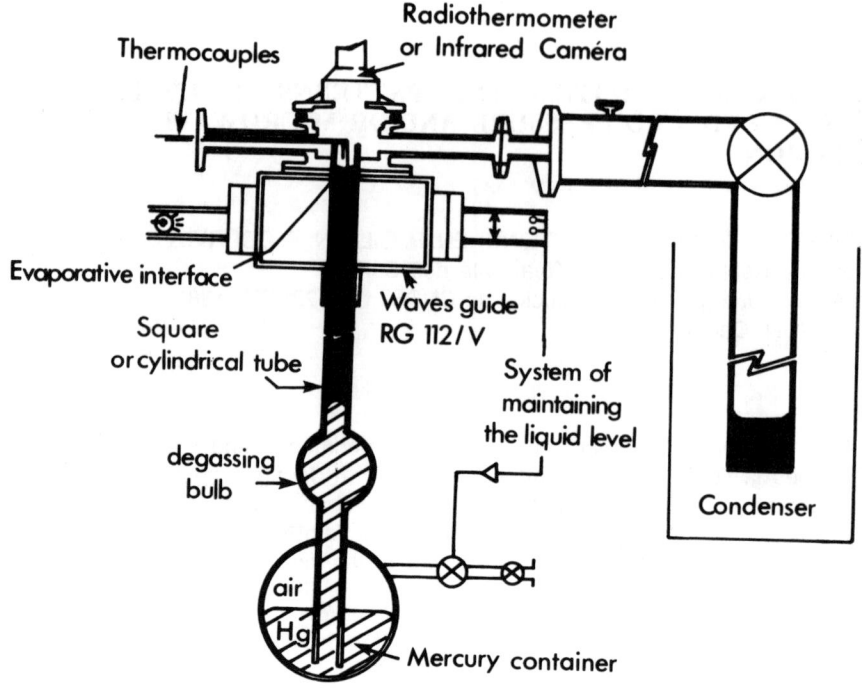

Figure 1 - Schematic view of the experimental device.

Figure 2 summarizes some results which were obtained, when the outside temperature is imposed at 20° C. At this temperature, the equilibrium pressure P_{eq} is 5.73 KPa. The reported experiments correspond to more and more reduced pressure, each experiment running for a given pressure. $P_r = P/P_{eq}$ defines the relative reduced pressure.

If P_r is greater than 0.68, in the region near the equilibrium, the evaporation goes slowly and no particle motion is observed in the liquid. The I.R. Camera does not reveal any local variation of the interface temperature. Clearly, the driving regime is the conductive regime.

Below this value, convection occurs. Near the transition point, the volumic stationnary convection is observed, as in confined systems, with an axisymmetrical flow. The convective roll is stable in time.

For larger constraints, P_r being less than 0.45, the convection becomes unsteady ; the centre of the roll is slightly out of the axis and turns around the centre of the tube. While increasing disequilibrium the motion becomes more and more complex with cold thermal spots plunging into the liquid and modifying the roll motion. But it is remarkable, that an interface layer remains stable and undeformed, despite the bulk convection.

This situation changes when P_r reaches more or less 0.05. The Marangoni instability localized in the interfacial layer is clearly observed, as it is shown on the infra-red picture of the surface which presents small regular hot spots, and also on a video picture which exhibits regular wavelets along the interface. In the volume of the liquid, the convective cells continue to be more and more perturbed.

Finally, when the reduced vapour pressure is decreased below 0.014, an explosive increase of the evaporation rate is measured. At the same time, the mechanical stability of the surface is broken and violent deformations are observed.

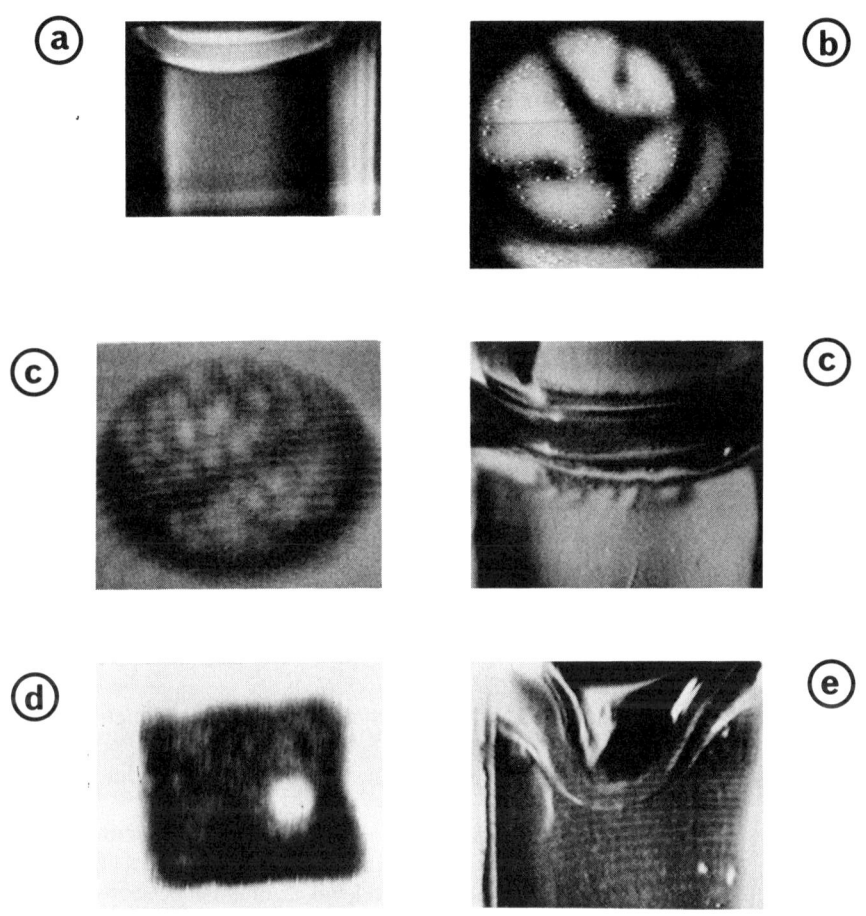

Figure 2 - Some infrared thermographs or optical views for increasing constraint
    a - the conductive regime   b - buoyancy driven convection cells
    c - Marangoni cells   d - hot spot of the surface
    e - deformation of the interface

If this instability is followed step by step, firstly a spot is observed, which is hotter than the others, after it is amplified. In the same place, the surface shows a crater, which grows and finally the surface breaks. After some swings, the surface becomes calm before a new identical scenario. The period, which depends on the vapour pressure can vary from 1 to 200 seconds.

It seems that this kind of instability is less known. Hickman, in 1952 [6], has given a first experimental description and the theory has been proposed by Palmer in 1976 [7] and more recently by Prosperetti and Plesset in 1984 [8] ; they have called it "the instability by vapour recoil".

So, when an evaporative system is removed by lowering the pressure further and further from its equilibrium, a succession of instabilities can emerge, first volumic and steady, then surfacic and unsteady. These instabilities are directly related to the increase of the evaporation rate.

3. THE OBJECTIVE OF THE MICROWAVE IRRADIATION AND THE RESULTS IN ETHANOL EVAPORATION.

The relation between instabilities and evaporation rate explains the interest to evaporate dipolar liquids under microwave irradiation.

Sure, this choise has had some applicative reasons. Evaporation is implied in numerous industrial processes ; it requires in general a very long time, because of the need for the heat energy transfer; it is heterogeneous. Microwaves irradiation leads to the local heating of the liquid, if it is a dielectric material, directly in the bulk, without the need of thermal conduction. So, the evaporation could be strongly activated, what is the aim in general for applications.

Beyond this concern, some new behaviours can be expected in the instabilities and patterns due to the new properties introduced : the dielectric property of the material, the matter-radiation interaction. The following questions arise :
How are the characteristic equations of the problem modified and what repercussions has the local supply of thermal energy on the phenomena ? How does the activation of the evaporation act ?
How are the instabilities modified ? What are their domain of existence ? Do new instabilities appear ?

Some results obtained on the evaporation of ethanol under microwave irradiation can be now described.

The evaporator shown before, is put in the centre of the microwave cavity, it crosses the waveguide. For maintaining the volume of the irradiated liquid and for avoiding discontinuities in the guide, the surface level is maintained constant, a few millimeters above the waveguide, by mean of a photosensitive device, which transduces the position information to the valve of a mercury piston. (figure 1).

The design of the microwave circuit is compatible with an electromagnetic wave which is of the transverse type $TE_{0.7.15}$ with a frequency of 2445 MHz, it guarantees a precisely stabilized resonant one mode cavity [5]. The intensity of the electric field is about 200 to 700 $V.m^{-1}$.

The diagram (figure 3) in which the reduced pressure P_r and the incident microwave power are the axis was studied systematically, point by point, so that the limits of the domains of existence of the different hydrodynamic regimes can be approximatively plotted.

The domain of the conductive regime is very narrow and finishes for a very small irradiation power, about 0.04 W. The region of the steady Benard convection, of the unsteady convection and so one are also identified. No new instabilities seem to appear.

So one can observe a progressive evolution from bulk phenomena to surface phenomena, when the evaporative system is removed from equilibrium. The microwave irradiation squeezes the stationary processes and enlarges the domain of unsteady behaviours. This is clearly seen in this stability diagram in which the span of the domain of the unsteady convection is much larger under microwave. The same prevails for the domain of the nearly steady Marangoni cells which gets narrower and disappears in favour of the Hickman instability, which is characterized by localized hot spots at the surface and violent motions of the surface.

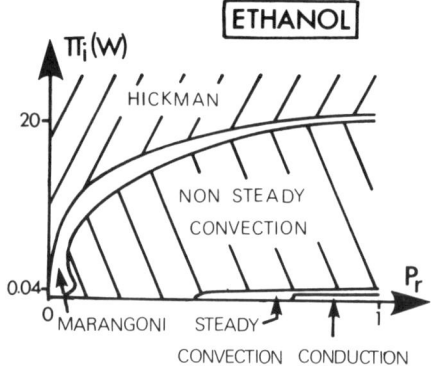
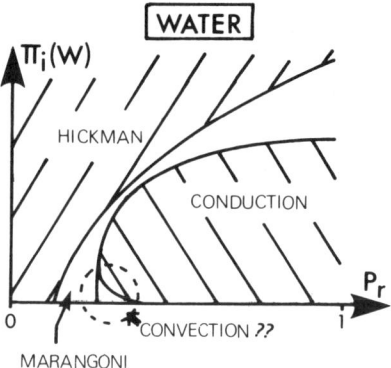

Figure 3 - Microwave irradiation vs depression plot - Ethanol evaporation. The different regimes of evaporation.

Figure 4 - Microwave irradiation vs depression plot - Water evaporation - The different regimes of evaporation.

4. EVAPORATION OF WATER.

Figure 4 shows directly the stability diagram of the different regimes, in the coordinates "reduced pressure-microwave power". It seems to be rather different from that drawn for ethanol evaporation either under reduced pressure or under irradiation beam.

- The domain of the conductive regime is wider. It persists for an irradiation power higher than some watts. For ethanol, it has disappeared for 0.04 Watt.

- The region of the Benard convection, steady or not, which was seen very easily in ethanol, is not underlined here. If it should exist, this would be only in a very narrow region near the abscissa, but no experiments were made, up to now, in this region. If it could appear, necessarily it would disappear for a few milliwatts.
- The Marangoni regime, which appeared in ethanol only in a restricted domain at low pressure is found in water also for a high microwave power. For lower irradiations, the conductive regime is stabilized.
- The Hickman instability is found again in a large domain of controlled parameters.

From this comparison of the behaviour of both liquids, we can say that a strong stabilization of the liquid bulk operates in water, whereas it is not effective in ethanol. We can say also that coupling between the non steady convection and the surface inhomogeneities cannot occur in water and this fact opens a new space for a pure Marangoni effect.

5. THEORETICAL APPROACH.

5.1. The stationary solutions and their limits.

The steady state solutions are deduced from a simplified model of evaporation. It consists of a one-dimensional model with the z axis perpendicular to the evaporative surface. The volume of the liquid is irradiated by the microwave beam, but the interface is out of the beam.

The following boundary conditions are assumed:

at $z = 0 \quad T = T_s$

and $D_{th}^{\ell} \left.\frac{\partial T}{\partial z}\right|^{\ell}_{z=0} - D_{th}^{v} \left.\frac{\partial T}{\partial z}\right|^{v}_{z=0} = LV_o$

L is the latent heat of evaporation and V_o the evaporation rate. D_{th} is the thermal diffusivity.

At $z = \infty \quad T = T_b$ which is a constant for a given experiment.

The balance equation for energy in a non convective state is given by the following relations:

$$\frac{\partial T}{\partial t} = D_{th}^{\ell} \frac{\partial^2 T}{\partial z^2} + V_o \frac{\partial T}{\partial z} + \varepsilon_{diss}$$

The three terms can be easily identified as the conductive term, the Stefan term and the dissipated energy term. This term can be written, taking into account the expression given by Böttcher and Bordewijk [9]:

$$\varepsilon_{diss} = \frac{\omega^2 E^2 \varepsilon_i}{8 \pi \rho^{\ell} c_p^{\ell}}$$

where $\omega = 2\pi f$

f being the frequency of the applied microwave radiations, E is the internal electric field intensity, which could be different from the applied one, ε_i the factor of dielectric loss, which comes from the expression of the dielectric constant :

$\varepsilon = \varepsilon_r + i\varepsilon_i$

ρ^ℓ and Cp^ℓ are respectively the density and the thermal capacity of the liquid phase.

E, ε_i, ρ and C_p can be temperature dependent ; but taking into account the amplitude of the variations of each of these terms with temperature in a small interval around the reference temperature T_o, one may only consider the variation of ε_i and suppose that the internal field is equal to the applied field E_0.

So, we have after linearization :

$$\varepsilon_i(T) = \varepsilon_{io}(1 + \alpha_{\varepsilon_i}(T - T_o))$$

in which $\alpha_{\varepsilon_i} = \dfrac{\partial \varepsilon_i}{\partial T} \dfrac{1}{\varepsilon_{io}}$

With all these assumptions, we can solve the stationary equation $\partial T/\partial t = 0$
Details of the calculation are given in Reference [10].
Two types of solution T(z) are found, depending on the sign of the parameter defined above :

α_{ε_i}

5.2. The limit of stabilities.

The length scales chosen for the definition of the non dimensional parameters, such as the Rayleigh number, are constraints dependent. So, the corresponding Rayleigh numbers with mobile boundary can be respectively defined by the following equations.

for reduced pressure experiments :

$$(Ra)_\ell = \dfrac{\alpha^* g}{D_{th}^\ell \nu^\ell} (T_b - T_s) \left(\dfrac{D_{th}^\ell}{V_o}\right)^3$$

for irradiation experiments, with $\alpha_{\varepsilon_i} > 0$

$$(Ra)_\pi = \dfrac{\alpha^* g}{D_{th}^\ell \nu^\ell} (T_b - T_s) \dfrac{\lambda^3}{4}$$

where λ is defined by :

$$\frac{2\pi}{\lambda} = \sqrt{\frac{\omega E_o^2 \varepsilon_i \alpha_{\varepsilon i}}{8\pi \rho^\ell c_p^\ell D_{th}^\ell} - \left(\frac{V_o}{D_{th}^\ell}\right)^2}$$

α^* is the thermal expansion coefficient.

These Rayleigh numbers vary non linearly with the applied constraints and this non classical Benard problem is close to some problems found in the directional growth of crystals or in oceanic and atmospheric layers.

We have been able to estimate the value of the Rayleigh number. Concerning the critical Rayleigh number Ra_c, it depends on the aspect ratio. Catton [11] performed the numerical calculations of this critical value for liquids which are confined in parallelipipedic containers with isothermal walls. His calculation can be a good approximation for us.

We have plotted (figure 5) the variation curves of Ra and Ra_c versus the deviation from equilibrium for the evaporation of ethanol under reduced pressure. The curves cross each other for $P_r = 0.7$, value in good agreement with the first bifurcation experimentally observed for the transition between the conductive regime and the convective one in ethanol evaporation. (Fig. 3).

Under an electric field, electroconvective instabilities could occur. For these, Turnbull has defined an electrical Rayleigh number $(Ra)_e$ [12], as:

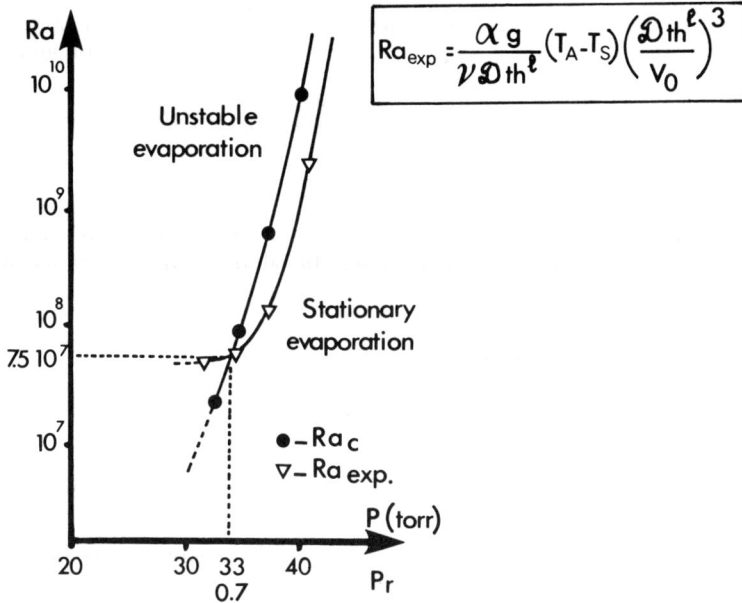

$$Ra_{exp} = \frac{\alpha g}{\nu \mathcal{D}_{th}^\ell}(T_A - T_S)\left(\frac{\mathcal{D}_{th}^\ell}{V_o}\right)^3$$

Figure 5 - Experimental Rayleigh number and critical Rayleigh number vs deviation from equilibrium pressure - Ethanol evaporation.

$$(Ra)_e = \frac{\alpha_{\varepsilon_r} \varepsilon_r \beta E^2}{\rho_0 g \alpha}$$

where β is the thermal gradient, and ε_r the real part of the dielectric constant, with :

$$\alpha_{\varepsilon_r} = \frac{\partial \varepsilon_r}{\partial T} \frac{1}{\varepsilon_{ro}}$$

In our case, the values of $(Ra)_e$ are about 10^{-9}. So, the electric contribution is clearly too small to induce convection.

If we consider the Rayleigh number (Ra_π) we have defined before, and if we calculate its value for ethanol evaporation under microwave, we can conclude that the stable steady conductive regime is broken when the bifurcation point, at about 0.05 W, is crossed over. Experimentally we have found this point near 0.04 W.

So, an exacerbation of the unsteady phenomena by microwave is obtained in ethanol evaporation. It seems qualitatively, that this effect could be linked to a feedback which is more effective under irradiation to amplify the temperature fluctuations.

The sign of the variation of the dielectric loss in function of temperature could be crucial. For ethanol [13], this sign is positive. For water [14], it is negative. If the expression given before of the dissipated energy in the dipolar liquid is considered, we have to pay attention to the properties of the local internal field E and of the dielectric loss ε_i.

Concerning the internal field E, which is the local electric field inside the material, we know that its intensity can be different to the intensity of the applied field ; but it cannot be directly measured. Considering some models which are able to take into account the effect of the dielectric sample in the specific conditions of a given assembly, this intensity has been estimated.

By a semi empirical method at the first order of perturbation, the calculation [5] has given the value of the mean maximum of the intensity E_{max} inside the material compared to the value in vacuum E_0 in function of the bulk temperature T_b, in an appropriate range of values (from 20° C to 60° C).

In ethanol, the internal field intensity remains high and shows a maximum near 40° C. In water, the internal field decreases continuously and tends to the intensity of the applied field. (Figure 6).

Qualitatively speaking, the dielectric losses vary similarly. What kind of consequences can be expected ? If a temperature fluctuation in the liquid is considered in this range of temperature, (for example $\delta T > 0$), the dissipated power, proportionnal to $E^2 \varepsilon_i$, rises if the liquid is ethanol, and on the contrary decreases strongly for water.

So, every hot perturbation is amplified in ethanol and becomes hotter. The result is that we observe large thermal heterogeneities in the bulk of liquid and consequently a strong convective regime.

In water, every perturbation is damped and the temperature homogeneity can be better guaranteed. Volumic convection does not find in water a favorable medium. We understand the predominance of convection in evaporative ethanol and its extinction in water under microwave irradiation.

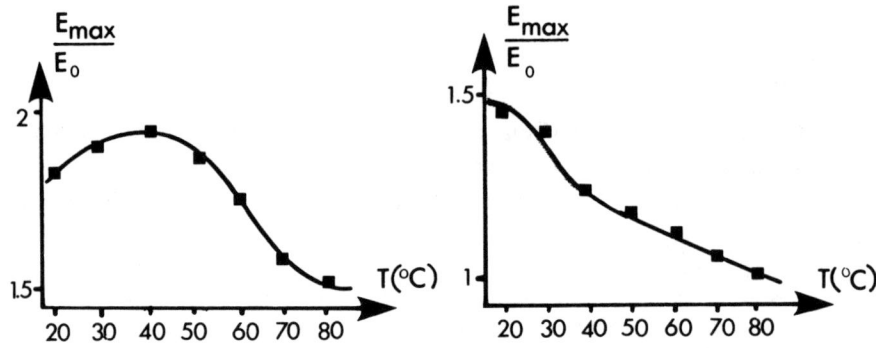

Figure 6 - Variation of E_{max}/E_o (relative internal field) in function of the temperature.

6. INTERFACIAL INSTABILITIES.

In the experiments described before, the interface is out of the irradiation beam and these instabilities concern the interfacial layer above the Benard zone, if it exists. The arguments of Palmer who studied such situations, are followed [7]. We have to take into account some properties of the layer :
W   the fluid velocities, in particular their vertical component,...
T   the temperature
P   the pressure
V_o the evaporation rate
z_o the position of the interface ; in the steady state of the layer, its value is $z_o = 0$
β   the vertical temperature gradient
δ   the depth of the layer
μ   the dynamic viscosity
σ   the surface tension

We assume that a steady state of the interfacial layer exists. (marked with the superscript r).

In this state, balance equations can be written, in particular the balance equation for momentum, which is :

$$P_\ell^r - P_v^r = \rho_v W_v^{r^2} - \rho_\ell W_\ell^{r^2}$$

and as
$$\rho_\ell W_\ell^r = \rho_v W_v^r = V_o^r \rho_\ell$$

we obtain the relation :

$$P_\ell^r - P_v^r = V_o^{r\,2} \rho_\ell \left(\frac{\rho_\ell}{\rho_v} - 1\right)$$

which defines what is called the <u>vapor recoil pressure</u>.
This reference state can be perturbed. The perturbed variables are defined by :
$(P', W', T', V'_0, z'_0)(x, y, z, t) = (P^r, W^r, T^r, V^r_0, z^r_0)\, z\, \exp(i(k_x x + k_y y) + \omega t)$
with the bidimensional wave number :

$$k = \sqrt{k_x^2 + k_y^2}$$

In the set of the balance equations for the perturbed state, the momentum balance equations are the most interesting.
For the tangential equation, we obtain :

$$\frac{\partial \sigma}{\partial T} \nabla_s^2 T' = \mu_\ell \left(\nabla_s^2 W'_\ell - \frac{\partial^2 W_\ell}{\partial z^2} \right) - \mu_v \left(\nabla_s^2 W'_v - \frac{\partial^2 W_v}{\partial z^2} \right)$$

The normal one writes as :

$$(P'_v - P'_\ell) + 2 V_o^r V'_o \rho_\ell \left(\frac{\rho_\ell}{\rho_v} - 1\right) + 2\left(\mu_\ell \frac{\partial W'_\ell}{\partial z} - \mu_v \frac{\partial W'_v}{\partial z}\right) + \ldots$$

$$\ldots + g(\rho_\ell - \rho_v) z'_o - \sigma \nabla_s^2 z'_o = 0$$

In this expression, the positive terms correspond to impulsions which push the surface down.

If we consider a perturbation, such as a hot crater, the first three terms destabilize the surface because they sink it. The first destabilizing term comes from the distorsion of the pressure profiles. The second one is the differential vapor recoil term. As δT is positive, V'_o is positive. The third corresponds to the contribution of the viscous forces. It is also destabilizing, because the velocity is increased in the crater.

The last two terms correspond to the contribution of gravity and surface tension ; they are stabilizing. The tangential balance equation contains terms which are specific to the Marangoni effects, which generally are destabilizing.
Now, some dimensionless numbers have to be introduced :

the well-known Marangoni number :

$$N_{Ma} = \frac{\partial \sigma}{\partial T} \frac{\beta \delta^2}{D_{th}^\ell \mu_\ell}$$

the so-called Hickman number :

$$N_H = \frac{\partial V_o^r}{\partial T} \frac{V_o^r \beta \delta^2 \mu_v}{D_{th}^\ell \sigma^r} \left(\frac{\rho_\ell}{\rho_v} - 1 \right)$$

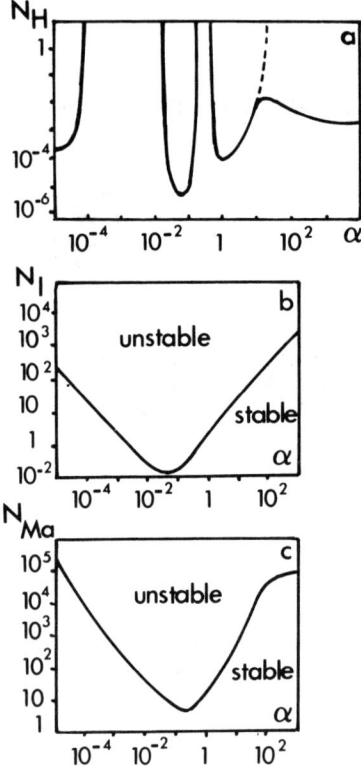

Figure 7 - Marginal curve of stability for different pure processes.
a - critical Hickman number variation
b - critical inertia number variation
c - critical Marangoni number variation

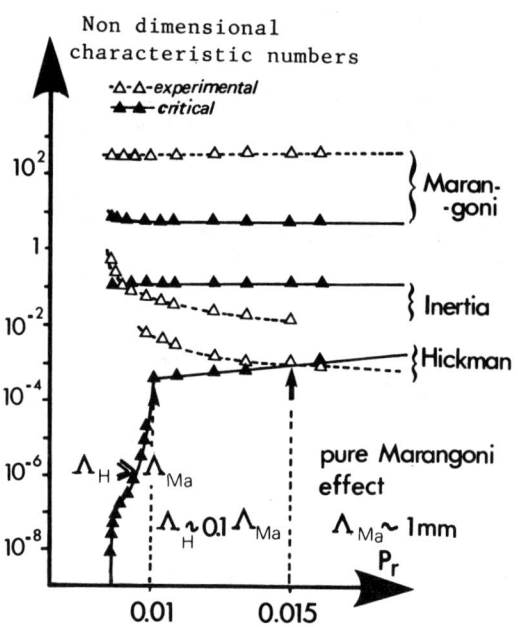

Figure 8 - Comparison of the variations of the theoretical or experimental nondimensionnal numbers
- Fonction of the reduced pressure
- Ethanol evaporation.

and the inertia number :

$$N_I = \frac{V_o^{r^2} \rho_l \delta}{\sigma^r} \left(\frac{\rho_l}{\rho_v} - 1 \right)$$

So, it is possible to compute the marginal curves of stability for each pure effect, in function of the reduced wave number $\alpha = k\delta$ and consequently to select the process which drives the instability. But, it should be difficult to discriminate between these processes because they could operate in the same range of values of the wave number (Fig. 7). So, couplings between processes, for example between Marangoni and Hickman effect, could be very crucial.

What is the dependence of the different instabilities versus the depression P_r ? On Fig. 8, are drawn the variations of the critical numbers and of the experimental ones. It exists a region where only the Marangoni instability occurs. The critical wavelength is about 1 mm, value which is experimentally almost verified.

Below $P_r = 0.015$, the Hickman number becomes supercritical, but the calculated wavelength is ten time smaller than the wavelength of the Marangoni instabilities. The Marangoni wavelength Λ_{Ma} diminishes when pressure decreases. It is the contrary for the Hickman wavelength Λ_H.

Below $P_r = 0.010$ Λ_H becomes greater than Λ_{Ma}. The size of the hot spot is bigger than the size of a Marangoni cell. The Hickman effect becomes dominant and violent motions of the surface occur (the experimental value is $P_r = 0.014$).

Under microwave, it is quite different. The coupling between the processes is more important in every condition and the region where the pure Marangoni effect can be observed is reduced in favor of the vapor recoil instability. But the precise theoretical analysis of this complicated situation is not yet achieved.

REFERENCES.

[1] H. BENARD, Rev. Gen. Sci. Pures Appl., 11 (1900) 1261, 11 (1900) 1309, Ann. Chim. Phys., 23 (1901) 62.

[2] Lord RAYLEIGH, Phil. Mag., 32 (1916) 529.

[3] C. MARANGONI, Nuovo Cimento, 16 (1871) 239.

[4] M. LALLEMANT, A. STEINCHEN-SANFELD, G. BERTRAND, R.G.E., (1988), n°5, 14.

[5] D. STUERGA, Doctoral Thesis, Université de Bourgogne, Dijon 1989.

[6] K. HICKMAN, Ind. Eng. Chem., 44 (1952) 1982, J. Vac. Sci. Technol., 9 (1972) 960, 13 (1976) 585.

[7] H.J. PALMER, J. Fluid. Mech., 75 (1976) 487.

[8] A. PROSPERETTI, M.S. PLESSET, Phys. Fluids, 27 (1984) 1590.

[9] C.F.F. BÖTTCHER, P. BORDEWIJK, Theory of Electric Polarization, Vol. II., Elsevier, Oxford (1978).

[10] P. COURVILLE, Doctoral Thesis, Université de Bourgogne, Dijon 1987.

[11] I. CATTON, Trans. of ASME, 2 (1970) 186.

[12] R.J. TURNBULL, Phys. Fluids, 12 (1969) 1809, 13 (1970) 2615.

[13] T.K. BOSE, R. CHAHINE, C. AKYEL, R.G. BOSISIO, J. Microwave Power, 19, (1984) 127.

[14] A. KUMAR, Int. J. Electronics, 47 (1979) 531.

DIRECTIONAL SOLIDIFICATION : THEORETICAL METHODS AND CURRENT UNDERSTANDING

HERBERT LEVINE
Department of Physics and Institute for Nonlinear Science
University of California, San Diego
La Jolla, CA 92093

The patterns formed via the directional solidification of binary mixtures are paradigmatic of the type of structures which can arise in non-equilibrium dissipative systems. This paper will review the current status of our attempts to understand these patterns. We will focus in turn on phenomena in shallow cell systems, the highly non-equilibrium deep cell regime and the transition to dendritic growth.

I. Introduction

Recent years have seen a great increase in research aimed towards understanding spatial patterns in non-equilibrium systems.[1] The goal of this research is to develop a conceptual framework for predicting dynamical behavior in systems driven far from thermodynamic equilibrium. The range of behavior considered ranges from unique, stable, coherent structures, to chaotic fluctuations to fully developed turbulence.

We are, in the end, most interested in concepts that have wide applicability. Nevertheless, it is often best to focus on a single, well chosen type of pattern forming system for an extended study. This is because one must necessarily have *quantitative* tests, via careful experiments, of any theoretical proposals. Quantitative tests invariably require comparison to a particular set of experiments. Of course, after completion of such a study, one must step back and sort out what has been learned into universal lessons and specific details.

It is in this spirit that these lectures will review current research into interfacial patterns formed by the directional solidification of binary mixtures.[2] The beginnings of the story here are rather familiar; stability, onset of a patially ordered state, secondary bifurcations, etc. However, there are several important advantages in this system as compared to some fluid mechanical alternatives. First, the equation for the pattern in thin samples is a one dimensional equation. Although there can be higher dimensional effects, careful control of the experimental system can lead to the ability to make a quantitative comparison with putatively exact methods. Next, it is easy to increase the driving to bring the system to the highly non-equilibrium state. Here we find new phenomena such as selection and subcritical wave instabilities (sidebranching) whose explanation lies well beyond the scope of the standard amplitude equation approach. Finally, an increase in the understanding of the solidification process will have technological benefits; that is, predicting microstructure is one of the primary ingredients in designing solidification methodologies for advanced materials.

The subject matter is naturally divided into three sections. First, we will discuss the basic experimental setup, typical patterns observed and the fundamental set of models that describe the solidification process. This leads to the Mullins-Sekerka instability[3] as the basic mechanism for driving pattern formation. Next, we will discuss the shallow cell regime, for which the tools of bifurcation theory and amplitude equations adequately describe the pattern. The typical structures seen in this regime are quite similar to the patterns found in convection and Taylor-Couette flow. The particular focus will be on showing that a (spatially) resonant interaction between modes at q and $2q$ can drive the system either into spatial period-doubling, intermittent or stable travelling waves or into oscillatory instabilities.[4] We can compare the predictions of this approach to some exact, numerical computations and to the experiments of Bechoefer, Simon and Libchaber.[5]

Finally, we turn to the deep cell and dendritic patterns. Here, a powerful solvability mechanism governs the actual pattern seen as well as its stability.[6] We comment on the poorly understood cell to dendrite transition within the context of a recently completed stability calculation.[7] At the end, we present a set of open issues, most of which are experimental.

The work I will be describing is the continuing effort of many scientists. My own efforts have been immensely ammplified by collaboration with W. Rappel, H. Riecke, E. Brener and especially D. Kessler. This work has been supported by in part by DARPA under the University Research Initiative, Grant no. N00014-86-K-0758.

II. Brief Review

In this section, we provide a quick overview of the equations describing directional solidification as well as some general experimental results.[2] The latter has been dealt with in some depth by C. Guthmann in these lectures and hence can be treated rather cursorily here.

The basic process to be considered is the solidification of a two component alloy from the melt. The alloy is placed in a thin region between two glass plates and a temperature gradient is imposed. Finally, the sample is pulled at some velocity v_0 towards the cooler region. This gives rise to a solid-liquid interface which remains approximately stationary in the lab frame, corresponding to the solidification front advancing at a velocity of v_0. The interface can be studied by visual inspection through a microscope.

The basic assumptions that govern the modeling of this process are as follows. First, we assume that there is no variation in the *thin* direction perpendicular to the sample plane. There are examples for which there is not sufficiently accurate, but these will not be considered here. Next, we assume that the relevant process for solidification is chemical diffusion of one of the species in a gradient set up by the miscibility gap between the liquid and solid phases. This idea requires that there be no nontrivial dynamics associated with heat transport. Finally, we assume that the solidification process occurs sufficiently slowly (on the time scale set by attachment kinetics) that the interface remains in local thermodynamic equilibrium.

Assuming standard linear relationships between the impurity concentration $C_{\ell,s}$ in the two phases and temperature, we immediately derive

$$D_\ell \nabla^2 C = \frac{\partial C}{\partial t} \qquad \text{inside liquid} \qquad (1)$$

$$D_s \nabla^2 C = \frac{\partial C}{\partial t} \qquad \text{inside solid} \qquad (2)$$

$$D_\ell \hat{n} \cdot \nabla C|_\ell - D_s \hat{n} \cdot \nabla C|_s = -v_n(C_\ell - C_s) \qquad (3)$$

where D_ℓ, D_s are the diffusion constant in two phases and the concentrations at the interface are given by

$$C_\ell = \frac{\tilde{T}_M - T}{m_\ell} \qquad (4)$$

$$C_s = \frac{\tilde{T}_M - T}{m_s} \qquad (5)$$

with $\tilde{T}_M$ the melting temperature and m_ℓ, m_s the slopes of the liquidus and solidus lines. Note that the melting temperature depends on the interface geometry through the Gibbs-Thomson effect

$$\tilde{T}_M = T_M(1 - d_0(\theta)\kappa) \qquad (6)$$

$$d_0(\theta) = d_0(1 - \epsilon \cos 4\theta) \qquad (7)$$

where θ is the angle made by the interface normal, κ is the curvature, T_M is the bulk melting point and ϵ is the coefficient of crystal anisotropy. The actual temperature T is given by $T_M + Gy$ where G is the imposed gradient.

There are basically two methodologies available for dealing with this type of system. The first approach solves the differential equation and boundary conditions by assuming small deviations from a planar interface. This approach predicts the onset of spatial structure where the Mullins-Sekerka instability first sets in. This instability corresponds to the point where the standard diffusional instability, corresponding to sharper tips having enhanced gradients, overcomes the stabilizing effects of the thermal gradient (at long wavelengths) and the surface tension (at short wavelengths). Subsequently, the interface is expanded in terms of a few modes and amplitude equations are derived for the expansion coefficients. This approach will be used in section III to discuss small amplitude cells.

As the driving velocity is increased past the onset of spatial structure, more complex patterns emerge. Experimentally, for materials where $k \equiv m_\ell/m_s$ is small ($\sim .16$ for Br_2 in CBr_4, e.g.[8]), there is a deep cell regime. Here the interface has a periodic structure of fingers seperated by extremely deep grooves. The wavelength of these structures is relatively well determined but does seem to vary in some allowed region. As we increase the driving still further, the cells turn into dendrites, replete with parabolic tips and sidebranching. Here the wavelength is less well determined.[9] Eventually, each dendrite becomes effectively decoupled from its neighbors and we approach the free dendrite problem studied in detail elsewhere.[6]

For all of the above phenomena, the small amplitude expansion methodology is inadequate. Instead, we use the by now common idea of rewriting the differential equation by a singular integro-differential equation for the solid-liquid interface. The derivation of this equation has been given elsewhere[10] and so we just quote the final results. Let G_ℓ and G_s be the Green's functions for the diffusion equation in the solid and liquid phases respectively. Then, the equations of motion for the interface and an auxillary unknown field Φ take the form

$$-\gamma\kappa - \frac{y - v_0 t}{l_T} = 1 - \int_{\text{liq}} ds' \hat{n}' \cdot \nabla' G_\ell(\mathbf{x}(t), \mathbf{x}'(s', t')) \left(-\gamma\kappa(s', t') - \frac{y(s', t') - v_0 t'}{l_T} \right) + \int ds' G_\ell(\mathbf{x}(t), \mathbf{x}'(s', t')) \Phi(s', t') \quad (8)$$

$$0 = \alpha k \int_{\text{liq}} ds' \hat{n}' \cdot \nabla' G_s(\mathbf{x}(t), \mathbf{x}'(s', t')) \left(-\gamma\kappa(s', t') - \frac{y(s', t') - v_0 t'}{l_T} \right) - \int G_s(\mathbf{x}(t), \mathbf{x}'(s', t')) \Phi(s', t'). \quad (9)$$

Here the integrals are evaluated on the liquid side of the interface. The dimensionless parameters which govern the process are

$$\gamma = \frac{v d_0}{2 D_\ell} \frac{T_M}{m_\ell C_\infty} \quad (10)$$

$$l_T = \frac{m_\ell C_\infty}{G} \frac{v}{2 D_\ell} \quad (11)$$

$$\alpha = \frac{D_s}{D_\ell}. \quad (12)$$

The basic advantage of this approach is that it allows us to do straightforward numerical computations of interface shape. These include steady-state calculations (via assuming $y(x,t) = y_0(x) + v_0 t$),[10] as well as simulations of the time evolution of the system.[11] These can be compared to approximate analytic methods and to experimental results. We will discuss the current state of knowledge of the deep cell and dendrite regime in section IV.

III. Small Amplitude Phenomena

The first set of interesting patterns in directional solidification occur as the system passes through the Mullins-Sekerka instability. This instability has been studied in great detail;

for reference we note the formula for the simple case of equal diffusivity

$$\omega = 2Q + \frac{k}{k-1}\left(\gamma q^2 + \frac{1}{l_T}\right)(k\tilde{Q} + Q + 2(k-1)) \qquad (13)$$

where ω is the growth rate in units of $v^2/4D$ and

$$Q = 1 + \sqrt{1 + \omega + q^2} \qquad (14)$$

$$\tilde{Q} = -1 + \sqrt{1 + \omega + q^2}. \qquad (15)$$

Note that both the surface tension term and the thermal term are stabilizing influences, limiting the range of the diffusive instability.

The next step beyond the linear calculation is the evaluation of the cubic coefficient governing whether the instability is subcritical or supercritical.[12] This calculation can then be compared to experiments, under the assumption that only in the case of a supercritical bifurcation will the wavelength of the pattern immediately above onset be given by the unstable mode of the linear analysis, λ_{MS}. This approach seems to work well in practice, with the measured wavelength significantly smaller than λ_{MS} in cases where the transition is predicted to be subcritical.

As we move away from onset, a new set of phenomena occur. As first seen in the numerical calculation of Brown and Ungar[13] and later expanded upon by others, there is a rather complex bifurcation structure at relatively small distances above the critical velocity. For example, there is a *fold* mechanism whereby solutions at some wavelength undergo a saddle node bifurcation and then merge with solutions at half the wavelength.[10,13] This mechanism provides a very sharp end to the allowed wavelength band, a feature that has also been noted in Taylor-Couette flow.[14]

A second set of suprising features has emerged from a recent set of experiments on the directional solidification of liquid crystals.[5] Here the partition coefficient k is large ($\sim .9$) and the diffusivities of the liquid and solid phases are approximately equal ($\alpha \sim .5$). This means that the system has no tendency to form deep cells which ultimately limit the applicability of any amplitude equation method. In these systems, there occur travelling waves in which the cells move systematically sideways as they grow; this type of cellular motion has also been seen in eutectics[15] and in a fluid mechanical analog of solidification.[16] Actually, it appears that the bifurcation to travelling wave is subcritical and so patches of travelling wave solutions can be observed in the steady-state region. At smaller wavelengths, there are spatially period doubled solutions and finally, at even smaller λ, an unsteady limit cycle.[17]

A recent paper by Coullet et al offers a natural explanation for travelling waves; these arise via a parity-breaking bifurcation of the original symmetric cellular pattern. There is a simple way of understanding why a parity-breaking bifurcation should give rise to travelling waves. Consider discretizing and solving numerically the integro-differential equation for a uniformly moving shape with overall velocity v_0 in the $\hat{y}$ direction and v_x in the $\hat{x}$ direction (the wave speed). The counting of equations versus unknowns for the reflection symmetric case has been given earlier; basically there are N angle variables θ_i and a tip position y_0. The angles are defined at the midpoints of an equal arclength grid running from $j = 0$ to N

as the interface goes from $x = 0$ to $x = \lambda/2$. There are $(N+1)$ values of Φ, giving $2(N+1)$ unknowns and exactly $2(N+1)$ equations; hence solutions generically exist without fixing v_0 or the wavelength.

Now extend this approach to non-reflection symmetric shapes. There are $2N$ angle variables, y_0 and $2N + 1$ Φ variables. Howver, translation invariance in the x direction allows as to set, say, $\theta_0 = 0$. This reduces the number of variables to $4N + 1$, which is one fewer than the $2(2N + 1)$ equations which must be satisfied. Hence solutions can only exist if an auxilliary variable, in this case, v_x, is also adjusted. Note that this approach naturally explains why v_x will start at zero right at the parity breaking bifurcation, a result also derivable from the amplitude equation approach.

This set of ideas connects travelling waves with parity breaking bifurcations but does not offer an immediate explanation for why this instability occurs. Recently, it has been pointed out[4] that there is a natural way to understand all of the aforementioned strucure by means of a degenerate bifurcation analysis where modes at wavevectors q and q go unstable. This type of analysis had already been suggested for the steady-state bifurcation diagram[13,18]; what is new is the idea that the new patterns seen in the liquid crystal experiments also follow from this approach. Finally it should be mentioned that the relevence of mode coupling is enhanced by the extreme shallowness of the marginal stability curve v vs. q.

More specifically, one can study the amplitude equations[19]

$$\frac{dz_1}{dt} = \epsilon_1 z_1 + (A_1|z_1|^2 + A_2|z_2|^2)z_1 + C_1 z_2 z_1^* \tag{16}$$

$$\frac{dz_2}{dt} = \epsilon_2 z_2 + (B_1|z_1|^2 + B_2|z_2|^2)z_2 + C_2 z_1^2. \tag{17}$$

Here ϵ_1 and ϵ_2 are unfolding parameters related to distance from the codimensiontwo bifurcation point v^*, q^*. The form of the coupling follows immediately from translation invariance once we recognize that the physical pattern takes the form (see talk by Coullet, these proceedings)

$$y(x) \simeq z_1 e^{iqx} + z_2 e^{2iqx} + \text{c.c.}. \tag{18}$$

These equation have been analyzed in detail by Jones and Proctor[19]; the analysis is too involved to be repeated here. What is worth mentioning though is the generic appearance of steady-state bifurcations from pure modes to mixed modes (spatial period doubling!) and from reflection symmetric state to parity broken travelling waves. A comparison of the exact prediction of this approach with numerical computation and experimental results is now in progress. Qualitatively, the comparison between this model system and the experimental findings is encouraging.

As mentioned in the introduction, directional solidification in this regime falls naturally into a branch of non-equilibrium physics populated by Taylor-Couette flow, convection near onset and, perhaps, driven surface waves. Here the emphasis is on a semi-systematic reduction of the full nonlinear equation to amplitude equations for the relevant modes of the spatial pattern. Not surprisingly, these systems exhibit universal mechanisms of wavelength selection (via ramps, e.g.) phase dynamics, defect motion, ... Each experimental realization offers distinct advantage and disadvantages and each is worthy of detailed study. The study of directional solidification (and its variants involving eutectics, phospholipid monolayers or fluid analogues) for this purpose is just beginning and there is quite a bit left to do and learn.

IV. Far From Equilibrium Patterns

The discussion in the last section focused on features of directional solidification which could in principle be explained by (small) amplitude equations. But while this regime is interesting, it is rather atypical. Most materials, upon being directionally solidified, form highly non-equilibrium nonstructures over most of the experimentally accessible parameter range. We therefore need to turn to a different approach in order to make progress.

Let us first imagine fixing the wavelength of the pattern and studying the shape of the single finger. THere are three distinct regions along the interface that can be discussed separately, at least in the usual limit where the spatial wavelength is much smaller than the thermal length. Over quite a long distance, there is a particular asymptotic solution to the diffusion equation, originally identified by Scheil. The basic approximations are that the interface is almost vertical with small curvature and that diffusion of impurities into the solid phase is unimportant. The latter condition will be satisfied whenever the partition coefficient k is small and the "gap" between neighboring cells is not too small. These together imply that

$$C(x,y) \sim -\frac{y}{l_T} + \frac{x^2}{2l_T}. \tag{19}$$

Substituting this into the Stefan condition results in the profile

$$y \sim y_0 x^{k-1}. \tag{20}$$

This result is independent of surface tension. This result has been verified in computer calculations based on the integro-differential equation approach and seems to be valid experimentally. Note that this is the same result as one finds in the pure one-sided model[20] where there is by definition no diffusion into the solid; here diffusion into the solid is suppressed both by small α and by small k.

At the end of the tail region, there is often found a small liquid bubble. In our model, this bubble must exist since the above asymptotic result cannot persist indefinitely if there is *any* solid diffusivity. The bubble region depends on solid diffusivity and surface tension; also, it is not clear whether experimentally observed bubble are really in steady-state. Although there is much to be said about this region and its possible instabilities, it is somewhat outside our main line of interest and will not be discussed further.

The remaining region of the finger is the tip region. Here, the problem can be stated as follows: given the above asymptotic form (at fixed λ and v) how exactly is the finger tip determined. It is important to realize that this is a selection problem completely analogous to the ones posed for free dendrites (fixed undercooling determines an asymptotic parabolic profile) and the Saffman-Taylor finger.[6] In these systems, finite surface tension causes the selection of a *unique* shape which can solve the steady-state equations at fixed driving. This has been analyzed in great detail and the theory of this *microscopic solvability* mechanism is on quite a secure footing. It will therefore not be surprising if we claim that essentially the same mechanism works here to determine the tip region of the cellular pattern.

The above picture can be verified numerically for all relevant parameter values. For illustrative purposes, though, it is more useful to discuss two limiting cases where we can understand the functioning of this mechanism analytically. First, if the Peclet number is small, the tip region problem can be mapped to that of a Saffman-Taylor (ST) finger. This was first done by Pelce and Pumir[21]; see also the closely related problem of growth in a channel.[22] We will present the simplest case of tha mapping which occurs for $k \simeq 1$ (constant miscibility gap) and negligible thermal gradient but; this approach can easily can be extended to physical values of k and moderate l_T (see Maashall et al[23]). The mapping identities the dimensionless surface tension of the ST problem γ as

$$\gamma \simeq \frac{2d_0}{\lambda} \frac{4D}{v\lambda}. \tag{21}$$

This then allows one to predict that the tip region looks like a Saffman-Taylor solution

$$y(x) = \frac{2(1-\beta)}{\pi} \log \cos \frac{\pi x}{2\beta} + y_{\text{tip}} \tag{22}$$

where the Saffman-Taylor width β is a known function of γ. Dombre and Hakim[24] have shown conclusively how to match this solution to the Scheil asymptotic law. The final step consists in fixing the absolute location of the tip in the temperature field. Again for small temperature gradient and $k = 1$ this is simply determined by the will known relationship equating the asymptotic ST width β and the effective undercooling[22]; this therefore implies

$$\beta = \Delta = 1 - \frac{y_{\text{tip}}}{l_T} \tag{23}$$

where y_{tip} is measured from the position of the planar interface solution. Again, this result can be extended to include thermal gradient effects and to $k \neq 1$.

So, for small pe the ST solvability mechansim selects a unique shape. This analytic approach based on this idea has been tested numerically for the relevant limit of directional solidification and has been shown to be quite accurate. Note that as driving velocity is increased (at fixed wavelength), γ decreases, β decreases and hence y_{tip} gets bigger.

Let us now turn to very large pe. Here, we expect the tip to behave as a free dendrite with the scaling $v \sim \Delta^4 \sigma^*(\epsilon)$, where ϵ is the crystal anisotropy. As the velocity increases, the undercooling Δ increases and the tip retracts. Again, there is not expected to be any problem matching the dendrite asymptotics with the Scheil law which sets in at length scales of order the wavelength λ. Again, the tip region is determined by a solvability mechanism, this time involving a parabolic profile and the crystal anisotropy. For intermediate values of pe, we get a smooth transition, necessarily finding a minimum in the Δ vs. v curve.

As we experimentally approach the free dendrite shape, we also see the emergence of sidebranching. There are currently two suggested explanations for this phenomena. Karma and Pelce,[25] in an approximate calculation based on the ST picture, suggest that there is an oscillatory instability of deep cells. Unfortunately, this does not seem to agree with recent numerical calculations[7] of the stability of the pattern in this regime. Barring some trivial reason for the disagreement (inappropriate parameter regime, programming error, etc.), this would suggest at the least that the KP picture could only apply for finite perturbation.

A more natural explanation is in terms of the noise generated sidebranching currently thought to hold for free dendrites. Here, noise of the tip would be amplified as soon as the tip pattern had a sufficiently long region of parabolic-like free dendrite structure.[26] A first guess is that this will take place around the Peclet number for which the curve Δ vs. v has its minimum. A calculation based on this idea is currently in progress. Clearly, this approach can be tested by measuring the correlation of branching on neighboring cells. The noise induction scenario surely predicts very little coherence.

All of the above discussion concerns patterns of fixed wavelength. Experimentally, there is a band of allowed wavelengths which is rather narrow in the cellular regime. Heuristically, $\lambda \sim v^{-1/2}$ which probably has something to do with the fact that $\lambda^2 v$ enters into the ST γ (see eq. 23). Presumably, there is some set of processes (tip-splitting, tip combination) which allows for wavelength adjustment. Once the dendrite regime is reached, the band becomes wider; that is, at some critical velocity, dendrite tips (with sidebranches) appear with wavelengths that are significantly larger than the extrapolated cellular band would allow.[9] Cells do exist, at the expected smaller λ. Somehow, the dendritic tip seems more reluctant to tip split and hence λ does not tend to decrease. This extra stabilization is undoubtedly connected with the importance of crystal anisotropy for the larger *pe* solutions; after all, Saffman-Taylor fingers will split under sufficient noise[27] but dendrites tend to sidebranch. Further investigation of mechanisms of wavelength selection and their interplay with the single finger solvability process is an exciting area for future endeavors.

In this work, I have attempted to summarize our theoretical methods and the understanding they have brought to directional solidification patterns. What the reader should marvel at is the amazing breadth of possible behaviors and lessons to be learned, all from a single system. And, if the past few years are a harbinger of things to come there will be additional phenomena and more puzzles left to study.

References

[1] *Systems Far From Equilibrium*, J.E. Wesfried, ed., Springer-Verlar, 1988.

[2] For an introduction, see Langer, J.S., *Rev. Mod. Phys.* **52**, 1 (1980).

[3] Mullins, W.W. and Sekerka, R.F., *J. Appl. Phys.* **34**, 323 (1964).

[4] Levine, H., Rappel, W.J. and Riecke, H., 'Resonant interactions and travelling solidification cells', preprint.

[5] Simon, A., Bechhoefer, J. and Libchaber, A., *Phys. Rev. Lett.* **61**, 2574 (1988).

[6] Kessler, D., Koplik, J. and Levine, H., *Adv. Phys.* **37**, 255 (1988); Langer, J.S., *Chance and Matter*, J. Souletie, ed., (North-Holland, 1987); Brener, E. and Melnikov, V.I., preprint.

[7] Kessler, D. and Levine, H., 'Linear stability of directional solidification cells', preprint.

[8] de Cheveigne, S., Guthmann, C. and Lebrun, M.M., *J. de Physique* **47**, 2095 (1986).

[9] See C. Guthmann, this volume.

[10] Kessler, D. and Levine, H., *Phys. Rev. A* **39**, 3041 (1989).

[11] Misbah, C., Saito, Y. and Muller-Krumbhaar, H., preprint.

[12] Wollkind, D.J., Oulton, D.B. and Sriranganthan, R., *J. de Physique* **45**, 505 (9184) and references therein.

[13] See Brown, R., et al in *Supercomputer Research in Chemistry and Chemical Engineering*, K. Jensen, ed., (ACS, Washington, 1987).

[14] Riecke, H. and Papp, H.G., *Phys. Rev. A* **33**, 547 (1987).

[15] See G. Faivre, this volume.

[16] Rabaud, M., Michalland, S. and Couder, Y., preprint.

[17] Coullet, P., Gunratne, G. and Goldstein, R., preprint (1989).

[18] Haug, P., preprint (1989).

[19] Jones, G.A. and Proctor, M.R.E., *Phys. Lett. A* **121**, 224 (1987).

[20] Ben-Amar, M. and Moussallam, B., *Phys. Rev. Lett.* **60**, 317 (1988).

[21] Pelce, P. and Pumir, A., *J. Cryst. Growth* **73**, 357 (1985).

[22] Kessler, D., Koplik, J. and Levine, H., *Phys. Rev. A* **34**, 4980 (1986).

[23] Maashall, M., Ben-Amar, M. and Hakim, V. (preprint, 1989).

[24] Dombre, T. and Hakim, V., *Phys. Rev. A* **36**, 2811 (1987); Ben-Amar, M. and Moussallam, B., *Phys. Rev. Lett.* **60**, 317 (1988).

[25] Karma, A. and Pelce, P., preprint (1989).

[26] Pieters, R. and Langer, J.S., *Phys. Rev. Lett.* **56**, 1948 (1986); Kessler, D. and Levine, H., *Europhys. Lett.* **4**, 215 (1987).

[27] Kessler, D. and Levine, H., *Phys. Rev. A* **33**, 2621, 2634 (1986); Bensimon, D., *Phys. Rev. A* **33**, 1302 (1986).

NEW INSTABILITIES IN DIRECTIONAL SOLIDIFICATION OF SUCCINONITRILE

P.E. Cladis,* J.T. Gleeson,†* and P.L. Finn*

* AT&T Bell Laboratories, Murray Hill, NJ 07974, USA
† Physics Department, Kent State University, Kent, OH 44242, USA

ABSTRACT. We report comprehensive measurements of interfacial properties as well as observations of new behavior at the cellular interface in directionally solidifying succinonitrile. Observations of the bifurcation from the planar state in samples ≥ 100μm thick show that the initial instability occurs *first* along the shortest dimension of the interface. This raises questions about the applicability of many conventional descriptions of patterns that rely on two-dimensional diffusive fields. We also describe new instabilities related to wavelength selection in cells and droplet formation at their roots.

We have accurately measured the position in the lab frame and the temperature of the planar interface and found, in general, that it does not attain a steady state while the cellular interface does. We speculate this is because the standard picture of a binary phase diagram is not always realized.

1. Introduction

Visual observation of the solid – liquid interface of a transparent material as the solid grows in a temperature gradient (directional solidification) was pioneered by Jackson and Hunt.[1] This opened a window on a technologically important process[2] as well as introduced a pattern forming system that has recently been the subject of much interest.[3-10]

Our objective is to understand better the physics behind the interface morphology that influences the quality of technologically important crystals grown for industry. To achieve this goal, we require quantitative information from patterns formed by moving interfaces such as can be obtained using high speed image processing and computer controlled data acquisition. The problem is particularly challenging for the following reasons. The parameter space is large: many parameters are easily controlled but some are not. Furthermore, a *single* pattern in these systems contains an enormous quantity of information. What is physically significant as well as realistically possible to measure? We shall present both qualitative and quantitative results that reflect on current theory as well as pose future challenges to both theorists and experimentalists.

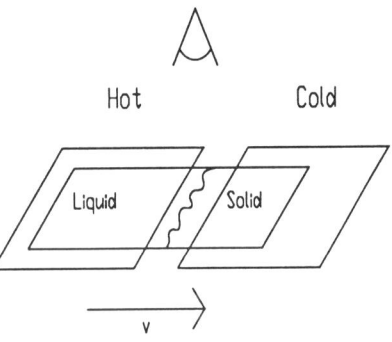

Fig. 1. Directional Solidification

Fig. 1 is a schematic of a directional solidification experiment. The "Hot" and "Cold" temperatures are chosen so that the solid-liquid transition temp̱ ature (and hence, the interface, shown as a wavy line) is in the field of view. In this way, temperature is mapped onto a spatial dimension and patterns formed by a moving interface, created by pulling the material at a speed, v, towards "Cold", may be observed for arbitrarily long times.

1.1 CONSTITUTIONAL SUPERCOOLING

Fig. 2 shows a temperature-concentration phase diagram for a binary mixture. The dashed line represents the concentration profile at zero speed, the horizontal portion being the discontinuity in concentration at the interface. Since impurity diffusion is faster in the liquid than in the solid (where the diffusion constant is assumed infinite), the concentration everywhere in the liquid is taken to be the equilibrium concentration, C_∞. Thus, at rest and in equilibrium, the position of the solid-liquid interface is at T_L.

As material freezes, it expels impurity leading to impurity build-up ahead of the solidification front. The dot-dash line represents the uniform speed, steady state concentration profile, assumed to be constant, C_∞, in the solid near the interface, and exponentially decaying to C_∞ over a length $\ell_D = D/v$ in the liquid. D is the diffusion constant in the liquid ($\sim 10^{-5} cm^2/sec$ for succinonitrile) and v is the pulling speed. Thus, the steady state interface position at constant speed is at T_S, a distance $\ell_T = \Delta T/G$ from its rest position (G is the temperature gradient and ΔT is the width of the two phase region). Since ℓ_D decreases with increasing speed, the concentration gradient at the interface will dip below the liquidus above a critical speed, where $\ell_D < \ell_T$, and a planar interface may become unstable (the Mullins-Sekerka instability (MS)).[10] This is known as *constitutional supercooling*.[11]

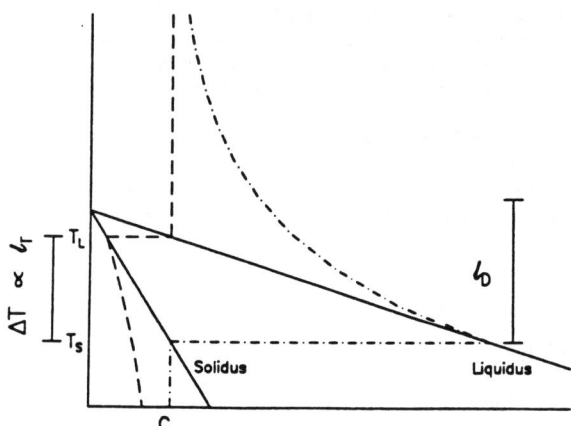

Fig. 2. Binary Mixture Phase Diagram.

Thus, the important parameters to know are the temperature gradient, the speed of the interface, ΔT and the partitition coefficient, k, defined as the ratio of the concentration in the solid to that in the liquid at the interface.

1.2 OUTLINE

In this paper, we first describe our experimental set-up. We follow this with some novel, qualitative observations of interfacial instabilities, in particular two that we call "mitten" and "cell fission", as well as some observations on droplets formed at cell roots. Next, we briefly mention a technique to measure the critical pulling speed, v_c, using high speed image analysis. Finally, we give some quantitative details of the propagating planar interface and our attempts to measure ℓ_T in nominally pure succinonitrile where the binary mixture scenario described above may not necessarily apply.

2. Experimental Setup

Our temperature gradient stage consists of two lower plates from Mettler FP 1 hot stages. Each of these is independently controlled by a Mettler FP 5 controller. In addition, the cold plate is thermally sunk to a thermo-electric device enabling control at or below room temperature. The surface temperature of each plate is monitored by a platinum resistance thermometer (2mm × 2.3mm × 1 mm) connected to a Fluke 1280A digital thermometer with a Y2000 multipoint selector. The Mettler controllers have been modified so that their set points can be controlled by a microcomputer (AT&T pc6300) to which the digital thermometer is also interfaced. With this system, we verified that the temperature of both plates is constant to within 20 mK over a nine hour period. The plates are mounted with ceramic standoffs on a phenol fiber board. The hot plate stand-offs are movable allowing adjustment of the gap between the plates. We have worked chiefly with gaps of 4 and 7 mm.

Our samples are placed on top of these plates and moved through the temperature gradient by a jig attached to a Klinger UT100 stepping motor driven translation stage with a KE 30002-N controller and a shaft decoder. The stage has a step size of 0.1µm and a travel of ~ 2.5cm. The motion is computer controlled directly, or, indirectly through an HP3312A function generator. Direct computer control has the advantage of precise control over the distance traveled while somewhat poorer control of the speed. The function generator can control the stepping rate to ~ 1µHz. Furthermore, with both types of control, the number of steps taken can be monitored by the computer.

This apparatus rests on the stage of a Leitz polarizing microscope. The interface between the solid and liquid is recorded in transmission by a high resolution video camera. The signal can be recorded with a micro-computer controlled video recorder as well as digitized by image processing equipment (Matrox MVP-AT in a Compaq Deskpro 386/20 with Image-Pro software).

2.1 SAMPLE PREPARATION

The experiments we discuss here are on the interface between the cubic plastic crystal and liquid phases of succinonitrile.[12] The motivation for this choice is that it has been extensively studied by others[13] as a pattern forming system as well as a useful model system of the solid-liquid interface of metals and semiconductors.

The succinonitrile as received (from Fluka Chemical) was judged to be too impure for our experiments. Moreover, the impurity(ies) were not known. Our purification procedure was:

1. Pumping on the material while it is freezing to remove dissolved gas. This was repeated until no more bubbles were seen.

2. Evaporation or sublimation and then condensation onto a cold finger while under vacuum. Sublimation usually resulted in purer material but took much more time.

These procedures resulted in an increase of ~ 2.2°C in the melting temperature and a decrease of ~ 1°C in the width of the two phase region. Both quantities were determined by differential scanning calorimetry.

After purification, the material is vacuum loaded into either 50mm long, rectangular (100µm×2mm or 200µm×4mm) glass capillaries (Vitro Dynamics) or flat containers made by sealing two cover slips together with Mylar spacers sandwiched in between. Both types of containers are sealed with Torr-Seal epoxy (Varian). Our experience has been that Torr-Seal does not contribute to material degradation evidenced by drifts in transition temperatures.

3. Observations and Experiments

3.1 THE INTERFACE CURVES BEFORE THE ONSET OF THE CELL PATTERNS

When $v < v_c$, the planar interface is observed as a sharp straight line (Fig. 3a). Above v_c, the first instability is a "thickening" of the interface (Fig. 3b) that develops into the appearance of several straight lines (Fig. 3c). The size of the region shown in each photograph is 0.41 mm wide and 0.37 mm high and the sample is 200μm thick. By focusing, we determine that the line furthest to the right is in the middle of the sample.

We interpret these lines to mean that the interface is no longer planar but now curved in a corrugated manner. In other words, an instability reminiscent of the MS instability has occurred but in the vertical direction (the shorter dimension of the interface) and *not* the longer, horizontal one. The important, novel feature here is that the interface becomes curved before the onset of the MS instability.

Once the interface has curved, the usual MS scenario is observed in the long direction (see Fig. 3d). However, the final pattern depends on the wavelength of the initial corrugation relative to the sample thickness. When there are several periods to the corrugation, a nearly regular cell pattern may occur through a 3-dimensional coarsening that we describe in 3.1.1 When there is only a single corrugation, the cell tips frequently split resulting in poor wavelength selection, described in 3.1.2

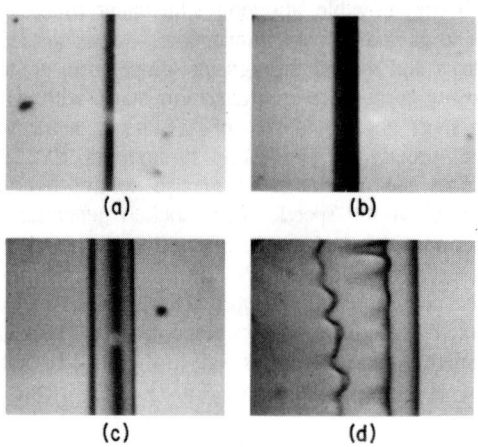

Fig. 3. Interface transition from planar to cellular

3.1.1. Three-dimensional Coarsening of the MS Instability. Fig. 4a shows several corrugations. Each undergoes the MS instability independently of the others (Fig. 4b). Eventually, as they ripen, cells on different levels coalesce leading to the final pattern (Fig. 4f). The cells are nearly regular and stayed so for another 3 mm pulling distance. Thus, this route to cell formation may have a wavelength selection mechanism. On the other hand, it must be stressed that these cells are 3-dimensional objects.

The sample thickness shown in the figure is ~ 140μm and the field of view is 0.51 mm wide and 0.68 mm high. The speed is about 1μm/sec ~ 1.5v_c. The material has no added impurity.

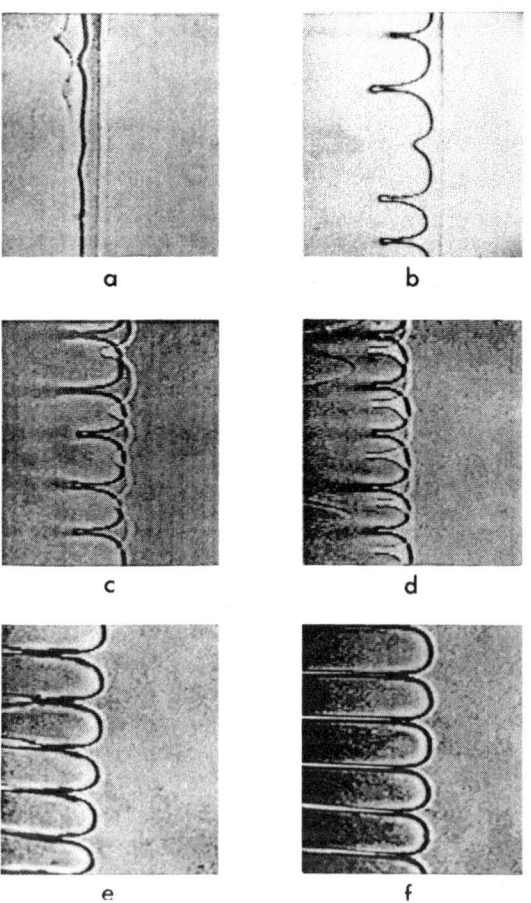

Fig. 4. 3-d coarsening

3.1.2. Tip Splitting: Cell Fission and Mitten Instabilities. In thinner samples or purer materials, the initial instability is a simple curvature of the interface in the short direction. The cell tip curvature is time dependent (Figs. 5 and 6) and the cell spacing is irregular. The photographs represent the entire field of view and show that the tips of the cells are stationary in the lab frame. The sequence of photographs is given by the inset showing the distance solidified (in tenths of a micron) at a pulling rate of 2μm/sec ~ 1.6v_c). The sample thickness here is 100μm. The succinonitrile has been doped with nominally 0.8 wt % acetone.

Two kinds of instabilities that are typical are shown in Figs. 5 and 6. We call them *cell fission* (Fig. 5) and *mitten* (Fig. 6) instability. Both the mitten and the cell fission instabilities arise because the curvature of the cell tip decreases with time. Once the curvature is zero, a depression appears somewhere along the cell tip and propagates to lower temperatures. For fission, this results in two daughter cells each having comparable (but not necessarily identical) curvatures at their tips and nearly the same widths as their mother had (cf. Fig. 5, first and last frames). For the mitten instability, the depression results in two cells where one has a curvature considerably larger than the other. The daughter cell with the larger curvature grows at a slower speed and becomes attenuated. In Fig. 6, an original mitten is seen to decay into a kink on the cell boundary. The width of the surviving daughter has not changed appreciably from that of its parent. Thus, from two similar initial conditions (first frames of Figs. 5 and 6), the number of cells either increases by one *or* remains the same (last frames of Figs. 5 and 6).

As the pulling speed increases, the frequency at which these instabilities occur increases and more thumbs and fingers are generated (and some thumbs may grow fast enough to stay even with the cell tips). This may be a precursor of the cell to dendrite transition. An open question is the time dependence of the occurrence of these instabilities in this regime.

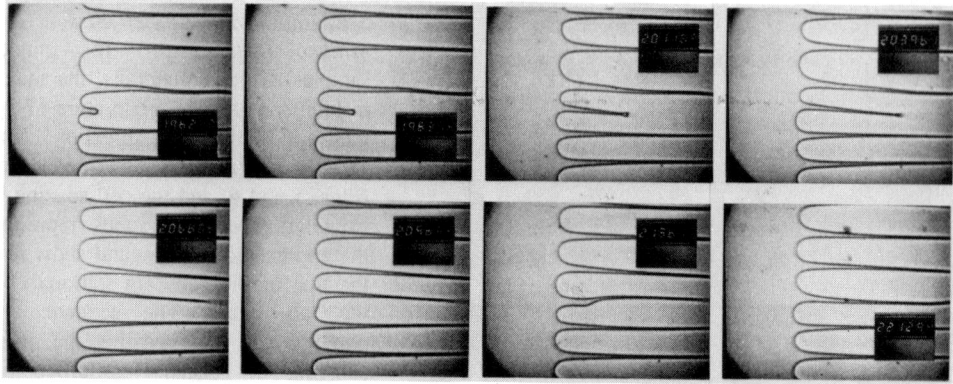

Fig. 5. Cell fission. The field of view is 0.78 mm wide by 0.58 mm high.

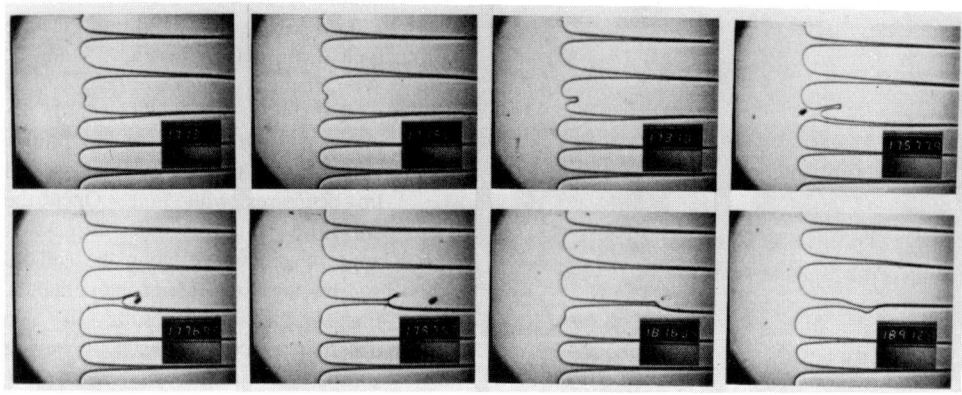

Fig. 6. Mitten Instability. Same conditions as Fig. 5.

3.2 CELL ROOTS – WHERE CELLS FUSE

In Figs. 5 and 6, initially, the cell amplitudes grow in time. In general, droplets do not form at the roots (where neighboring cells meet), however, sometimes they do but their origin is not understood.

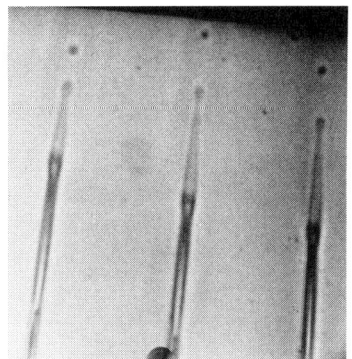

Fig. 7a. Plateau Instability at cell roots.

In Fig. 7a, droplets are observed to form from a necking instability (the Plateau instability).[14] Here the "grooves" are probably cylindrical jets (widest dimension ~ 10μm) of impurity rich liquid. Since the cylinder is not a minimum energy surface for a constant volume, it breaks into spherical drops.

In Fig. 7b, two photographs are shown of a cell root: the first taken just before a droplet separated from the root and the second, just after. The interesting feature is that as the droplet separated, the root shortened. This behavior is reminiscent of a dripping faucet. Both effects are similar to what is expected for behavior controlled by the surface energy of a 3-dimensional object.

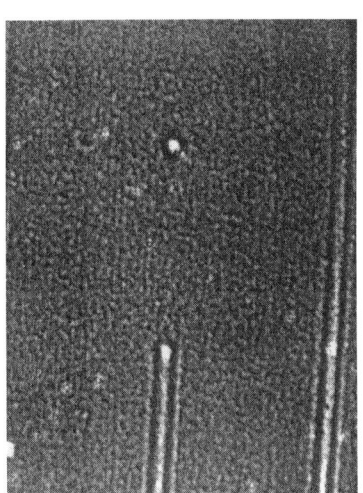 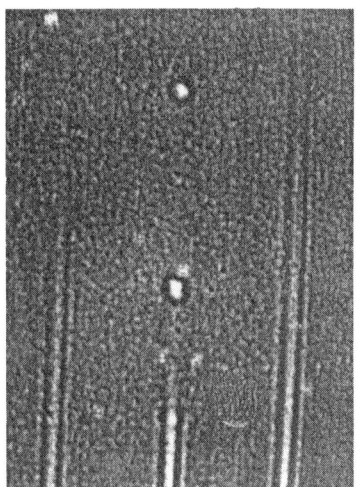

Fig. 7b. Root shrinking after droplet formation

3.3 DETERMINATION OF THE CRITICAL PULLING SPEED

A crucial quantity to measure in directional solidification is the critical speed for the planar to cellular bifurcation. We have devised a method for automatically determining the onset of interfacial instabilities. This method may be used to determine the critical speed in directional solidification.

First, we perform a spatial convolution of a digitized image that accentuates (makes brighter) regions where the intensity gradients are large. After performing this convolution, we count the number of pixels having an intensity greater than a preset cutoff value. This number should be roughly proportional to the arc length of the visible portion of the interface. In Fig. 8, the transition from a planar interface is plainly visible as a sharp eightfold increase in pixel count. There is also a sharp, but smaller, pre-transitional decrease resulting from the loss in contrast as the interface blurs during the formation of the instability described in 3.1 (Fig. 4).

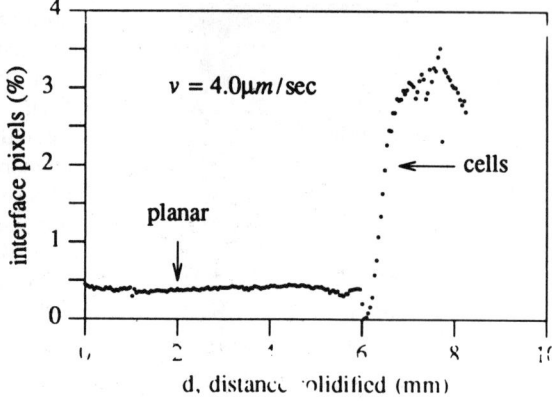

Fig. 8. Interface pixels show planar-cellular bifurcation.

Our strategy is to step down the pulling speed when the interface pixel count changes from the constant value measured for the planar interface. When the constant value is recovered, the pulling speed is increased. In short, the interface pixel count is used to servo the stepping motor speed around v_c.

3.4 THE PLANAR INTERFACE

Initial transients are expected when solidification first starts at a constant rate. They result from the evolution of latent heat and advection of heat that alter the temperature gradient and the changes in impurity concentration at the interface.[15] The over-all effect is a net displacement of the planar interface towards the cold plate that is, in principle, related to the thermal length, ℓ_T.

We have automated the measurement of the displacement (in the lab frame), z, of the planar interface from its zero speed position after the speed was quickly raised to a fixed value. In this experiment the sample is left to equilibrate for longer than its *relaxation time* (which will be discussed later), then, its zero position is recorded. After starting the stepping motor, frames are acquired from the video camera and digitized at regular intervals at which time the position of the interface and the stepping motor displacement d, (which is also the distance solidified), are recorded. This is continued for the entire travel of the translation stage (typically 25 mm).

The most striking result of this experiment is that the planar interface does not fall back a finite distance and then come to rest (*lock*) but continues to travel *in the lab frame* as long as the sample moves (See Fig. 9). In contrast, the cell tips of the non-planar interface are stationary in the lab frame (see e.g. Figs. 5 and 6). If one assumes the temperature gradient is unaltered, the observed displacement of the planar interface corresponds to a change in the interface temperature of ~ 5°C.

We have observed this phenomenon in samples of many different purities, thicknesses and temperature gradients. Furthermore, when the interface between the nematic and isotropic phases is studied, locking is always observed.

To verify that the interface falling back is a real effect and not because of nonlinearities in the temperature gradient, we measure the *background*

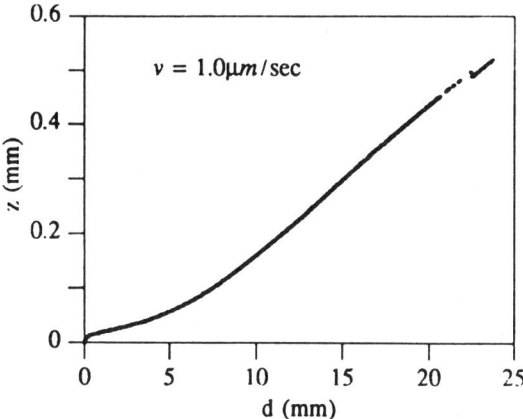

Fig. 9. Unlocked planar interface.

position of the interface. That is, we measure its position in the lab frame at zero speed for different displacements of the sample across the gap. This experiment has also been automated and that has led to new observations about transient behavior of the interface. Principally, the relaxation times (referred to above) after the sample motion has stopped can be up to 3 hours or more (see Fig. 10).

An obvious explanation for why the interface does not appear to reach a steady state is that the two phase region is too wide, i.e. the partition coefficient, k, is smaller than expected (from DSC measurements for example). A simple model for the redistribution of impurities at the start of solidification shows that steady state growth distribution is not attained until a distance ~ D/kv has been solidified and this can be long (~ cm) when k is small (~ 0.01).

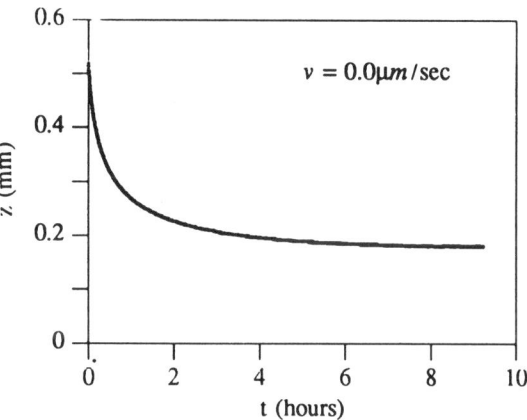

Fig. 10. Relaxation of the planar interface

However, while the interface goes through this transient, its temperature decays from T_L to T_S, i.e. through the two phase region (see Fig. 2) which becomes larger for smaller k (at fixed total concentration). This suggests an experiment to test the explanation.

First, we measured the temperature gradient across our stage. This was done by constructing a sample with a hole drilled in the center and gluing a small (25–50μm) copper – constantin thermocouple in it. It was then vacuum filled with succinonitrile and placed in the apparatus. We displaced the sample a distance, d, either at constant speed or by small increments (~100 μm) to determine the zero speed gradient and monitored the thermocouple temperature, T, with an Omega DP10 digital thermometer interfaced to the computer controlling sample motion.

The results are shown in Fig. 11 where the vertical lines represent the edges of the hot and cold plates. The zero-speed gradient is not significantly different from that measured at constant speed when $v < 10 \mu m/s$. Furthermore, the measured profile is independent of where the thermocouple starts meaning that transients in this profile are small.

When constructing this sample, we always take care that the thermocouple weld will be visible in the microscope. This is necessary to observe visually at what point the thermocouple is directly over the edges of the plates, and also for the next experiment in which we monitored the interface temperature.

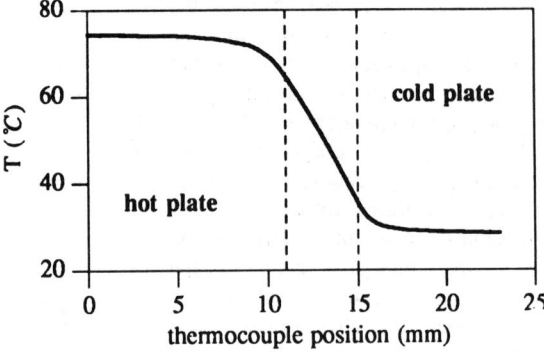

Fig. 11. Observed temperature gradient.

Knowing that one explanation for the unlocked planar interface was the presence of a small k impurity so that the length associated with the decay in interface temperature from T_L to T_S was unusually large, we measured the interface temperature with a thermocouple for different solidification distances, d, and speeds, v. In the simplest model, the transient in interface temperature, $T_I(d)$, is:

$$T_I(d) = T_S + \Delta T \exp(-kvd/D) \quad .$$

The exact transient has been calculated[15] and this approximation is accurate to ~ 10%. Fig. 12 represents preliminary data on our purest material in a 200μm thick capillary. The best fit line corresponds to $\Delta T \sim 0.9°C$ and $k \sim 0.2 \pm 0.05$. The important point is that the decrease in temperature at the interface, while measurable, does not adequately account for the apparent displacement of the interface of ~5°C in the gradient. Thus, we exclude the possibility of a wide two phase region or a k < 0.1 impurity.

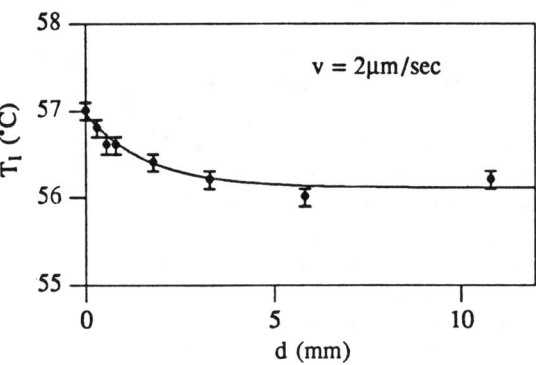

Fig. 12. Interface temperature vs distance solidified

3.5 STATIONARY PLANAR INTERFACE IN DOPED SUCCINONITRILE

We stress, however, the following points. We have been working with not a perfect but only nominally pure succinonitrile. There are probably tiny amounts of several different impurities still present in even our most refined compounds. Thus, its phase diagram may not correspond to a simple binary mixture. When we doped the purest grade of succinonitrile with about 0.8% acetone (k ~ 0.1), we did indeed observe the planar interface lock (Fig. 13) for several different speeds.

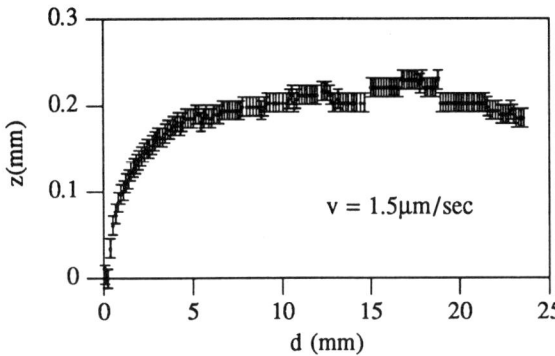

Fig. 13. Stationary Planar Interface

The lifetime of a sample was approximately a day after which we no longer observed the stationary planar interface. We speculate this was because we had zone-refined enough acetone out of the succinonitrile into the hot end of the sample where it presumably remained. Thus, the material was effectively no longer doped with a significant amount of a single impurity.

4. Conclusions

The apparent lack of a steady state planar interface in succinonitrile and its presence in the cellular interface and the planar interface between the isotropic and nematic liquid crystal phases (where the latent heat is 20 times smaller) suggest that latent heat and temperature effects are more important in our experiments than expected. However, since we observed stationary planar interfaces in a material that had been significantly doped, we speculate that departures from a binary phase diagram even in the limit of almost pure materials, could be important for this dynamical process.

The observation that the onset of cell patterns occurs differently in *each* direction perpendicular to pulling leads us to conclude that current theories that consider only *one* may not adequately describe pattern formation at the solid-liquid interface during directional solidification. Cell fission and mitten instabilities appear to be inefficient wavelength selection processes because of the stochastic way they arise. Whether the cell spacing is significantly adjusted depends strongly on the apparently randomly chosen initial position of the depression of the cell tip. This suggests that wavelength selection might be elusive. We know of no theory that explains droplet formation in cell roots or under what circumstances it occurs. The droplet formation from a necking instability suggests that it, also, is a three dimensional phenomenon.

Note Added:
During this conference we were pleased to learn of the work being done[16,17] on the droplet forming instabilities of Sec. 3.2.

ACKNOWLEDGEMENTS We thank W. van Saarloos, J. D. Weeks, K. A. Jackson, P. Palffy-Muhoray and B. Shraiman for useful discussions as well as N. Bhargava and R.C. Dunne for assistance preparing figures. We also thank the conference organizers for supporting our participation at this ASI. J.G. acknowledges support from the Natural Sciences and Engineering Research Council of Canada and the Liquid Crystal Institute at Kent State University.

REFERENCES

1. K.A. Jackson and J.D. Hunt, *Acta Metall.*, **13**, 1212 (1965).
2. *Crystal Growth*, edited by B.R. Pamplin (Pergamon, Oxford, 1980).
3. J.S. Langer, *Rev. Mod. Phys.*, **52**, 1 (1980) and Science **243**, 1150 (1989).
4. J.D. Weeks & W. van Saarloos, *Phys. Rev. A.*, **39**; 2772 (1989).
5. G.J. Merchant & S.H. Davis, *Phys. Rev. Lett.*, **63**; 573 (1989)
6. S. de Cheveigné, C. Guthmann & M.M. Lebrun, *J. Phys. (Paris)*, **47**, 2095 (1986)
7. B. Caroli, C. Caroli & B. Roulet, *J. Phys. (Paris)*, **43**, 1767 (1982).
8. S. de Cheveigné and C. Guthmann in *Propagation in Systems Far from Equilibrium*, J. E. Wesfried, H. R. Brand, P. Manneville, G. Albinet and N. Boccara (eds.), (Springer-Verlag, New York, 1988) p. 44.
9. M. Ben-Amar and B. Moussallam, *Phys. Rev. Lett.*, **60**, 318 (1988).
10. J. Bechhoefer, A.J. Simon, A. Libchaber & P. Oswald, *Phys. Rev. A.*, **40**, 2042 (1989).
11. W.W. Mullins and R.F. Sekerka, *J. Appl. Phys.*, **35**, 444 (1964).
12. W. A. Tiller, K. A. Jackson, J. W. Rutter, B. Chalmers, Acta. Metall **1**, 428, 1953.
13. *The Merck Index* edited by M. Windholz, entry no. 8746.
14. See for example: S.C. Huang & M.E. Glicksman, *Acta Metall.*, **29**, 701 (1981); M.A. Eshelman, V. Seetharaman & R. Trivedi, *Acta Metall.*, **36**, 1165 (1988); H. Chou & H.Z. Cummins, *Phys. Rev. Lett.*, **61**, 173 (1988).
15. J.A.F. Plateau in *Statique Expérimentale et Théorique des Liquides soumis aux Seules Forces Moléculaire*, (Gauthier-Villars, Paris, 1873), Vol. II, pp. 228-307, also Lord Rayleigh, *Proc. R. Soc. Lond.*, **29**, 71 (1879).
16. V.G. Smith, W.A. Tiller and J.W. Rutter, *Can. J. Phys.* **33**, 723 (1955).
17. K. Brattkus, *preprint*.
18. P. Kurowski, S. de Cheveigné, G. Faivre and C. Guthmann, *J. Phys. France*, **50**, 93 (1989).

STATIONNARY CELLS IN DIRECTIONAL SOLIDIFICATION

M. Mashaal and M. Ben Amar
Laboratoire de Physique Statistique, Ecole Normale Supérieure
24 rue Lhomond
F-75231 Paris Cedex 05, France

1. Introduction

The directional solidification of a binary alloy constitutes a typical example of a pattern forming system [1, 2, 3]. In this kind of experiments, a complex structure (cellular or dendritic) appears as a result of the underlying Mullins and Sekerka's instability [4] of crystal growth. The problem of wavelength selection in the cellular regime is still an open one: in a steady state analysis, a well defined shape for the solid-liquid interface is always found once the wavelength is fixed. Leaving this problem aside, we focus here on the shape of the interface and report on numerical results obtained with the one-sided model of directional solidification, where the impurity diffusion is neglected in the solid phase. For every set of the relevant physical parameters, our numerical code was able to find the corresponding interface shape. A useful analysis of these results consists in considering them in the light of a Saffman-Taylor analogy at low Péclet numbers; this analogy gives approximate relations and an effective parameter which can help in comparing the different experimental results. Note that analogous effective relations have been derived by Billia *et al.* [5] on phenomenological grounds, using an extended compilation of published experimental data.

In the framework of the one-sided model of directional solidification of a binary alloy, one can define the following four typical lengths: the wavelength of the pattern α, the diffusion length $l_d = 2D/U$ where D is the diffusion constant and U the pulling velocity, the thermal length $l_t = m \Delta C/G$ where m is the liquidus slope, ΔC the miscibility gap and G the thermal gradient, and the capillary length $d_0 = T_m \gamma/(m \Delta C L)$ where γ is the surface tension, L the latent heat and T_m the melting temperature.

With these four lengths, we can form the following three independent parameters: the Péclet number $P = 2\alpha/l_d$, $\nu = l_t/l_d$ ($\nu = 1/2$ corresponding to the planar front threshold) and the capillary constant $\sigma = d_0 l_t/\alpha^2$.

In the following, we explain first how to obtain, by a Green's function method, an integro-differential equation for the interface; we solved this equation numerically for different values of the relevant parameters. Before presenting and discussing the numerical results, we will recall the Saffman-Taylor analogy at low Péclet numbers and its main consequences.

2. Derivation of the Integral Equation for the Interface

In terms of the dimensionless field $u(x,y) = [\,C(x,y) - C_\infty\,]/\Delta C$, the basic equations of solidification in the stationnary case [2] consist of:
. the stationnary diffusion equation in the liquid phase,

$$\Delta u - P\frac{\partial u}{\partial x} = 0 \tag{1a}$$

. the conservation of impurities at the interface (Stefan law),

$$[K + (1 - K)u_{int}]\, P \cos\theta = -(n.\nabla u)_{int} \tag{1b}$$

. the local thermodynamical equilibrium at the interface (Gibbs-Thomson law),

$$u_{int} = 1 + \frac{P}{2\nu}\,(x_{int} + \sigma\Omega) \tag{1c}$$

where K is the partition coefficient, θ the angle between the normal at the interface and the growth direction, and $\Omega = y''/(1+y'^2)^{3/2}$ is the curvature, if the interface is $y = y(x)$. Lengths have been scaled by the cell spacing α, and our choice of coordinates is sketched in Figure 1.

To obtain an equation for the stationnary half-interface defined by $y(x)$, one uses the Green's function defined by

$$\Delta G(q_0, q) + P\frac{\partial G}{\partial x}(q_0, q) = -\delta(x_0 - x)\,\delta(y_0 - y)$$

where $q = (x, y)$ and $q_0 = (x_0, y_0)$ are arbitrary points of the liquid phase. Fixing the point q_0 on the interface, integrating the quantity $u\,\Delta G - G\,\Delta u$ over the liquid phase domain Σ and using the Green's theorem and the diffusion equation, one obtains

$$\frac{u(q_0) - 1}{2} = -\int_0^\infty dx\,\frac{ds}{dx}\,\widetilde{G}(q_0,q)\,\hat{n}.\vec{\nabla}_q u\;-$$

$$\int_0^\infty dx\,[u(q) - 1]\,[\,y'\,P\widetilde{G}(q_0, q) - \hat{n}.\vec{\nabla}_q\widetilde{G}(q_0, q)\sqrt{1 + y'^2}\,] \tag{2}$$

where q_0 and q are interface points, and

$$\widetilde{G}(q_0, q) = \frac{2}{P}[\,e^{-\frac{P}{2}(x-x_0+|x-x_0|)} - 1\,] - \frac{1}{4\pi}e^{\frac{P}{2}(x_0-x)}(\ln\lambda_+ + \ln\lambda_-) + c.c. \tag{3}$$

is built from an approximation [6] of G when $(P/4\pi)^2 \ll 1$, with

$$\lambda_\pm = 1 - \exp 2\pi [i(y_0 \pm y) - |x_0 - x|]$$

Using (1b) and (1c), we obtain

$$\frac{w(q_0)}{2} = 2\nu \int_0^\infty dx\, y'\, \widetilde{G}(q_0, q) - \int_0^\infty dx\, w(q) \left[y'\, P\widetilde{G}(q_0, q) - \hat{n}.\vec{\nabla}_q \widetilde{G}(q_0, q)\sqrt{1 + y'^2} \right]$$

$$+ (1 - K) \int_0^\infty dx\, w(q)\, y'\, P\widetilde{G}(q_0, q)$$

(4)

with $w(q) = \frac{2\nu}{P}(u-1) = x + x_{tip} + \sigma\Omega$, x_{tip} explicitly appearing because the origin $x = 0$ is chosen at the cell tip. To soften the singularities coming from the term $(x+x_{tip})\nabla_q G$ at $x = x_0$ large, an exact identity was established [7], valid for any function describing a symmetric cell:

$$(x_0 + x_{tip})/2 = \int_0^\infty dx\, y'\, \widetilde{G} - \int_0^\infty dx\, (x + x_{tip})\left[y'\, P\widetilde{G} - \sqrt{1+y'^2}\, \hat{n}.\vec{\nabla}\widetilde{G} \right] + P\iint_\Sigma dx\, dy\, G$$

(5)

Substracting (4) from (5), we get

$$-\frac{1}{2}\sigma\Omega(x_0) = (1 - 2\nu)\int_0^\infty dx\, y'\, \widetilde{G} + \sigma\int_0^\infty dx\left[y'\, P\widetilde{G} - \hat{n}.\vec{\nabla}\widetilde{G}\sqrt{1+y'^2} \right]\Omega(x)$$

$$+ P\iint_\Sigma dx\, dy\, G + (K-1)\int_0^\infty dx\, y'\, P\widetilde{G}\, (x + x_{tip} + \sigma\Omega)$$

(6)

This last integro-differential equation can be worked out further by using expression (3) for the Green's function, giving an equation composed of terms suitable for numerical computation; for details, see [7, 8]. This integro-differential equation can then be solved numerically by discretizing the integrals and approximating the derivatives by finite differences, resulting in a system of non-linear equations where the unknowns are the values $y_i = y(x_i)$ for a suitable mesh. This system is then solved by Newton's method.

Equation (6) for the half-cell profile $y(x)$ can thus be solved numerically for every set of the parameters K, P, ν, σ, the Péclet number not being too large, otherwise the approximation (3) of G looses its validity. Although standing by their own, it is much more instructive to analyse the numerical results in the light of the Saffman-Taylor viscous fingering analogy [9, 10], which is summarized below.

3. Analogy with Saffman-Taylor Fingering, at Low Peclet Numbers

If one introduces the diffusion field $\phi = \frac{2\nu}{P}(u-1) - x$, the basic equations (1a,b,c) become

$$\Delta\phi - P\frac{\partial\phi}{\partial x} = P \tag{7 a}$$

$$\left[(2\nu - 1) + (1-K)P(\phi + x)_{int}\right]\cos\theta = -(n.\nabla\phi)_{int} \tag{7 b}$$

$$\phi_{int} = \sigma\,\Omega \tag{7 c}$$

One can then see that, at length scales of order 1 around the tip and for small enough P, these equations are identical to those describing viscous fingering in a Hele-Shaw cell, where the field ϕ corresponds to the velocity potential (or pressure) in S.T. fingering.

Saffman-Taylor fingers are characterized by their relative width λ, which is a function of a surface tension parameter $\sigma_{S.T.}$ given by the ratio $b^2T/(48\mu U a^2)$ where b is the Hele-Shaw plates spacing, a its half-width, μ the viscosity of the driven fluid, U the constant velocity of the interface and T the surface tension. For viscous fingering, we have thus

$$\lambda = \mathfrak{F}(\sigma_{S.T.}) \tag{8}$$

where the function $\mathfrak{F}$ is numerically known [11].

In directional solidification at small P, the rôle of $\sigma_{S.T.}$ is played by the effective parameter

$$\sigma_{eff} = \frac{\sigma}{(2\nu - 1)[1 + (1-K)\mu x_{tip}]} \tag{9}$$

where $\mu = P/(2\nu - 1)$. Thus, when the analogy is valid, the tip of a solidification cell is identical to that of a Saffman-Taylor finger whose relative width is given by

$$\lambda = \mathfrak{F}(\sigma_{eff}) \tag{10}$$

This is illustrated in Figure 1. The asymptotic shape of the solidification cell (exponential for K=1, algebraic for K<1) can then be related to λ by matching the tail of the cell to the corresponding S.T. finger in the tip region. This matching gives, at dominant order in P:

$$y(x) \sim \frac{1}{2} - \frac{(1-\lambda)}{2}e^{-\mu x} \tag{11a}$$

for K = 1, and

$$y(x) \sim \frac{1}{2} - \frac{(1-\lambda)}{2}(1+\alpha x)^{-\beta} \tag{11b}$$

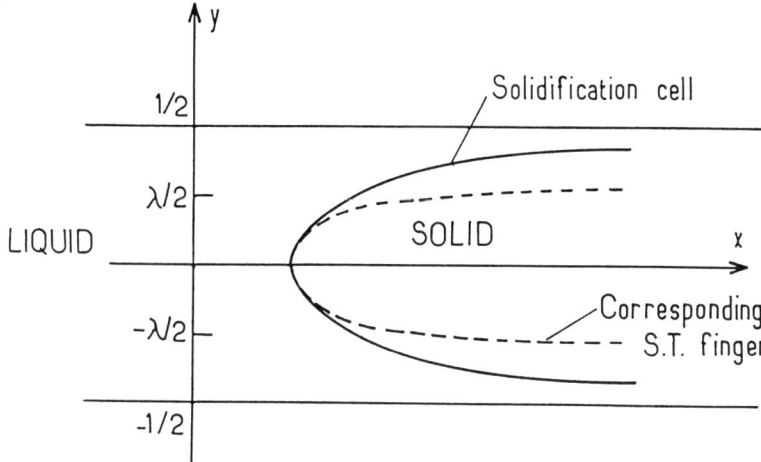

Figure 1: Sketch of a cell showing the coordinates system used.

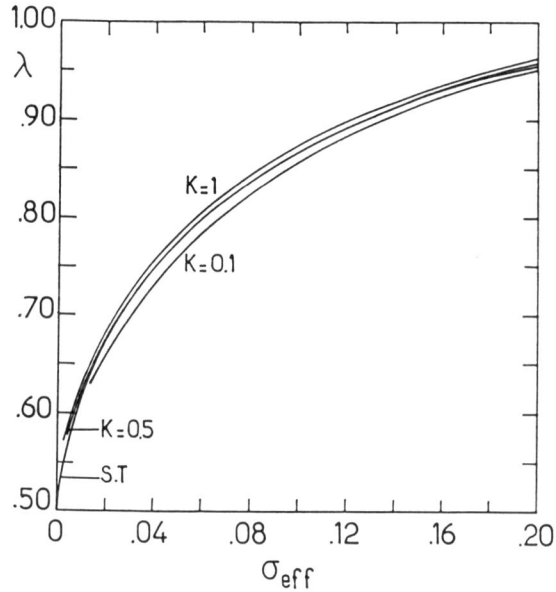

Figure 2: λ as a function of σ_{eff} for viscous fingering and for directional solidification ($P = 0.1$, $\nu = 1$) at different values of K.

for K<1, with $\beta = 1/(1 - K)$ and $\alpha = \dfrac{\mu}{\beta + \mu x_{tip}}$. The position x_{tip} is determined by the global conservation of impurities at the interface, giving

$$x_{tip} = 2 \left[\sigma + \int_0^\infty dx \, (y - \tfrac{1}{2}) \right] \tag{12}$$

which can be computed once the profile is known. The parameter λ corresponding to a given solidification cell can then be numerically extracted from the cusp by Eqns (11) and (12). If the S.T. analogy is valid, this extracted parameter is also the relative width of the S.T. finger fitting the cell tip.

With the help of Eqns(11), an estimation [8] of x_{tip} at dominant order in P can be obtained:

$$\mu \, x_{tip} = -\dfrac{1-\lambda}{1-(1-K)\lambda} + O(P) \tag{13}$$

and, replacing in (9):

$$\sigma_{eff} = \dfrac{1-(1-K)\lambda}{K} \dfrac{\sigma}{2\nu - 1} \tag{14}$$

We can now examine the numerical results and compare them to the predictions (10), (13), (14) above, remembering that the λ parameter is numerically obtained from the cusp and not from the tip.

4. Numerical Results and Discussion

At small Péclet number ($P = 0.1$), Figure 2 shows the $\mathfrak{F}$ function (relation (10)) for the Saffman-Taylor case and for some values of K. The different curves almost overlap, the gaps with the S.T. curve being due to an effect linear in P which is beyond the reach of the analysis above, made at dominant order in P. This P effect can be seen in Figure 3, where the gap increases roughly linearly with increasing P.

Figure 4 shows the curvature at the tip as a function of the λ parameter which is extracted from the cusp, for $P = 0.1$. Here too, the different curves nicely overlap, showing that the λ parameter extracted from the tail corresponds effectively to the relative width of the S.T. finger fitting the cell tip.

How much small must be the Péclet number in order the viscous fingering analogy to remain useful is illustrated in Figures 5 (for K=0.5) and 6 (for K=0.1). These figures show that the tip curvature is less and less the one of a Saffman-Taylor finger of relative width λ as P grows. The viscous fingering analogy is thus valid until a maximal P value of the order of 1/2. One can remark that the domain of "small" P is smaller as K becomes smaller. This can be understood, at least qualitatively, by looking at Eqn (7b): for the Saffman-Taylor case, the coefficient of $\cos \theta$ is a constant, whereas for directional solidification, this coefficient contains a variable quantity multiplied by $P(1 - K)$.

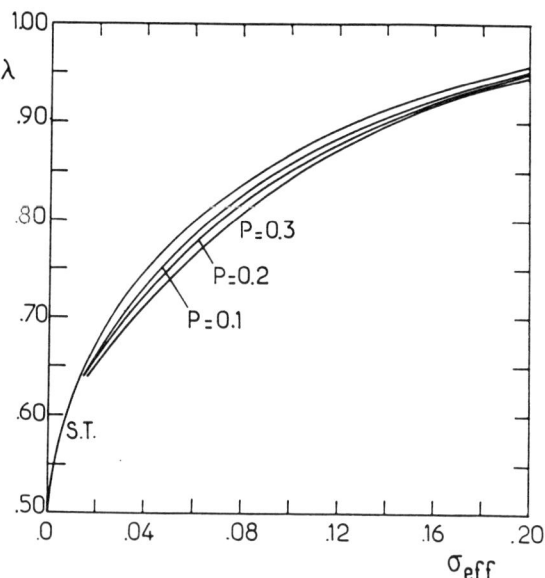

Figure 3: λ as a function of σ_{eff} for P = 0 (Saffman-Taylor fingering) and P = 0.1, 0.2, 0.3 (K = 0.1, ν = 1).

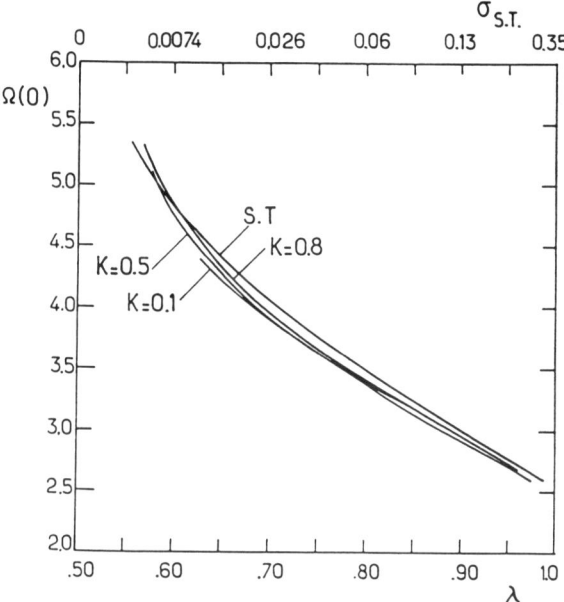

Figure 4: The curvature at the tip as a function of λ for Saffman-Taylor fingers and for directional solidification (P = 0.1, ν = 1) at different values of K.

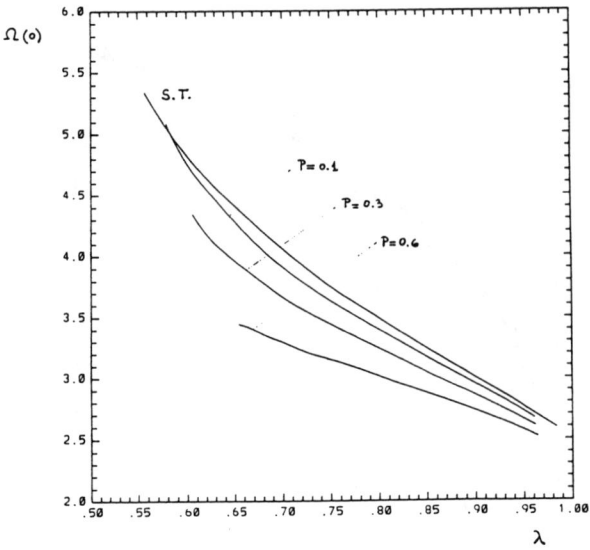

Figure 5: The curvature at the tip as a function of λ for Saffman-Taylor fingers and for directional solidification ($K = 0.5, \nu = 1$) at different values of P.

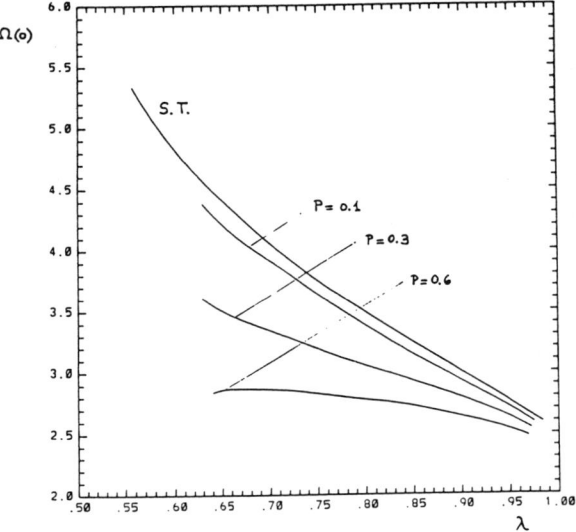

Figure 6: The curvature at the tip as a function of λ for Saffman-Taylor fingers and for directional solidification ($K = 0.1, \nu = 1$) at different values of P.

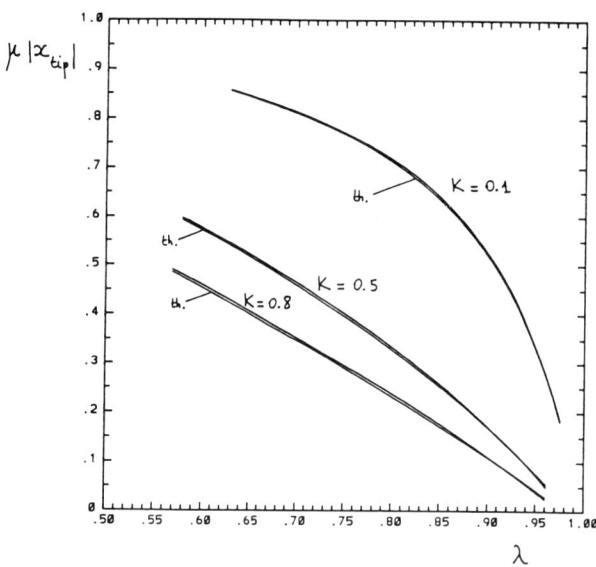

Figure 7: $\mu |x_{tip}|$ as a function of λ for K = 0.1, 0.5 and 0.8 (P = 0.1, ν = 1). The curves indicated by "th." are those given by Eqn (15).

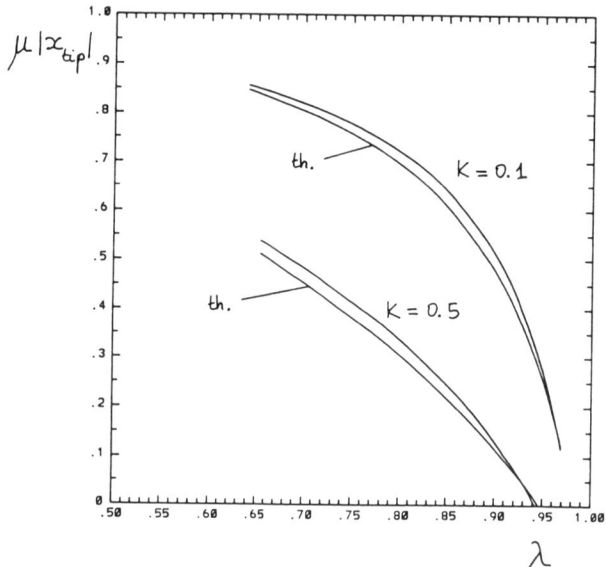

Figure 8: $\mu |x_{tip}|$ as a function of λ for K = 0.1 and 0.5 (P = 0.6, ν = 1). The curves indicated by "th." are those given by Eqn (15).

The numerical results concerning x_{tip} are shown in Figure 7 (for P=0.1) and Figure 8 (for P=0.6); the numerical x_{tip} is there compared to the phenomenological relation

$$x_{tip} = -\frac{1}{\mu}\frac{1-\lambda}{1-(1-K)\lambda} + \frac{2\nu-1}{\nu}\sigma \qquad (15)$$

This relation differs from Eqn (13) above by a term proportionnal to σ. Although not accessible by the analysis at dominant order in P, it has been included since it corrects very nicely Eqn (13) for high surface tension parameters. The agreement of the numerical results with (15) is very good not only for P=0.1, but also, as Figure 8 shows, for relatively high Péclet numbers where the parameter λ looses its physical meaning.

5. Conclusion

We have thus presented numerical results concerning the shape of steady-state cells in directional solidification. These results can be described in terms of an analogy with viscous fingering. The Saffman-Taylor analogy provides approximate relations which are well satisfied by the numerical results, at least for low Péclet numbers. This should help to compare the different experimental results in a meaningful way. An extension of this work to higher Péclet numbers is the object of future studies, hoping to shed light on the wavelength selection and cell-dendrite transition problems.

Acknowledgments

We are grateful to Vincent Hakim and Bachir Moussallam for many valuable discussions and advices.This research was supported by the Centre National d'Etudes Spatiales, Grant: "Incitation à la Recherche 1988, N01236" and by the Direction des Recherches Etudes et Techniques, Grant : "88 C 0219".

References

[1] C. Guthmann, this Conference.

[2] H. Levine, these Proceedings.

[3] S. de Chevigné, C. Guthmann and M.M. Lebrun, J. Physique **47**, 2095 (1986)
    H. Jamgotchian, B. Billia and L. Capella, J.Crystal Growth 82,342 (1987); **85**,318 (1987)
        M.A. Eshelman, V. Seetharaman and R. Trivedi, Acta metall. **36**, 1165 (1988);
        V. Seetharaman, M.A. Eshelman and R. Trivedi, Acta metall. **36**, 1175 (1988).
        A. Simon, J. Bechhoefer and A. Libchaber, Phys. Rev. Lett. **61**, 2574 (1988).

[4] W.W. Mullins and R.F. Sekerka, J. Appl. Phys. **35**, 444 (1964).

[5] B. Billia , H. Jamgotchian and L. Capella, J. Crystal Growth **82**, 747(1987) and **94**,987 (1989).

[6] A. Karma, Phys. Rev. Lett. **57**, 858 (1986).

[7] M. Ben Amar and B.Moussallam, Phys. Rev. Lett. **60**, 317 (1988).

[8] M. Mashaal, M. Ben Amar and V. Hakim, "Directional solidification cells at low velocities" (September 1989), submitted to Phys. Rev. A.

[9] T. Dombre and V. Hakim, Phys. Rev. **A36**, 2811 (1987)
M. Ben Amar, T. Dombre and V. Hakim, in "Propagation in Systems Far From Equilibrium", Les Houches 1987 Proceedings (Springer-Verlag 1988)

[10] P. Pelcé and A. Pumir, J. Cryst. Growth **73**, 357 (1985).

[11] J.W. McLean and P.G. Saffman, J. Fluid Mech. **102**, 455 (1981);
J.M. Van den Broeck, Phys. Fluids **26**, 2033 (1983).

RECENT PROGRESS IN THE THEORY OF THE GROWTH OF NEEDLE CRYSTALS

M. BEN AMAR and Y. POMEAU
*Laboratoire de Physique Statistique
Ecole Normale Supérieure,
24 rue Lhomond,
F-75231 Paris Cedex 05
France*

ABSTRACT. This presents a review of recent progress made in the theory of growth of needle crystal. We consider first the low undercooling limit and then the finite undercooling limit, where the result of a detailed calculation is compared with recent experiments. We review first the now classical Ivantsov solution and explain how to derive the proper scaling laws from simple boundary layer estimates. This allows us to consider then the effect of an imposed axial flow on the growth. We find by using scaling arguments several regimes of growth for a needle crystal under those conditions. The growth velocity and tip radius are derived, depending on the fluid velocity and undercooling. Finally we present results relevant for a finite dimensionless undercooling, including in particular kinetic effects, and we compare this with recent experiments on rapid solidification of metals and alloys.

1. Introduction

The purpose of this paper is to review recent progress made in the theoretical study of the growth of needle crystals. The foundation of this subject were laid many years ago by Ivantsov, when he derived his exact solution of the coupled diffusion and Stefan equations. As it is well known too and although this solution is correct, it does not say everything on the real solution. Actually a continuum of solutions is so described simply because the physical data lack a length scale, so the solution is invariant under a continuous one parameter group. It was noticed quite early too that by adding the so-called Gibbs-Thomson condition to the mathematical data, this spurious invariance is lost and unique or at least discretely many solutions are selected. Even though this idea was true, a lot of effort has been needed to prove it. The tale may be found in the recent review by Kessler et al. (1989), and we shall not try to reproduce it here.

Below, we explain first by means of simple scaling argument how to understand the Ivantsov solution in the low undercooling limit and then how the introduction of the Gibbs-Thomson relation leads to a complete selection of the needle crystal parameters, satisfying the famous relationship between the radius of curvature ρ at the tip and the the growth velocity u_c $\rho^2 u_c$ = constant, independent of the undercooling. However it must be recalled that this scaling approach is not everything, since it leaves to solve a difficult mathematical nonlinear eigenvalue problem for computing dimensionless constants, and it turns out precisely that this one has no solution in the absence of anisotropy in surface tension between the solid and the melt. We shall not consider this specific question and merely refers to the original publications [Ben Amar and Pomeau (1986), Barbieri et al.(1987)].

However it turns out that even without going into the details of this mathematical analysis, the situation of a needle crystal growing in an axial flow presents already nontrivial question of scalings. The difference with the standard case (no flow) comes from the fact that the concentration field (or the excess of latent heat) now is both spread by molecular diffusion and advected. The balance between those two phenomena depends in principle on a dimensionless number, the so-called Péclet number: Pe=$\frac{l_0 u_\infty}{D}$ where u_∞ is the fluid velocity far from the crystal and l_0 a typical length. When Pe is large (and this can happen quite easily because the molecular diffusion coefficients in liquids-D is usually very small), the convection takes over diffusion. However things are more complicated than that because the length l_0 in the Péclet number depends itself on the efficiency of the convection-diffusion through the Stefan-Lamé and the Gibbs-Thomson relations. Thus we consider, in the second section below, how all those effects combine to produce various well defined regimes where different physical phenomena do not balance the same way. This leads to three different regimes, delineated by the value of a single dimensionless quantity, called N that combines the imposed undercooling, the flow velocity at infinity and various molecular parameters, as the capillary length and the molecular transport coefficients.

In section three we expose in some details the general theory based on the Nash and Glicksman integrodifferential equation that has to be solved to obtain the shape of a needle crystal, when the growth is limited by diffusion only. We explain too what happens to this equation in the small undercooling limit as well as how to solve it outside of this limit. We finally present some numerical results in the finite undercooling limit and with kinetic terms added, relevant for describing the experimental findings.

2. Scaling Laws without and with an Axial Flow

Ivantsov (1947) has found a remarkable solution of both the Stéfan-Lamé and diffusion equation. This describes the parabolic needle crystal growth when the heat (or the impurities) released by this growth process is evacuated by molecular diffusion only in the undercooled liquid. Through the Gibbs-Thomson relation, the capillary effects determine [Pelcé and Pomeau(1986), Ben Amar and Pomeau (1986), Barbieri et al. (1987)] all the parameters of the solution, although in a not obvious way in the mathematical sense. Below we study the same question (i.e. scaling laws for the tip radius and for growth rate) when an axial flow in the liquid phase is imposed from the outside. It is assumed uniform at large distances and opposite to the growth direction (i.e along the y direction). This question has already been examined [Ben Amar et al. (1988), Ananth and Gill (1988)], in particular a nice parabolic solution was found for a perfect fluid flow. We plan here to consider an axial and steady flow of viscous fluid which creates a Blasius boundary layer along the crystal. In a first step, we recover the known scaling laws [Kessler et al.(1988)] for both the growth rate and the tip radius of the free needle crystal as a function of the undercooling and in the absence of flow, then we extend the results to the case when a forced flow makes an important perturbation in a sense which has to be precised. The calculations are presented for the bidimensionnal case; the results for the tridimensional case are indicated without any detailed derivation when they differ from the 2D case. Since we aim at establishing scaling laws, the equalities have not to be understood rigorously but indicate order of magnitude or equivalence only.

2.1. SCALING LAW WITHOUT AXIAL FLOW :

Let us determine, in a first step, scaling laws for the boundary layer in the asymptotic part of the growing needle crystal. In the stationary frame of the crystal, the diffusion equation for temperature field (or the impurity concentration field) reduces to:

$$u_c \frac{\partial \theta}{\partial y} = D \frac{\partial^2 \theta}{\partial x^2} \quad (2.1),$$

where u_c is the growth rate, D is the molecular diffusion coefficient (either thermal or solutal), and where x (/y) is the direction perpendicular(/ parallel) to the crystal axis. For small undercooling (this limit will be precised later on), the transverse width of the crystal is negligible compared to the thickness of the thermal boundary layer. Then, in the asymptotic part of the crystal, one can replace the crystal by a straight line (x=0, y<0) to compute the diffusion field. This boundary layer correspond to a similarity solution of (2.1) such that θ is a function of the dimensionless quantity $\frac{Dy}{u_c x^2}$ only. For large arguments, or far away from the crystal, this function tends to Δ, the dimensionless undercooling given by $\frac{(T-T_{eq})c_L}{L}$, where T_{eq} is the melting temperature, c_L the heat capacity and L the latent heat.

The typical length of the thermal boundary layer is then the local thickness $e_{th}(y)=(\frac{Dy}{u_c})^{1/2}$, while the crystal width $e_{cr}(y)$ at the distance y from the tip is found by considering the growth rate as given by the Stéfan-Lamé law during a typical time of $\frac{y}{u_c}$: as the crystal grows at the speed u_c, this is the time elapsed since the crystal began to grow at a given location "downstream". Assuming the local temperature gradient along the x axis to be of order $\frac{\Delta}{e_{th}(y)}$, one gets:

$$e_{cr}(y) = \Delta\, e_{th}(y)$$

For Δ small, the crystal width is small compared to the thermal length. Near the tip, length scales along x or y become the same. The thickness of the diffusion boundary layer is fixed there by $e_{th}(y)=y=e_{th}$, so

$$e_{th} = \frac{D}{u_c}, \qquad (2.2).$$

To get both the growth rate u_c and the tip radius, one has to consider the Stéfan-Lamé and Gibbs-Thomson relations [Langer(1987), Kessler et al.(1988)]. A first condition can be found by noticing that, near the tip, the convection term becomes negligible in the diffusion equation: the characteristic length scale is the tip radius assumed to be small compared to e_{th}. As a consequence the temperature field satisfies there Laplace's equation with a Dirichlet condition on the crystal. For a half straight line, this field has a square-root singularity near the tip and behaves locally as

$$\theta = \Delta \left(\frac{r}{e_{th}}\right)^{1/2} \qquad \left[3D:\ \theta = \frac{\Delta\ \mathrm{Ln}(r/\rho)}{\mathrm{Ln}(e_{th}/\rho)}\right]$$

with r the distance scale near the tip. The denominator e_{th} allows to match the diffusion field with the outer boundary layer. The gradient of θ near the tip is of order:

$$\Delta\left(\frac{1}{\rho e_{th}}\right)^{1/2} \qquad \left[3D:\ \frac{\Delta}{\rho \mathrm{Ln}(e_{th}/\rho)}\right]$$

Combining this with the Stéfan-Lamé law, one finds the laws for the growth rate u_c:

$$u_c = \frac{D\Delta}{(\rho\ e_{th})^{1/2}} \qquad \left[3D:\ u_c = \frac{D\ \Delta}{\rho\ \mathrm{Ln}(e_{th}/\rho)}\right] \qquad (2.3).$$

equivalent to the Ivantsov relation between the Péclet number ($Pe = \frac{u_c \rho}{D}$) and Δ:

$$Pe = \Delta^2 \qquad \left[3D:\ Pe = -\frac{\Delta}{\mathrm{Ln}\Delta}\right] \qquad (2\text{-}4)$$

The Gibbs-Thomson law provides the second relation needed to fix u_c and ρ independently:

$$\theta = \frac{\Gamma}{\rho},$$

where Γ is the capillary length that is proportional to the surface energy. Then:

$$\Delta \left(\frac{\rho}{e_{th}}\right)^{1/2} = \frac{\Gamma}{\rho} \qquad [\ 3D: \ \frac{\Gamma}{\rho} = -\frac{\Delta}{\text{LnPe}}\ ] \qquad (2.5)$$

When combined with (2.3), Equation (2.5) gives, both in the 2D and 3D case, the well-known relation

$$\rho^2 \, u_c = D\Gamma. \qquad (2.6)$$

This relation eliminates the undercooling which is often difficult to measure experimentally. It does not exhibit too the thermal length e_{th} at the tip. To get (2.6), we used the hypothesis of small undercooling and assumed a parabolic tip. In particular (2.6) does not require the crystal to behave like a parabola in its asymptotic part. This is why it is observed to hold in experiments wherein the crystal sides exhibit secondary sidebranchings. Eqn (2.6) is no more valid as soon as the undercooling is not small anymore, as shown by Ben Amar (1989).

Below we shall use the same method as above to establish the relevant scaling laws when a steady and axial flow is imposed in the melt assumed to be a viscous fluid. As shown later on, it is necessary to distinguish several limits depending on the relative order of magnitude between ρ, the tip radius of the crystal, e_{th} and e_{hy}: thermal and hydrodynamical length estimated near the tip. Nevertheless, some relations which describe the asymptotic part of the crystal are the same in different cases, so they will be derived in the last subsection. On the other hand, in all subsections, we will use the same notations for the same physical quantities although the algebraic expression of this quantity differ most of the time according to the considered approximation.

2.2. SCALING LAWS FOR THE GROWTH IN A FORCED FLOW WHEN $\rho < e_{th} < e_{hy}$:

Here, we shall not consider the case where the flow is a small perturbation: then, obviously, the results of the first subsection remain valid and can be applied in a first approximation. This subsection is devoted to the case where the flow is an important perturbation in a sense to be precised later. The convection-diffusion equation (2.1) has now to be changed into:

$$u_c \frac{\partial \theta}{\partial y} + u_h \nabla \theta = D \frac{\partial^2 \theta}{\partial x^2} \qquad (2.7)$$

where u_h is the flow field in the vicinity of the needle crystal measured in its moving frame. Contrary to a recent work which assumes inviscid flow [Ben Amar et al. (1988)], we shall consider the case - which is a priori realistic if the u_h value at infinity is large enough - where the first term on the left side of (2.7) is negligible compared to the second one and where the Blasius boundary layer is much thicker than the thermal one. This last condition physically means that the vorticity diffuses more quickly than the heat (or the scalar field θ). It then implies that the Prandtl number (or the Schmidt number for impurities) is large. The first condition depends on the dimensionless ratio: $\frac{u_c}{u_\infty}$ where u_∞ is the speed of the flow far away from the solid. Evaluating the viscous boundary layer thickness in the same way as done for the thermal one, we get

$$e_{hy}(y) = (\frac{\upsilon y}{u_\infty})^{1/2} \qquad (2.8),$$

where υ is the fluid kinematic viscosity. This estimate is consistent with the Stokes approximation, but also with the Blasius one where the Reynolds number is arbitrary. The following calculation based on this hypothesis would become inconsistent if, near the tip, the hydrodynamic boundary layer characterized by $e_{hy} = \frac{\upsilon}{u_\infty}$ is slimmer than the tip radius. We will come back later to this condition: $\frac{\upsilon}{u_\infty} \gg \rho$. In the asymptotic branches of the crystal, the axial component of the hydrodynamic speed is given by $u_y = \frac{u_\infty x}{e_{hy}(y)}$, as far as x $< e_{hy}(y)$. By substitution of this estimate into the convection-diffusion equation near the parabola:

$$u_y \frac{\partial \theta}{\partial y} + u_x \frac{\partial \theta}{\partial x} = D \frac{\partial^2 \theta}{\partial x^2} \qquad (2.9),$$

where u_y and u_x are the Cartesian components of the flow velocity, and by noticing that, due to the incompressibility of the liquid phase, both terms on the left side of (2.9) have the same order of magnitude, we get the estimate:

$$e_{th}(y) = \left(D \sqrt{\upsilon}\right)^{1/3} (\frac{y}{u_\infty})^{1/2} \qquad (2.10),$$

for the thermal length along the x axis. This is consistent with the assumptions if $e_{th}(y) \ll e_{hy}(y)$, or $\frac{\upsilon}{D} \gg 1$ as mentionned above. We can neglect the first term in (2.7), as done before if

$$\frac{u_c}{u_\infty} \ll (\frac{D}{\upsilon})^{1/3} \qquad (2.11).$$

Let us study now the vicinity of the tip. The thickness of the thermal boundary layer is then e_{th} such that $e_{th}(e_{th}) = e_{th}$ where we use the expression of $e_{th}(y)$ as given by (2.10). It reads:

$$e_{th} = \frac{(D^2 \upsilon)^{1/3}}{u_\infty}$$

which defines for this case the typical length around the tip analogous to the length e_{th} already introduced for the case without flow. The following calculation for various estimates in the vicinity of the tip use the same strategy as used before, but in the absence of axial flow. On one hand it leads to the relation (2.6) between the tip radius and the growth rate and, on the other hand, it leads to the following dependence of these two quantities in the undercooling.

$$u_c = \frac{u_\infty^{2/3} D^{5/9} \Delta^{4/3}}{\Gamma^{1/3} \cdot \upsilon^{2/9}} \quad [\ 3D: \ u_c = \frac{D \Delta^2}{\Gamma Ln[(D^2\upsilon)^{1/3}\Delta/u_\infty\Gamma]} \ ] \qquad (2.12)$$

The condition (2.11) and the initial assumption ($\rho < e_{th}$) can be expressed in term of the control parameters: the undercooling, and the hydrodynamical velocity. This is equivalent to the following inequalities:

$$1 < N < \frac{1}{\Delta^3} \quad \text{in 2D} \quad \text{and} \quad 1 < N < \frac{1}{\Delta} \quad \text{in 3D}$$

$$\text{with N defined by} \qquad N = \frac{\Delta \ (D^2\upsilon)^{1/3}}{u_\infty \ \Gamma}$$

2.3. SCALING LAWS FOR THE GROWTH WITH AN AXIAL FLOW SUCH THAT $e_{th} < \rho < e_{hy}$:

When the velocity u_∞ is such that the three typical lengths of the problem, that are ρ, e_{th} and e_{hy} are such that $e_{th} < \rho < e_{hy}$, (contrary to the former case where they are ordered as $\rho < e_{th} < e_{hy}$) it is necessary to modify the estimates which have just been presented. The inequality $e_{th} < \rho$ means that the thermal boundary layer is much thinner than the characteristic width of the crystal (measured for example by the radius of curvature at the tip). The length scale e_{th} cannot be found as previously, where we considered the needle crystal as a simple half line, at least for the determination of the thermal field in its vicinity. Our starting point is again the boundary layer approximation for the hydrodynamical velocity field which keeps the following dependence in x and y :

$$u = u_\infty \ F[\frac{x \ u_\infty^{1/2}}{(\upsilon \ y)^{1/2}}],$$

where the function F is the solution of the classical Blasius similarity equation. Due to the boundary conditions, this function must go to 1 for large arguments and vanish for small arguments. Near zero, its Taylor expansion begins with a linear term. Before, this kind of argument lead us to an estimate of the thermal boundary layer thickness since we always assumed that this one was nearly parabolic:

$$x = (\ e_{th} \ y)^{1/2}.$$

When this estimate is put in the above expression for the velocity u, we conclude that this parabolic dependence holds for small arguments of the F function and that :

$$u = u_\infty \left(\frac{e_{th}}{e_{hy}}\right)^{1/2},$$

This estimate, when put into the convection-diffusion equation leads to the previous result for the thermal boundary layer. In the case under consideration, the thermal boundary layer is a priori thinner than the tip radius. The order of magnitude of the velocity field is always given by the boundary layer approximation but now it has to be evaluated near the parabola which physically represents the crystal given by:

$$x = (\rho y)^{1/2},$$

So the new estimate for the magnitude of the velocity field is :

$$u = u_\infty \left(\frac{\rho}{e_{hy}}\right)^{1/2}$$

In the thermal boundary layer and near the tip, we need an estimate for the gradient of the velocity field since the velocity must vanish on the crystal. Then, along the crystal surface and in the vicinity of the tip, the gradient of the velocity, within the boundary layer approximation is given by :

$$\mathrm{grad}\, u = \frac{u_\infty}{(\rho\, e_{hy})^{1/2}}$$

From this, we deduce an estimate of u in the thermal boundary layer of width e_{th} (unknown for the moment):

$$u = \frac{u_\infty}{(\rho\, e_{hy})^{1/2}} e_{th} \qquad (2.13).$$

When this estimate for the velocity is introduced in the convection-diffusion equation, we get:

$$e_{th} = \frac{D^{1/2} \rho^{1/4} e_{hy}^{1/4}}{u_\infty^{1/2}} \qquad (2.14).$$

It now remains to fix the needle crystal parameters, by use of the Stefan-Lamé and Gibbs-Thomson relations and to determine the range of validity of the present calculation:

$$u_c = \frac{D\Delta}{e_{th}} \quad \text{and} \quad \rho = \frac{\Gamma}{\Delta}$$

so
$$u_c = \Delta^{5/4} u_\infty (\frac{D}{\upsilon})^{1/2} (\frac{\upsilon}{\Gamma u_\infty})^{1/4}, \quad (2.15).$$

Equation (2.15) is written in order to point out dimensionless quantities. This second case is relevant if $(\frac{D}{\upsilon})^{2/3} < N < 1$,

The formula for the 3D case are identical if x means then the positive distance from the axis.

2.4. SCALING LAWS FOR THE GROWTH WITH AN AXIAL FLOW SUCH THAT $e_{th} < e_{hy} < \rho$:

Let us now consider the last case, where the hydrodynamical velocity is large enough to make the thickness of both the thermal and hydrodynamical boundary layer much less than the tip radius as given by the Gibbs-Thomson law :

$$\rho = \frac{\Gamma}{\Delta}$$

Near the tip the boundary layer does not obey the Stokes approximation, since the Reynolds number calculated with the tip radius is $Re = \frac{\rho}{e_{hy}}$ which is much larger than 1 by hypothesis. We are concerned with the classical situation of a viscous boundary layer assumed to be attached, which is certainly true in the vicinity of the stagnation point in front of the crystal. The external flow when calculated at distances of order of the tip radius ρ is the stationnary Euler flow around the tip, the no slip condition being ensured by the boundary layer. The classical calculation of hydrodynamical boundary layers gives for the vicinity of the stagnation point a typical length of order :

$$e_{hy} = (\frac{\rho^2 \upsilon}{u_\infty})^{1/3},$$

If the Péclet-(or Schmidt-) number which describes the diffusion of the scalar quantity θ is small, one can neglect e_{hy} compared to the thermal length. In this case, it is necessary to apply the method of Ben Amar et al. (1988) to solve this problem but one needs to apply moreover the Gibbs-Thomson condition. Anyway, since the beginning, we are interested in the other limit, where the diffusion of θ is much slower than the one of vorticity. Then, the thermal boundary layer is inside the hydrodynamic boundary layer where we can evaluate the velocity (tangential) by $u_\infty (\frac{e_{th}}{e_{hy}})^2$. The square comes from the Taylor expansion of the flow field near a stagnation point that begins by second order terms. By inserting this approximation into the convection-diffusion equation one gets:

$$e_{th} = \frac{D^{1/3} \rho^{4/9} \upsilon^{2/9}}{u_\infty^{5/9}} \qquad (2.16)$$

which, with the relation $u_c = \frac{\Delta D}{e_{th}}$, leads to the approximate value of the growth rate:

$$u_c = \frac{\Delta^{13/9} D^{2/3} u_\infty^{5/9}}{\upsilon^{2/9} \Gamma^{4/9}} \qquad (2.17)$$

Knowing the growth rate, this applies to flows such that

$$N < (\frac{D}{\upsilon})^{2/3}$$

2.5 SCALING LAWS FOR THE ASYMPTOTIC BRANCHES :

As suggested by the geometry of the problem itself, the asymptotic parts of the crystal have little chance to depend in great details on the mechanism which fixes the curvature of the tip. As shown below, in a way, these parts depend on local parameters and also on the growth rate u_c, which is no more precised in the following part and has to be replaced by the convenient expressions found above. The main result of this section will be that the asymptotic part are always parabolic and have for Cartesian equation:

$$y = \frac{x^2}{E_c},$$

where E_c gives the length scale. If the needle crystal were defined by an unique length scale E_c, this one should be of the order of magnitude of the tip radius. This is what happens in the first two cases considered above : growth in 2D with $\rho < e_{th} < e_{hy}$ or without flow. On the other hand, in the other cases, E_c has not the same order of magnitude as ρ: if the inner (=near tip) problem has an asymptot for $x \approx \rho$, we expect that the matching is achieved along this parabola for $y \approx \frac{\rho^2}{E_c}$, which is consistent whatever the value of ρ and E_c. At the distance y from the tip, the width of the viscous boundary layer is :

$$e_{hy}(y) = (\frac{y\upsilon}{u_\infty})^{1/2},$$

from which we deduce by the same argument as before that:

$$e_{th}(y) = (yE_{th})^{1/2},$$

with $\quad E_{th} = \dfrac{(D\upsilon^{1/2})^{2/3}}{u_\infty}.$

The crystal width $e_c(y)$, at the y distance from the tip, is calculated by noticing that the growth rate in the x-direction is given by the Stéfan-Lamé law, with a gradient of temperature of order $\dfrac{\Delta}{e_{th}(y)}$, that is:

$$u_c(y) = \dfrac{D\Delta}{e_{th}(y)}.$$

We get $e_c(y)$ now by integration of the growth rate during the growth time at the distance y, that is $\dfrac{y}{u_c}$. This yields:

$$e_c(y) = (y\, E_c)^{1/2} \quad \text{with} \quad E_c = \dfrac{D^2\Delta^2}{E_{th}u_c^2}$$

which is the most general expression for this length scale and apply also to the case without flow, in 2D. In the 3D case, because of curvature effects, it is only valid if $E_c > E_{th}$. In 2D and with an external flow, one has the following general approximation for E_c:

$$E_c = \dfrac{\Delta^2 D^{4/3} u_\infty}{\upsilon^{1/3} u_c^2} \qquad (2.18).$$

We have shown here that three regimes of dendritic growth occur when a forced flow is imposed far away from the tip. More precisely, we have pointed out a dimensionless parameter N, a combination of both the control parameters as the undercooling Δ and the flow velocity u_∞ and of the molecular parameters as the diffusion coefficient D, the viscosity ν and the surface tension Γ. Depending on the values of N, we have found different scaling laws for the growth rate which can in principle be checked experimentally [Glicksman et al.(1986), Bouissou et al.(1989)]. Obviously, precise comparison with experimental measurement would require the determination of numerical constants, out of reach of the present analysis. To get them, it is necessary to solve rather complicated free boundary problems. The present analysis is the first step to solve them since it points out useful and necessary approximations specific of each regimes of growth. In the third part, we expose the solution of one of them: he 2D solidification case without any axial flow, but at arbitrary undercooling.

3. Growth Rate at Arbitrary Undercooling

3.1. THE FREE BOUNDARY PROBLEM :

We want to determine the growth rate of a needle crystal growing in its pure melt, without any deformation. Contrary to the second section of this paper, we focus on the existence of solutions, and we plan to calculate any numerical constants missing in the previous part. It is why now we recall the basic equations in the bidimensionnal case only. Far at infinity, the bulk temperature T_∞ is fixed at a lower value than the melting one T_m. For sake of simplicity, we assume equality of the diffusion coefficients in the solid and liquid phases. The shape of the needle crystal and its velocity is solution of a free boundary problem: one has to solve the heat diffusion equations in both phases, with two boundary conditions to apply to the unknown interface:

i) the Stefan law relates the jump in the temperature gradient across the interface to the normal velocity:

$$v_n = D\,(\nabla\theta_s - \nabla\theta_l)$$

Here, θ_s (resp. θ_l) is a dimensionless quantity which measures the departure from the equilibrium temperature in the solid phase (resp. the liquid phase), it is given by:

$$\theta = c\,\frac{(T-T_m)}{L}$$

with c the heat capacity and L the latent heat, released at the interface.

2) The Gibbs-Thomson law gives the temperature of a moving curved interface:

$$\theta_{int} = \frac{\Gamma}{R} - \beta\,v_n \qquad (3.1)$$

Here $\frac{1}{R}$ is the curvature, Γ the capillary length. When local equilibrium is assumed, the second term in (3-1) is usually neglected [Ben Amar and Pomeau (1986), Barbieri et al (1987)]. Here we add it to indicate that the interface presents a small departure from thermal equilibrium. Its expression, assumed to be linear in v is valid only for rough interfaces. In case of facetting, it has to be modified (Ben Amar and Pomeau (1988)). We will come back in the last part of this paper to estimate its importance for the growth rate. For the moment, we neglect it.

Assuming that the interface is an isothermal surface (i.e neglecting completely (3-1)), Ivantsov (1947) succeeded in solving this free boundary problem : he found a continuum of solutions with parabolic (resp paraboloid) shape:

$$\xi_{iv}(x) = -a\,x^2$$

However, only the Péclet number Pe was given in term of the undercooling

$$\Delta = c \frac{(T_m - T_\infty)}{L}$$

by his analytical treatment.

If we define Pe as $Pe = \frac{a^{-1} v}{D} = 2\frac{\rho v}{D}$ (with ρ is the tip curvature of the parabola) then

$$\Delta = \frac{(\pi Pe)^{1/2}}{2} \exp(\frac{Pe}{4}) \, \text{Erf}(\frac{Pe^{1/2}}{2}) \tag{3.2}$$

The interested reader can check that, when the Péclet number vanishes, the relation (3-2) restores (2-4). Hereafter, we choose the growth direction along the y axis.

3.2. THE GREEN'S FUNCTION TECHNIQUE APPLIED TO THE SYMMETRIC MODEL OF SOLIDIFICATION

3.2.1. *The Nash and Glicksman equation*

The Ivantsov theory explains the experimentally observed shape, but not the selection of the growth rates. In order to introduce the capillary effects, Nash and Glicksman (1974) transformed the free-boundary problem by using Green's function techniques. They established an equation for the sought profile, in 2 D:

$$\Delta + \frac{\xi_{xx}}{(1+\xi_x^2)^{3/2}} (1 - \varepsilon \cos(4\Theta)) = \frac{U}{2\pi} \int_{-\infty}^{\infty} \exp[-U/2(\xi(x) - \xi(t))] K_0(U/2 \, |R(x) - R(t)|) \tag{3.3}$$

Here, the capillary length Γ is our choice of length units, while velocities are scaled by D/Γ, so $v = U \, D/\Gamma$. ξ_x (resp. ξ_{xx}) means the first (resp. the second) derivative with respect to the variable x. $|R(x) - R(t)|$ means the distance between two points of the interface: $[(x-t)^2 + (\xi(x) - \xi(t))^2]^{1/2}$. K_0 is the modified Bessel function of first order. We have also included four-fold anisotropy of surface tension and Θ is the angle between the normal interface and the growth axis. If one assumes that capillary effects are negligible far away from the tip, we can seek solutions to (3-3) with parabolic shape at infinity:

$$\xi(x) \propto \xi_{iv}(x) \propto -a x^2 \quad \text{when} \quad x \to \pm\infty \tag{3.4}$$

Recalling that ξ_{iv} is solution to (3-3) without curvature effect in the left hand side, by obvious subtraction, we derive an equivalent equation more suitable for any analytical but also numerical treatment:

$$\frac{\xi_{xx}}{(1+\xi_x^2)^{3/2}} (1-\varepsilon \cos(4\Theta)) =$$

$$\frac{C}{2\pi} \int_{-\infty}^{\infty} dt \, \exp[-Pe/2 \, (\xi(x)-\xi(t))] K_0(Pe/2 \, |R(x) - R(t)|)$$

$$- \frac{C}{2\pi} \int_{-\infty}^{\infty} dt \, \exp[-Pe/2 \, (\xi_{iv}(x)-\xi_{iv}(t))] K_0(Pe/2 \, |R_{iv}(x) - R_{iv}(t)|) \tag{3.5}$$

All lengths are scaled by $a^{-1}=2\rho$ (twice the tip radius of the asymptotic parabola ξ_{iv} which has the same velocity than the needle crystal) so C equals a^{-2} U. This rather complicated equation has C as a nonlinear eigenvalue and the undercooling Δ is hidden in the continuous parameter Pe. In order to calculate the growth rate, one has to solve (3-5) and find the Péclet dependence of the eigenvalue spectrum, since

$$v = U \frac{D}{\Gamma} = \frac{Pe^2}{C(Pe)} \frac{D}{\Gamma} \tag{3.6}$$

Much progress has been made in the last three years by looking at two different limiting values of the undercooling Δ : $\Delta \to 0$ [Ben Amar and Pomeau (1986), Ben Amar and Moussallam (1988), Kessler et al. (1986)] or $\Delta \to 1$ [Caroli et al. (1986), Barbieri (1987)]. The first limit (vanishing Péclet number limit) seems relevant in many experiments of solidification induced by thermal diffusion, except the very recent experiment on Nickel by Willnecker et al., whose results will be discussed in the last part of the paper. In case of chemical diffusion, one can expect to get larger values for Pe. Let us sum up the known results.

3.2.2. The vanishing Péclet number limit:

If Pe is small, the diffusion length as given by D/v is quasi-infinite. Since the capillary effects act only on distances of order of the tip radius, Pelcé and Pomeau (1986) have proposed a matched asymptotic expansion to get a simplified equation for the profile. They prove that it is only necessary to consider vanishing arguments of the modified Bessel function K_0 in (3-3) or (3-5). Since

$$K_0(x) \propto -\text{Ln}(\tfrac{1}{2}x) \quad \text{as} \quad x \to 0^+$$

they obtain:

$$\frac{\xi_{xx}}{(1+\xi_x^2)^{3/2}}(1-\varepsilon \cos(4\Theta)) = -\frac{C}{4\pi}\int_{-\infty}^{+\infty} dt \; \text{Ln} \; \frac{|R(x)-R(t)|}{|R_{iv}(x)-R_{iv}(t)|} \quad (3.7)$$

The dependence in Pe has disappeared. If one assumes isotropic surface tension (ε equal zero), it has no solution due to a mismatch at the tip measured by ξ_x (0), whatever the C value. In the limit of large C, Ben Amar and Pomeau (1986) evaluate this one by means of a nonlinear W.K.B. analysis, the details of which will be given hereafter (Section 3-3). They find:

$$\xi_x(0) \propto A_0 \; C^{-27/28} \exp(-S(1/2) \; C^{1/2})$$

with $\quad S(1/2) = \int_0^{1/2} dt \; S'(t) \approx 0.2176 \quad$ and $\quad S'(t) = (1-4 \; t^2)^{1/4}(1/2-t)^{1/2} \quad (3.8)$

A_0 is a pure constant which can be evaluated numerically.

So, it is impossible to impose simultaneously vanishing first derivative at the tip and asymptotic parabolic behaviour far from the tip, in case of isotropic surface tension. If ε is different from zero but less than one, Ben Amar and Pomeau (1986) and Ben Amar and Moussallam (1987) have found a discrete set of possible profiles. In the limit of large eigenvalues C_n, they establish that:

$$C_n(\varepsilon) = \frac{(\theta_0 + n \pi)^2}{T(\varepsilon)^2} \quad (3.9 \text{ a})$$

θ_0 is a nonlinear phase shift, $T(\varepsilon)$ is given by a path integral between two singularities so:

$$T(\varepsilon) = \int_{t_0}^{1/2} dt \; \frac{S'(t)}{(4 \; \alpha \; t^2/(1-4t^2)^2-1)^{1/2}} \quad (3.9\text{b})$$

with $\alpha = 8\epsilon/(1-\epsilon)$ and $t_0 = (-(\alpha)^{1/2} + (\alpha+4)^{1/2})/4$ (3.9c)

Although valid for small ϵ, Eqn (3-9) reproduces conveniently the numerical set of eigenvalues. By a numerical stability analysis, Kessler and Levine (1986) show that only the smallest one (which gives the quickest needle crystal) can be observed in an experiment so we deduce the growth rate:

$$v = \frac{16 \Delta^4}{\pi^2 C_0(\epsilon)} \frac{D}{\Gamma}$$

while $a^{-1} = 2\rho = \frac{\pi C_0(\epsilon)}{4} \frac{1}{\Delta^2} \Gamma$ (3.10)

3.3. W.K.B. ANALYSIS FOR ARBITRARY PECLET NUMBER.

3.3.1. *Analytical continuation in the complex plane*

We will follow exactly the various steps which have been extensively described in Ben Amar and Pomeau (1986) and Ben Amar (1988) The analysis rests on an analytical continuation of (3-5) to the complex plane. If C is large enough, (3-5) can be linearized around the Ivantsov solution in part of the complex z plane where ξ is not singular. On the real axis, we get:

if $\quad \xi(x) = -x^2 + \frac{1}{C'} u(x) \quad$ with $\quad C' = \frac{C}{1-\epsilon}$ (3.11)

then

$$\frac{-2}{(1+4x^2)^{3/2}} + \frac{1}{C'} \frac{d}{dx} \frac{u_x G(x)}{(1+4x^2)^{3/2}} =$$

$$-\frac{Pe}{4\pi} \int_{-\infty}^{+\infty} [u(x) - u(t)] H(x,t) \exp[Pe/2(x^2 - t^2)] dt \quad (3.12a)$$

with $\quad H(x,t) = K_0[Pe/2 \, |R_{iv}(x) - R_{iv}(t)|]$

$$- \frac{(x^2 - t^2)}{|R_{iv}(x) - R_{iv}(t)|} K_1[Pe/2 \, |R_{iv}(x) - R_{iv}(t)|] \quad (3.12b)$$

and
$$G(x) = 1 + \frac{4\alpha x^2}{(1+4x^2)^2} \quad (3.12c)$$

Written in this way, (3-12) is not suitable for analytical continuation, mainly because of the absolute value. Recalling that

$$K_0(x \pm i a) = K_0|x| - \pm i\pi I_0(x)\theta(-x) \quad \text{if } a \to 0^+$$

and $K_1(x \pm i a) = K_1|x| \pm i\pi I_1(x)\theta(-x) \quad \text{if } a \to 0^+$ (3.12d)

I_0 and I_1 are the modified Bessel functions, regular at the origin, $\theta(x)$ is the Heaviside distribution, equal to 1 when x is positive and zero elsewhere. We have chosen the Riemann sheet which makes symmetric the upper and lower half complex plane. Hence, we will focus on the behaviour of u along the positive imaginary axis ($z = iv$), we write only the analytical continuation of (3-12) on this axis, for sake of simplicity:

$$\frac{-2}{(1-4v^2)^{3/2}} - \frac{1}{C'}\frac{d}{dv}\frac{vG(iv)}{(1-4v^2)^{3/2}} = \frac{u(iv)}{2(1-4v^2)}$$

$$+ \frac{Pe}{4\pi} PP \int_{-\infty}^{+\infty} u(t)L(iv,t)\exp[-Pe/2(v^2+t^2)]dt$$

$$+ \frac{iPe}{4} \int_\Gamma u(z)l(iv,z)\exp[-Pe/2(v^2+z^2)]dz \quad (3.13a)$$

with:

$L(iv,t) = K_0[Pe/2(iv-t)(1+(iv+t)^2)^{1/2}] -$

$$\frac{(iv+t)}{(1+(iv+t)^2)^{1/2}} K_1[Pe/2(iv-t)(1+(iv+t)^2)^{1/2}] \quad (3.13b)$$

and

$l(iv,t) = I_0[Pe/2(iv-t)(1+(iv+t)^2)^{1/2}] -$

$$\frac{(iv+t)}{(1+(iv+t)^2)^{1/2}} I_1[Pe/2(iv-t)(1+(iv+t)^2)^{1/2}] \quad (3.13c)$$

The first integral has to be performed on the real axis, while the Γ contour involves two parts :Γ_1 and Γ_2. Γ_1 begins at iv on the imaginary axis, goes to zero along this axis. Γ_2

takes place on the positive real axis. PP means the Cauchy principal part. Notice that l(iv,z) is purely real on Γ_1. Note that the second integral in (3-13a) has been pointed out first by Barbieri and Langer (1989) and Tanveer (1989). As shown later, it is responsible of the Péclet dependance of the eigenvalues and cannot be neglected as assumed in (et al (1987)

The Kruskal and Segur method (1985) rests on a solution of (3-13) in the complex plane: it assumes the validity of the ordinary pertubation expansion of $\xi(x)$ when x goes to $+\infty$ The first term of this expension u_1 is solution of (3-13) if one formally makes C' = ∞. Contrary to the vanishing Péclet number limit [Ben Amar and Moussallam(1987)], we did succeed to invert this equation analytically, but after numerical investigation which will be explained later, we can claim that u_1 has no derivative at the tip (with or without any cristalline anisotropy). So as before, the ordinary perturbation expansion does not give any information on the selection mechanism. The linearization of (3-5) breaks down in the part of the complex plane where each term of (3-13) become singular : that is $\pm i/2$. However, in the close neighborood of $\pm i/2$, we can take into account the most singular terms to simplify a little the original nonlinear equation.

3.3.2. *The internal region and the W.K.B. analysis on the imaginary axis.*

We assume the validity of the scaling laws found at Pe equal zero, since we are interested in an arbitrary but finite Péclet number, contrary to the limit involved by Barbieri (1987). So:

If ε equal zero

$$\frac{1}{C} u(i/2 - iv) = C^{-4/7} F(C^{2/7} v) \tag{3.14}$$

F obeys the following local but nonlinear equation:

$$4(2 + F_{tt}) = [F + P_1 \int F](4t + 2F_t)^{3/2} \quad \text{with} \quad P_1 = Pe\, C^{-2/7} \tag{3.15}$$

t is the natural variable of F, that is : $C^{2/7}$ v. After inspection of the second integral in (3-13a), one can deduce that the main contribution comes from $x \approx t$, so l(iv,z) is equal to one, and (3-15) involves only an integral of F denoted: $P_1 \int F$.

If α is not equal to zero: let us define $C_1 = \dfrac{C}{(1-\varepsilon)\alpha}$ and a similar definition of F

$$\frac{1}{C} u(i/2 - v) = C_1^{-4/11} F(C_1^{2/11} v) \quad \text{if } t = C_1^{2/11} v \tag{3.16}$$

$$(2 + F_{tt}) 16 = -[F + P_2 \int dt\, F](4t + 2F_t)^{7/2} \quad \text{with} \quad P_2 = Pe\, C_1^{-2/11} \tag{3.17}$$

Equations (3-15) and (3-17) are valid in a region of the complex plane around i/2 of very small extent in the first case: $C^{-2/7}$, in the second: $C_1^{-2/11}$. So we need the asymptotic behaviour of F as soon as we leave this area, in order to match to the ordinary W.K.B. solution, valid on the imaginary axis between i/2 and zero. At first, let us focus on the case without anisotropy:

i) isotropic case

When t is large, the leading behaviour of F is purely real and given by:

$$\int F \approx \frac{1}{P_1 t^{3/2}} + \ldots \ldots \quad \text{when } t \to +\infty \tag{3.18}$$

Let us recall that only odd terms contribute to the first derivative at zero. They are purely imaginary on the imaginary axis. So we focus only on the asymptotic expansion of the imaginary part of F. Its main contribution is obtained by linearization of (3-17) around the leading real behaviour (3-18) which gives

$$h_{tt} - \frac{3}{2} \frac{h_t}{t} - 2 h(t) t^{3/2} - 2 P_1 t^{3/2} \int h = 0 \tag{3-19a}$$

so we obtain:

$$h(t) = 2^{1/2} A(P_1) t^{3/8} \exp(-2^{5/2} t^{7/4}/7) \exp(P_1 t/2) \tag{3-19b}$$

A is a nonlinear eigenvalue which obviously is a function of P_1. Equation (3-19b) gives in variable v:

$$\text{Im}[u (i/2 - iv)] = u^* (i/2 - iv) =$$

$$2^{1/2} A (P_1) C^{-13/28} v^{3/8} \exp(-C^{1/2} 2^{5/2} v^{1/4}/7) \exp(\frac{Pe}{2}v)$$

On the imaginary axis, in the outer region, Im(u) obeys an ordinary differential equation:

$$\frac{u^*_{vv}}{(1-4v^2)^{3/2}} + \frac{12 u^*_v}{(1-4v^2)^{5/2}} - \frac{C}{2} \frac{u^*(v)}{(1+2v)} + \frac{Pe\ C}{4} \int u^* = 0 \tag{3.20}$$

We can solve this equation by a W.K.B. approximation since the small parameter 1/C is in front of the highest derivative u^*_{vv}. Assuming this approximation for u^*, we get simply a integral of $\int u^*$ by a simple asymptotic evaluation of the second integral in (3-13a) and we derive

$$u^*(v) = B(1-4v^2)^{3/8}(1/2+v)^{1/4}[\exp(-\frac{Pe}{4}(v+v^2))\exp(C^{1/2}S(v)) -$$

$$\exp(\frac{Pe}{4}(v+v^2))\exp(-C^{1/2}S(v))] \quad (3.21)$$

with $S(v) = \int_0^v S'(t)\,dt$ as in (3-9). By matching the two expressions of u in their common domain of validity, we can calculate B in term of the nonlinear eigenvalue A and finally obtain for the first derivative at zero:

$$u_v(0) = 2^{1/4} B\, C^{1/2} = A(PeC^{-2/7})\, C^{1/28}\exp(3\,Pe/16)\exp(-C^{1/2}S(1/2)) \quad (3.22)$$

We shall assume that A is not a singular function of P_1, so it admits a regular series expansion in term of P_1:

$$A(PeC^{-2/7}) = A_0 + A_1(PeC^{-2/7}) + A_2(PeC^{-2/7})^2 \ldots \quad (3.23)$$

When compared to (3-8), Eqn.(3-22) proves that the mismatch at the tip is larger than at non zero Péclet number. Although derived at rather small Péclet number (Pe< $C^{2/7}$), it indicates that there is no chance to observe a needle crystal without anisotropy, at any Péclet number.

2) Anisotropic case:

Let us come back to the local nonlinear equation with anisotropy (3-17). As in the former case, we do not need to solve completely this equation, valid only around i/2, but we need its asymptotic expansion. The leading behaviour is real :

$$\int F \approx -\frac{1}{P_2\, t^{7/2}} + \ldots\ldots \quad \text{when } t \to +\infty \quad (3.24)$$

and the imaginary part is solution to:

$$h_{tt} - \frac{7}{2}\frac{h_t}{t} + 8\,h(t)\,t^{7/2} + 8\,P_2\,t^{7/2}\int h = 0 \quad (3.25)$$

Choosing $h(t) \approx e^{iS(t)}$ we finally find :

$$h(t) = D(P_2)\, t^{7/8} \exp(i\,2^{9/2}\,t^{11/14}/11 + i\Phi(P_2))\exp(P_2\,t/2) + \text{C.C.} \quad (3.26)$$

Φ (P_2) is a nonlinear eigenphase-shift and for it, we will assume a regular series expansion in powers of P_2. So we get

$$\Phi(P_2) = \Phi_0 + \Phi_1 P_2 + \Phi_2 P_2^2 \ldots\ldots\ldots \quad \text{with} \quad P_2 = Pe\, C_1^{-2/11}$$

Tanver (1989) does not make explicitly this expansion for Φ called in its paper Arg(A_1), Eqn 66. h(t) has to be matched to the W.K.B. outer solution u^* which is given by:

$$u^*(v) = r(v)\, \exp[C'^{1/2} S_1(v)]\, \exp[-\frac{Pe}{4}(v+v^2)] \qquad \text{with} \quad C' = \frac{C}{(1-\varepsilon)}$$

and

$$r(v) = E\, \frac{(1-4v^2)^{3/8}(1/2+v)^{1/4}}{(1 - 4\alpha\, v^2/(1-4v^2)^2)^{1/2}}$$

$$S_1(v) = \int_0^v dt\, \frac{S'(t)}{(1 - 4\alpha\, t^2/(1-4t^2)^2)^{1/2}} \tag{3.27b}$$

Notice that the Péclet contribution to u^* is real along the imaginary axis, so does not modify the selection rule. This W.K.B. expansion breaks down at t_0 which has been defined earlier (3-9c) where a new internal region has to be defined and a new local equation to be established, taking care only of the most singular terms. If we neglect terms of order PeC^{-1}, we derive the same equation as for Pe equal zero [see Ben Amar and Pomeau (1986)]. So the eigenvalues are fixed by a path integral between t_0 and $i/2$:

$$C'_n(\varepsilon, Pe)^{1/2}\, T(\varepsilon) = n\, \pi + \theta_0 + \Phi_1 P_2$$

θ_0 takes into account the two eigenphase shifts arising from the treatment of the two singularities. We can estimate it numerically by solving either the local boundary layer equations around t_0 and $i/2$ or by solving the original Nash and Glicksman equation. Note that both θ_0 and Φ_1 do not depend on the anisotropy. If P_2 is small, that is Pe smaller than $C_1^{2/11}$, then we get for the Péclet dependence of the discrete set of eigenvalues:

$$C_n(\varepsilon, Pe) = C_n(\varepsilon) + \frac{2\, Pe\, \Phi_1}{T(\varepsilon)}(1-\varepsilon)^{15/22}\, \alpha^{2/11}\, C_n(\varepsilon)^{7/22} \tag{3-28}$$

$C_n(\varepsilon)$ is the eigenvalue at Pe equal to zero, found earlier. Notice that the slope of the linear effect in (3-28) is not small: it is given by $\varepsilon^{-5/4}$ while the C eigenvalues (at Pe=0) are scaled by

$\varepsilon^{-7/4}$ at small anisotropy. It is important to realize that the Péclet contribution to the eigenvalues $C_n(\varepsilon, Pe)$ arises from the internal region around $\pm i/2$ and not from the W.K.B. expansion along the imaginary axis.The relation (3-28) is valid within the W.K.B. approximation, i.e for large C. To do so, we can choose either very small anisotropy ε or unphysical eigenvalues C_n corresponding to high n values.They are unphysical since they cannot been observed in an experiment. For usual anisotropy values [Bouissou (1989)], we can compare this asymptotic relation to numerical eigenvalues, solution of the real nonlinear, nonlocal integro-differential equation (3-5). By direct comparison with the numerics (see Ben Amar 1989), estimation of the validity of our treatment can be made. In any case, the W.K.B. treatment explains the existence of the eigenvalues but a precise comparison with experimental data requires a numerical solution of the integro-differential equation (3-5). This is achieved in the next section.

3.4. IMPORTANCE OF KINETIC EFFECTS AND COMPARISON WITH THE EXPERIMENT.

To our knowledge, the experiment of Willnecker et al. (1989) is the only one which covers such a wide domain of undercooling between 0. and 0.75, that is, in Péclet between 0. and 10. The authors have measured the growth rate of pure Nickel and of an alloy ($Cu_{70}Ni_{30}$) using a rapid solidification technique. We can discuss here only the experiment where solidification is induced by undercooling. We have treated only the symmetric model of solidification with equality of the diffusion coefficients between the liquid and solid phase. For the alloy solidification, the one-sided model (with vanishing diffusion coefficient in the solid phase) is more adapted and results will be published elsewhere. The authors insist on a break in the experimental data around $\Delta = 0.440$. It is clear that the curve representing the growh rate versus the undercooling shows a break in curvature for this value. They explain this by kinetic effects and indicate a coefficient μ of order 1.6 m/sK. Unfortunately, they do not indicate the anisotropy of surface tension which is so important for selection. It is time to discuss now the importance of the kinetic effects which indicate that the interface presents a small departure from thermodynamical equilibrium. To measure their strength, it is necessary to write down the proper equation with dimensionless parameters that is for us C and Pe. If one restricts to the linear expansion, in (I-1), one must add on the left hand side of (3-4) the following term:

$$- \beta \, Pe \, \frac{1}{(1+\xi_r^2)^{1/2}}$$

Here, we assume dendritic growth in 3 dimensions and axisymmetry. Note that β is also anisotropic if surface tension is chosen anisotropic. Kinetic effects are not important at low undercooling for two reasons: first in the equation to solve, they are multiplied by Pe so they modify the growth rate only for $Pe > \beta^{-1}$. Second, they include only first derivative, so within the W.K.B. approximation, they modify only the prefactor also called the Van-Vleck factor which is not important for selection.The phase factor (called S in the previous section) depends on the highest derivative, that is, the curvature and so the selection is mostly induced by surface tension. When the Péclet number increases, as shown by [Lemieux et al.(1989)], it is necessary to incorporate these effects, at least for calculating numerically

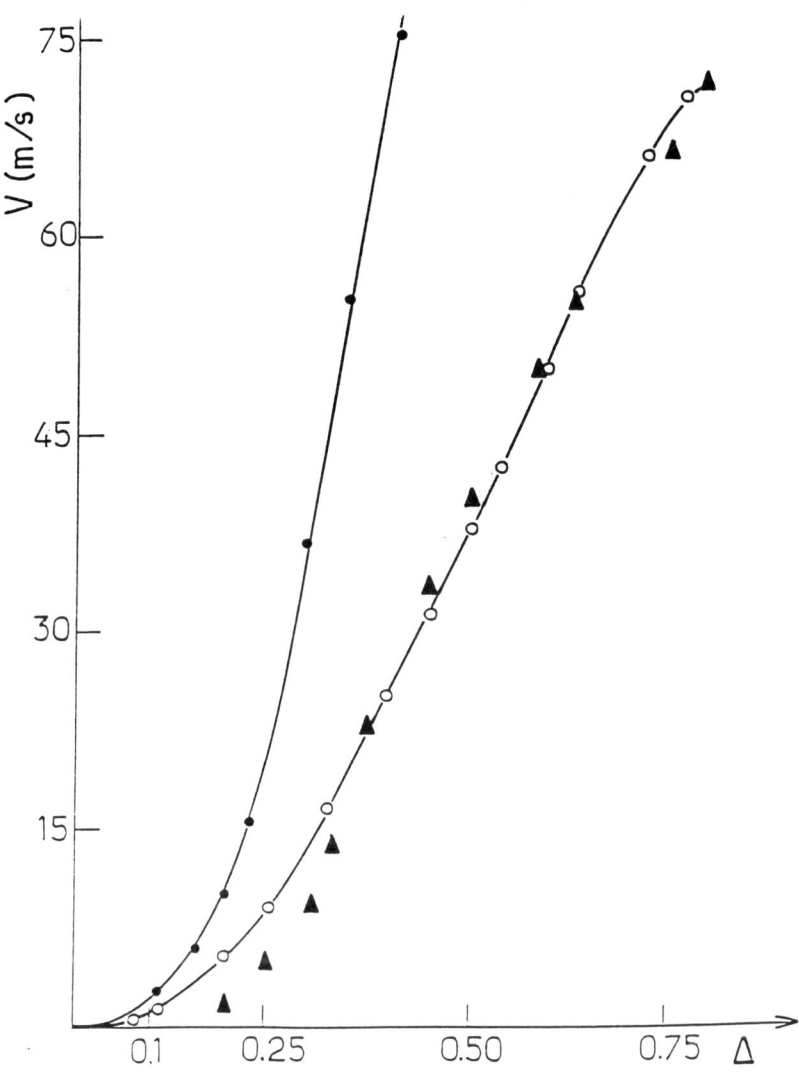

FIGURE :   DENDRITIC GROWTH RATE OF NICKEL

• • • •   Calculated growth rate assuming thermodynamical equilibrium for the interface
o o o o   Calculated growth rate with anisotropy of surface tension and kinetic effects included
▲   Experimental points of Willnecker et al.

the growth rates. We want to stress the especially high value of the β coefficient for Nickel when written with our units

$$\beta = \frac{c}{L\,\mu} \qquad \frac{D}{\Gamma} \approx 15.65 \qquad \Gamma = \frac{c\,\gamma}{L^2 \rho^*}\, T_m \approx 6.53$$

For the definition of each thermodynamical constant, see section I. We have taken into account experimental data collected in Willnecker et al. (1989). Note that γ is the interface energy and ρ^* the density of nickel that is 8.88 10^3 Kg/m^3. After some trials, we decide to calculate the 3D growth rate for a four fold anisotopy coefficient δ_{3D} of 0.0266. In fact here, we have two adjustable parameters anisotropy of free energy and the kinetic coefficient . We fix the last one to the numerical value indicated above. Figure 1 proves that the agreement between experimental and numerical data is rather good as soon as Pe is greater than 1.4 (Δ >0.4) . For Δ < 0.4, our calculation seems to underestimate the C eigenvalues so to overestimate the growth rate. We recall that the growth rate is an increasing function of the anisotropy and a decreasing function of the β coefficient. However, surface tension acts more at low undercooling and the kinetic effects at large undercooling. To obtain a better agreement , it will be necessary to reduce both the anisotropy of surface tension and the β coefficient or to sophisticate the calculation by introducing anisotropy in β. When β is neglected, the growth rate is too high and never fits the experimental data, as shown in Figure 1.

4. Conclusion

Important scaling laws have been established in the second part of this paper, concerning the growth of a needle crystal.By handling physical arguments, we are able to recover known relations proved a long time ago [Ivantsov (1947)] by solving the Stefan problem. We also give some physical insight to the so-called stability constant: we show that it is the result of both the singularity of the Laplace field at the tip of the needle and the effect of surface tension. Since it assumes the Laplace field, it is not a constant as soon as the undercooling and so the Péclet number increases. This is shown in the third part of this paper by solving a complicated integro-differential equation. Such a sophisticated analysis can not or has not been made for the growth in an external flow. This is why we restrict ourselves to the physical analysis of the second part to establish useful scaling laws which can be checked experimentally.

Acknowledgements

We would like to thank Vincent Hakim, Bachir Moussallam for many useful discus and suggestions during the realization of this work. This research was supported in part b Centre National d'Etudes Spatiales, Grant : Incitation à la Recherche 1988, N^0 1236.

References

R. Ananth and W.N. Gill, Chem. Eng. Comm. **68**,1 (1988); R. Ananth and W.N. Gill, Journ. of Cryst. Growth **91**,587 (1988).
A. Barbieri, D. Hong and J. Langer, Phys. Rev.A**36**, 3340 (1987).
A. Barbieri Phys. Rev.A **36**,5353 (1987).
A. Barbieri and J. Langer , to be published in Phys.Rev.A (1989).
M. Ben Amar and Y. Pomeau Euro. Phys. Lett.**2** ,307 (1986).
M. Ben Amar and B. Moussallam, Physica **25D**, 155 (1987).
M. Ben Amar Physica **31D**,409 (1988).
M. Ben Amar, P. Bouissou and P. Pelcé, J. of Crystal Growth **92**,97 (1988).
M. Ben Amar and Y. Pomeau, Euro. Phys. Lett. **6**, 609 (1988).
M. Ben Amar, Phys. Rev. A (1989) in press.
P. Bouissou, B. Perrin and P. Tabeling, Phys. Rev.A **40**,509 (1989).
P.Bouissou Université P. et M. Curie Phd Thesis:"Influence d'un écoulement en croissance dendritique: aspects experimentaux et théorique." (1989).
B. Caroli, C. Caroli, B. Roulet and J.S. Langer Phys. Rev. A**33**, 442 (1986).
M.E. Glicksman, S.R. Coriell and G.B. McFadden, Ann. Rev. Fluid Mech., 307 (1986).
G.P. Ivantsov, Dokl.Akad. Nauk SSSR **58**,567 (1947).
D. Kessler, J. Koplik and H. Levine, Phys.Rev. A **33** ,3352 (1986).
D. Kessler and H. Levine , Phys. Rev. Lett. **57**,3069 (1986)
D. Kessler, J. Koplik and H. Levine, Adv. in Phys. **37**, 255 (1988).
M. Kruskal and H. Segur "Asymptotics beyond all orders in a model of dendritic crystals, Aero. Res. Ass. of Princeton Tech. Memo (1985).
J.Langer :"Chance and Matter", Compte-Rendus des Houches (North Holland, 1987).
M.A. Lemieux, J. Liu and G. Kotliar Phys.Rev. **36**,1849 (1987).
G.E. Nash and M.E. Glicksman Acta Metall. **22**,1283 (1974).
P.Pelcé and Y.Pomeau, Stud. in Appl. Math. **74**,245 (1986).
S. Tanveer , preprint (1989).
R. Willnecker, D.M. Herlach and B. Feuerbacher, Phys. Rev. Lett. **62**,2707 (1989).

STRUCTURAL INVARIANTS AND THE DESCRIPTION OF THE LOCAL STRUCTURE OF CONDENSED MATTER

ANTONI C. MITUS[1] and ALEXANDER Z. PATASHINSKII[2]
[1] University of Saarland, Dept. of Theoretical Physics,
6600-Saarbrucken, West Germany,
and Institute of Physics, Technical University,
50-378 Wroclaw, Poland
[2] Institute for New Materials, University of Saarland,
6600-Saarbrucken, West Germany,
and Institute of Nuclear Physics, Novosibirsk, USSR

ABSTRACT. We discuss the methodological aspects of the problem of the choice of the parameters intended for the description of local structure of condensed matter. The commonly used approaches are critically analyzed. The main difficulty reflects the lack of well-defined concept of degree of similarity of fluctuating structures. Consequently, some conclusions of the analysis of type of local order in "computer" model liquids are shown to be meaningless. We introduce the parameters -structural invariants- which are free from these difficulties.

1. INTRODUCTION

The problem of the type of local arrangement of atoms in liquids near the melting temperature T_c is open. The direct observation of the local structure of liquids remains a challenge to experiment.

The intensive study of the structure of melts is done via computer simulation methods (Monte Carlo, molecular dynamics). The ensemble of typical configurations of atoms contains the full information about the local and global structures. The success of the analysis of this information depends critically on the choice of appropriate set of parameters which should ensure the detailed description of both topological and metrical properties of local structures.

The aim of this lecture is to discuss the methodological aspects of the problem of the choice of such the set of parameters, to review critically the commonly used parameters and to discuss briefly the formalism of structural invariants (Mitus and Patashinskii [1]) for the study of local and global structures of "computer" liquids.

The arrangement of atoms in volume δV can be described, e.g., via the multipole moments $T_{\alpha_1 \ldots \alpha_n}$ (n=0,1,...) of the density of matter $\rho(\underline{R})$ [2]:

$$T_{\alpha_1 \ldots \alpha_n} = \frac{1}{\delta V} \int_{\delta V} \rho(\underline{R}) \, x_{\alpha_1} \ldots x_{\alpha_n} \, d^3 R, \qquad (1)$$

where $\underline{R}=(x_\alpha)$. The equivalent set of local order-parameters can be constructed with the help of spherical harmonics Y_{nm} [3]:

$$T_{nm} = \frac{1}{\delta V} \int_{\delta V} \rho(\underline{R}) \, w(\underline{R}) \, Y_{nm}(\underline{R}/|\underline{R}|) \, d^3R, \qquad (1')$$

where $w(\underline{R})$ denotes some weight function. The fourth-rank tensor field $T_{ijkl}(\underline{r})$ plays the central role in the phenomenological theory of orientational melting of simple crystals [2].

2. THE LANGUAGE OF STRUCTURAL INVARIANTS

2.1 GENERAL FORMULATION OF THE PROBLEM

The local order in melts manifests itself in the presence of thermal fluctuations, characterized by parameter $\xi=\xi'/a$. Here ξ' denotes the mean-root-square displacement of an atom from its supposed "ideal" position and a denotes the mean interatomic distance. At $T=T_c$ $\xi=\xi_c=$ 0.07-0.17 for most of the elements. The intuitive concepts of the resemblance of structures are insufficient for quantitative analysis; it is necessary to work out the mathematical formalism for the comparison and recognition of fluctuating structures.

The main difficulty can be illustrated with the simple example. Consider two deformations of the structure of a cluster of atoms - one corresponding to small displacements of all the atoms, the other to the noticeably bigger displacement of only one atom. The question of principal importance is: which of the resulting structures is "closer" to the initial one ? In order to look for the answer one has to introduce a concept of "distance" (metrics) in the space of structures. Careless choice of the metrics may lead to totally wrong interpretation of results of computer simulation (see next Section).

In order to study the similarity of structures one introduces the concept of feature space - the space of parameters which identify these structures. The parameters can be defined in a large variety of ways. Each of them must meet the following requirements:

(I) it must provide the quantitative characteristics of the degree of similarity of structures; the concept of similarity should correspond to some clear physical intuition;

(II) it should be a complete one in the sense that it should contain the full information about the structure of the cluster (i.e. should be equivalent to the set of coordinates of the atoms). In particular, it should be known what type of information is lost when only a subset of parameters is used. This requirement makes possible the control of the degree of detail of the description;

(III) it should yield satisfactory distinction between the local structures of crystals at the presence of fluctuations at $T=T_c$.

When these requirements are not fulfilled (requirement (I) is the most critical) the concept of similarity and difference between the structures is meaningless.

2.2 METHODS OF STUDY OF LOCAL STRUCTURE OF CONDENSED MATTER: METHODOLOGICAL PROBLEMS.

In this Section we review the methods of description of local structure of "computer" liquid in light of the above requirements.

The structure of condensed matter can be studied in terms of the corelation functions of density $g_k(\underline{r}_1,\ldots,\underline{r}_N)$, k=1,..., which satisfy infinite set of BBGKY equations (see e.g. [4]). Mathematical difficulties restrict, as a rule, the study of the structure of the system to the simplest version, where only the radial distribution function (RDF) $g_2(r)$ is used. Such the description of the structure does not yield a detailed information, which is contained in the correlation functions of higher orders. Moreover, RDF seems to be very sensitive to irregularities of the structure (see [5]).

Another comonly used approach uses the concept of Voronoi polyhedron (Wigner-Seitz cell in real space) and the Delaunay symplex (DS) [6]. These geometrical objects provide the description of the topological properties of the structure in terms of the topological indices $|n_1, n_2, \ldots |$, where n_k denotes the number of polygons with k sides on Voronoi polyhedron. Unluckilly, topological indices turned out to be very sensitive to the small fluctuations of the coordinates of the atoms. Medvedev and Naberukhin [7] showed that small deformations of the Voronoi polyhedron of the ideal fcc lattice (with index $|0,12,0|$) can lead to the change of this index to $|0,0,12|$ which, in particular, corresponds to the icosahedrical ordering. Consequently, these indices characterize the details of the local arrangement of atoms and can be used for classification purposes only at very low temperatures.

In order to avoid these difficulties metrical properties of local structure must be studied. A large variety of metrical characteristics of DS was proposed : "irregularity" δ (Hiwatari, Saito and Ueda [8]) and generalized "irregularity" δ' (Medvedev, Voloshin and Naberukhin [9]):

$$\delta = (6\bar{1})^{-1} \sum_i |1_i - \bar{1}| \quad ; \quad \delta' = (15\bar{1}^2)^{-1} \sum_{i,j} (1_i - 1_j)^2 \quad ; \qquad (2)$$

where 1_i denotes the length of the i-th edge of DS and $\bar{1}$ - the average value of 1_i ; dispersion $(\Delta \Omega)^2$ of solid angles of DS (Kimura and Yonezawa [6]):

$$(\Delta \Omega)^2 = \frac{1}{6} \sum_{\mu,\nu} (\Omega_\mu - \Omega_\nu)^2 \qquad (3)$$

where Ω_ν - one of the solid angles of DS; or coefficient K_t of "tetrahedricity" (Medvedev and Naberukhin [10]):

$$K_t = 9\sqrt{3}\ V/(2R)^3, \qquad (4)$$

where V denotes the volume of the DS and R - the radius of the sphere containing the DS.

As a rule, the following approach to the problem of the description of local structure is used. Let some pattern of local ordering of atoms be described by the value $P=P_{id}$ of one of the parameters (2-4). The physical system is represented as the set of DS; the statistics of quantity P characterizes the structure of condensed matter in the presence of thermal fluctuations. In this approach the degree of similarity of the ideal and trial structures is determined via ΔP:

$$\Delta P = |P - P_{id}|/P_{id} \qquad (5)$$

Unluckilly, such a simple euclidean metrics do not provide any physical intuition about the degree of similarity between the structures described by the values P and P_{id}. For example, the result $\Delta P > 0.3$ does not imply, as sometimes argued, that the DS differs strongly from the pattern with $P=P_{id}$ - such the statement is meaningless.

Another important problem to be faced is that of the systematical choice of metrical parameters. Without some underlying physical or mathematical intuition, these parameters can be chosen only "at random". This leads, e.g. to the choice of the "new" parameters which are only slightly modified "old" ones (e.g. parameters δ and δ' (2)).

An illustrative example of methodological problems follows from recent MC simulation of rapidly supercooled one-component plasma (Ogata and Ichimaru ([11]) where, in particular, an attempt is made to classify the local structures of instantaneous configurations of charged atoms. The criterion for identification of type of local structure is quite formal and offers no physical intuition concerning the relevance of the identification. In particular, the cruicial problem of the "stability" of the results of the classification against the fluctuations remains open.

Analogous remarks concern the minimal spanning tree approach for studying order and disorder (Dussert et all. [12]).

It is worth stressing once again the main cause of the methodological problems of the analysis of local structures of model liquids:

neither the smallness of parameter ΔP (5) guarantees the similarity of structures nor the relative big value of this parameter proves difference between these structures, unless the requirements I-III from the preceeding Section are fulfilled.

In the next three Sections we argue how to introduce the characteristics of local structure which satisfy these requirements.

2.3 STRUCTURAL INVARIANTS

The spatial arrangement of N atoms in a cluster can be described by the set of vectors $\underline{r}_i$ (i=1,...,N) pointing to the centers of these atoms, or equivalently, by the set of local order-parameters (1), (1'). In either of the cases, these characteristics determine both the structure of the cluster (e.g. cube) and its orientation in the space (e.g. via three Euler angles determining the rotation of the three main crystallographic axes of the cube with respect to some fixed coordinate system). The latter information is of principal interest in the study of the semi-macroscopic and macroscopic spatial order in locally ordered condensed matter and, in particular, in phenomenological treatment of the melting of crystals with simple lattices [2]. On the contrary, this information is not essential for the description of the structure of the cluster because the structure is independent on the choice of the coordinate frames. In other words, the characteristics of the structure have to be both rotationally and translationally invariant. Translational invariance is ensured by chosing the origin of the coordinate system, in which the order-parameters (1),(1') are evaluated, to coincide with one of the atoms of the cluster. The rotationally invariant characteristics of the structure of the cluster can be chosen, e.g., as SO(3) scalars constructed from tensor quantities (1) by contraction of the indices of products of two or more parameters $T_{\alpha_1...\alpha_n}$ [2]. The SO(3) invariant characteristics of structure of the cluster can also be constructed from parameters T_{nm}. From the definition, T_{nn} transform themselves under the rotations of SO(3) group as the angular parts of the wave functions of particle with angular momentum n. Finding the scalar quantities means constructing quadratic, cubic or higher-order forms (of parameters T_{nm}) which are proportional to the wave function of system of two, three or more particles with total null angular momentum. This goal can be achieved by the use of the formalism of addition of angular momenta in quantum mechanics [13,14].
The parameters T_{nm} and the irreducible part of parameters $T_{\alpha_1...\alpha_n}$ have in general (i.e. when no point-symmetry of the cluster is assumed) 2n+1 independent components from which 2(n-1) independent invariants can be constructed. These invariants $\Psi_n^{(k)}$ (n=0,1,...; k=1,...,2(n-1)) will be referred to as structural invariants. By definition, a cartesian tensor is irreducible when it is fully symmetric and the contraction of any two of its indices yields null. As a simple example consider the case of n=2; irreducible tensor $T_{\alpha\beta}$ is represented by a traceless symmetric 3×3 matrix with 2n+1=5 independent components and 2(n-1)=2 independent invariants.

Properly chosen set $\left[\Psi_s\right]$ contains the full information about the structure of the cluster. Namely, the 3N equations

$$\Psi_s(\underline{r}_1,\ldots,\underline{r}_N) = \alpha_s, \qquad (6)$$

s=1,..,3N, can, in principle, be solved with respect to variables $\underline{r}_i$. It is assumed that all the invariants Ψ_s are independent. When one (or more) of them can be expressed via the others, it must be replaced by another.

Thus, any set $\left[\Psi_s\right]$ of 3N independent structural invariants can, a priori, be regarded as a candidate for the required set of parameters for the description of local structure. The choice of one of them depends on the fulfilment of the other requirements formulated above and is studied in the following sections.

The concept of the use of invariants to the study of the shape of structures was proposed by Steinhardt, Nelson and Ronchetti [14].

2.4 PROBABILISTIC INTERPRETATION OF THE SIMILARITY OF STRUCTURES

The structural invariants $\underline{\Psi} = \left[\Psi_s\right]$ offer the possibility of the introduction of the concept of similarity between the structures.

To each geometrical structure (set of atoms) corresponds a point in the feature space. The cluster undergoing the small thermal fluctuations (physical structure) is represented by a "cloud" in the feature space, or by the probability density function $\rho(\underline{\Psi})$ of random variable $\underline{\Psi}$. Consider the fluctuations of two ideal structures Γ_1, Γ_2 represented by $\underline{\Psi}_1^{(0)}$ and $\underline{\Psi}_2^{(0)}$ (Fig.1). A point $\underline{\Psi}'$ can correspond to

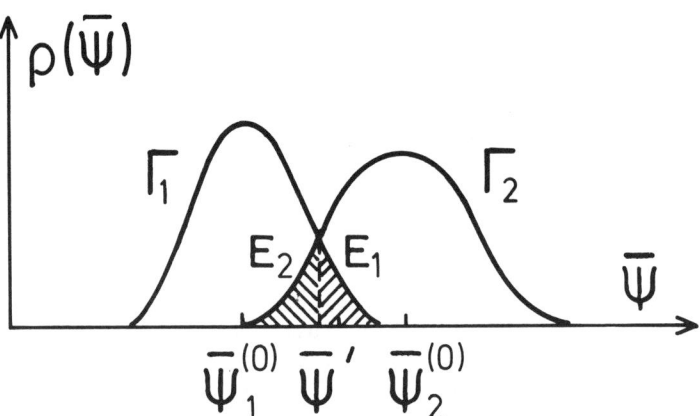

Figure 1. The statistical recognition of fluctuating structures.

deformed states of each of two patterns. In other words, the instantaneous configuration of the cluster's atoms can be treated as the fluctuation of any ideal pattern of N atoms. The classification of an instantaneous (geometrical) structure has probabilistic character: one can determine the probability density $\rho_i(\underline{\Psi}')$ that this structure corresponds to a fluctuation of some ideal pattern Γ_i. The concept of a "definite" type of geometrical structure is thus meaningless unless one of $\rho_i(\underline{\Psi}')$ is much bigger than any of the others.

The above physical picture suggests that the characteristics of the similarity of physical structures can be defined in the framework of the probability theory. We propose the following definition:

> *the degree of similarity of two physical structures is given by the probability E of erroneous identification of either of the structures on the basis of a "single measurement", i.e. with the help of only one geometrical structure.*

The above definition uses the concept of the probability of erroneous identification. The correspondence of the structure represented by $\underline{\Psi}'$ to Γ_1 or Γ_2 is settled on via the maximal-probability decision rule. It states that the fluctuating structure is the deformed state of this of two patterns Γ_1, Γ_2 for which the value of $\rho(\underline{\Psi}')$ is the bigger one. The probability E of the erroneous identification of either of the structures is:

$$E = E_1 + E_2 = \int \min\left[\rho_1(\underline{\Psi}); \rho_2(\underline{\Psi})\right] d\underline{\Psi} \qquad (7)$$

The integration in (7) is carried over the whole feature space.

The general concepts of the theory of thermodynamic fluctuations (see e.g. [15]) make possible the study of small fluctuations of the structure via normally distributed (gaussian) statistically independent invariants. One obtains [1]:

$$E(T) = 1 - \Phi\left[\sqrt{\frac{1}{8}\Delta(T)}\right], \qquad (8)$$

where $\Phi(x) = \frac{2}{\sqrt{\pi}}\int_0^x \exp(-t^2)dt$ is the error function and $\Delta(T)$ is proportional to the euclidean distance between the centers $\langle\underline{\Psi}\rangle$ of two Gaussian distributions: $\Delta(T) \propto |\langle\underline{\Psi}\rangle_1(T) - \langle\underline{\Psi}\rangle_2(T)|$.

More details of mathematical nature can be found in [1].

The quantity E has simple physical interpretation. Consider a system consisting of N=13 fcc clusters in the presence of thermal fluctuations at temperature T and fix an instantaneous configuration of the atoms. The concentration of clusters which will be identified (via the maximal-probability decision rule) as being fluctuations of some other 13-atom structure (e.g. hcp) is approximately 1/2E(T).

2.5 RELEVANT AND IRRELEVANT INVARIANTS

Structural invariants are constructed from tensor order-parameters (1), (1') which describe with various degrees of detail the local arrangement of the atoms. It can be presumed that the fluctuations of invariants depend strongly on the value of n. This hypothesis was checked for the case of the close-packed 13-atom clusters: fcc, hcp and icosahedron [1] using quadratic structural invariants Q_n:

$$Q_n^2 = \frac{4\pi}{2n+1} \sum_{m=-n}^{m=n} |T_{nm}|^2 \qquad (9)$$

The distance $\Delta^{(n)}(T)$ between the centers $\langle Q_n \rangle$ of 1-D gaussian distributions displays the systematical dependence on n. Namely, the bigger value of n, the lower is the temperature T_n at which $\Delta^{(n)}(T_n)=0$, i.e. the structures are completely undistinguishable: E=1. At any temperature the space of structural invariants can be splitted into two parts. One of them contains the relevant (informative) invariants, i.e. the ones that contribute noticeably into E(T). All the other invariants are irrelevant - they do not improve the conditions of the recognition. The high-rank invariants become relevant at low temperatures and can play some role in the description of the structure of crystals. For the study of the local arrangement of the atoms in liquids it is sufficient to use the first few structural invariants. For example, at the melting temperature T_c of many simple substances $\Delta^{(n)}(T_c)=0$ for n>10 [1]. The main contribution into $E(T_c)$ is given by 4-6 invariants with the smallest values of n.

Structural invariants constructed from parameters $T_{\alpha_1...\alpha_n}$ display analogous properties.

Let us make an important methodological comment. The recognition of the fluctuating structure via an irrelevant invariant may lead to meaningless results. For example, the use of Q_{11} at $T=T_c$ leads to ridiculous conclusion: there exists only one type of close-packed 3-D crystals ! In particular, the Voronoi-polyhedra approach should be re-examined from this point of view.

Finally, let us consider briefly the the problem of the distinction between the local structures of 3-D close-packed crystals at $T=T_c$. For crystal of inert gases $\xi_c \simeq 0.1$; the probability of erroneous recognition of 13-atom fcc and hcp clusters is $E \simeq 10^{-4}$ [1]. Above the melting temperature the value of E grows rapidly and the two clusters become undistinguishable (E=1) approximately at $\xi=0.5$.

Some examples of the analysis of local structure of instantaneous configurations of atoms of "computer" liquid via the use of structural invariants can be found in [16].

Acknowledgments

This work was done during our stay in Saarbrucken, West Germany, as Alexander von Humboldt Fellow (ACM) and as visiting Professor in Institute for New Materials (AZP). The warm hospitality and numerous discussions with Profs. H.Gleiter and A.Holz are gratefully acknowledged.

REFERENCES

[1] A.C.Mitus and A.Z.Patashinskii, Physica, A150, 371 (1988); A150, 383 (1988).
[2] A.C.Mitus and A.Z.Patashinskii, Sov.Phys. JETP, 53, 798 (1981); Phys.Lett., A87, 79 (1982).
[3] D.R.Nelson and J.Toner, Phys.Rev., B24, 363 (1981).
[4] J.P.Hansen and I.R.McDonald, Theory of simple liquids, Academic Press, 1976.
[5] A.C.Mitus and A.Z.Patashinskii, preprint 88-114, Institute of Nuclear Physics, Novosibirsk, USSR (1988).
[6] M.Kimura and F.Yonezawa, J.Non-Cryst.Solids, 61&62, 535 (1984).
[7] N.N.Medvedev and Yu.I.Naberukhin, J.Struct.Chemistry (USSR), 26, 59 (1985).
[8] Y.Hiwatari,T.Saito and A.Ueda, J.Chem.Phys., 81, 6044 (1984).
[9] N.N.Medvedev,W.P.Voloshin and Yu.I.Naberukhin, J.Struct. Chemistry (USSR), 28, 62 (1987).
[10] N.N.Medvedev and Yu.I.Naberukhin, DAN USSR, 288, 1104 (1986).
[11] S.Ogata and S.Ichimaru, Phys.Rev., A39, 1333 (1989).
[12] C.Dussert,G.Rasigni,M.Rasigni,J.Palmari and A.Llebaria, Phys. Rev., B34, 3528 (1986).
[13] L.D.Landau and E.M.Lifshits, Quantum Mechanics, Nauka, Moscov, 1984.
[14] P.J.Steinhardt,D.R.Nelson and M.Ronchetti, Phys.Rev., B28, 784 (1983).
[15] L.D.Landau and E.M.Lifshits, Statistical Physics, Nauka, Moscov, 1976.
[16] A.C.Mitus and A.Z.Patashinskii, Proceedings of the seventh international meeting on ferroelectricity, Saarbrucken, West Germany, 1989 (to be published in Ferroelectrics).

STRUCTURAL ASPECTS OF DOMAIN PATTERNS IN CERAMICS AND ALLOYS

CH. LEROUX, G. VAN TENDELOO and J. VAN LANDUYT
University of Antwerp (RUCA), Groenenborgerlaan 171, B-2020 Antwerp, Belgium

Abstract

The usefulness of electron microscopy observations of domain patterns in the frame of phase transitions studies is illustrated by means of two cases. An in-situ study of the $\alpha \rightarrow \beta$ phase transition in quartz establishes the existence of an incommensurate intermediate phase which could be fully characterized by TEM. An in-situ study of the $L1_2 \rightarrow A1$ transition and post-transformation observations of the variants in the Co-Pt system revealed the role of twodimensional defects in these order-disorder related transformations.

1. Introduction

The aim of this contribution is to illustrate by some examples, how electron microscopy can contribute, with the use of a set of techniques such as - diffraction contrast electron microscopy, - high resolution electron microscopy, - selected area electron diffraction, - optical diffraction of high resolution electron micrographs, to a better understanding of phase transitions. In fact, direct images are usefully complemented by very localized electron diffraction data which allow a cristallographic analysis of the observed images of phase fronts or defects involved with the transition. Phase transformations are almost invariably accompanied by changes in symmetry : a reduction of symmetry mostly occurs in going from the high temperature to the low temperature phase. The loss of elements of rotational symmetry gives rise to orientation variants separated by twin boundaries whereas the loss of translationnal elements gives rise to translational variants separated by antiphase boundaries. The interfaces between variants can be revealed in diffraction contrast by a characteristic fringe pattern and also by the presence or absence of intensity differences between the variants [1, 2, 3]. It is also possible to produce high resolution images exhibiting directly the relationship between the variants [4]. Thus, the imaging of these domain patterns, associated with selected area diffraction, often allows to propose a detailed model for the transition that produced them. For phase transitions occurring in a temperature range between -120°C and 900°C, any electron microscope equipped with the available heating and cooling stages can be used to obtain direct information on phase transitions, through diffraction contrast images which allow a direct observation of the transformation. The diffraction contrast mode (for low magnification up to ~10.000x) is usually applied in the bright field (Fig. 1.a) resp. dark field (Fig. 1.b) whereby the transmitted resp. one scattered beam is allowed to enter the magnifying lens system through the objective aperture. High resolution electron microscopy, yielding information down to atomic scale, makes use of interference of several beams in the multiple beam imaging mode (Fig. 1.c).

The two examples we have chosen are a displacive transformation (the $\alpha \rightarrow \beta$ transformation in quartz) observed in situ by diffraction contrast electron microscopy and the two-phase microstructures in Co-Pt alloys, studied by diffraction contrast electron microscopy and HREM.

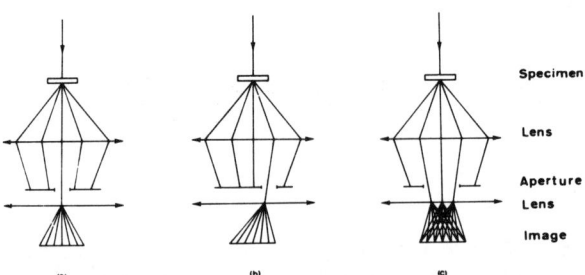

FIG. 1. The three main imaging modes in transmission electron microscopy. a) bright field. b) dark field. c) multiple beam

2. The α→ β transformation in quartz

It is known for more than 100 years that quartz (SiO_2) undergoes a transformation from the low quartz (α) to the high quartz (β) upon heating above 573°C. It is therefore quite surprising that only recently electron microscopy has been able to elucidate the structural aspects of this transition [5], leading to further studies using different techniques, in particular small angle neutron scattering [6, 7, 8]. These studies produce evidence for the occurrence of an intermediate incommensurate phase between the α and the β phase, in a temperature range of 1.3 °C. The high symmetry phase of quartz (β) is hexagonal with the point group 622 : on going through the transition from high to low quartz, the 180° symmetry rotation around the sixfold axis is lost, reducing the sixfold axis to a threefold one. The loss in symmetry is evident on a projection along the c-axis in Fig. 2. The structure consists of interlinked SiO_4 trapezia, the crystal breaks up in two types ($α_1$, $α_2$) of domains related by a 180° rotation around the c-axis, called Dauphiné twins. These domains can be observed by electron microscopy using dark field images made in reflections such as 301, for which the structure factor is significantly different for the two variants $α_1$ and $α_2$. In order to study the α→ β transformation, specimens were heated in the heating holder of the microscope to a temperature somewhat lower than the transition temperature, and a gradient over the transition region was produced by local heating with the electron beam. In the region of the temperature gradient, one can observe simultaneously the α and β phase, and in between a transition phase. The α phase breaks up into small triangular prims parallel with the c-axis, as it was deduced from observations on specimens cut parallel with the c-axis. In the transition region, these Dauphiné twin columnar domains form fairly regular networks of which the mesh size decreases with increasing temperature from 30 nm until it reaches the sizes below the resolution limit of the microscope which is about 10 nm, due to the fact that the boundaries are vibrating. This microstructure can be interpreted as an incommensurate phase, the incommensurability being associated with the irrational relation of the domain size with respect to the quartz unit cell. The sides of the "triangles" are rotated with respect to the $[100]_α$ directions over an angle φ which decreases with temperature. The small triangles form macrodomains of similarly oriented triangles (Fig. 3.a). This regular arrangement gives rise to diffraction effects in neutron diffraction as well as in X-ray [9] or electron diffraction [6], i.e. the Bragg spots acquire satellites along three equivalent directions. The satellites vary continuously with temperature and the corresponding q-vector was found to be consistent with the TEM observed domain sizes (Fig. 3.b).

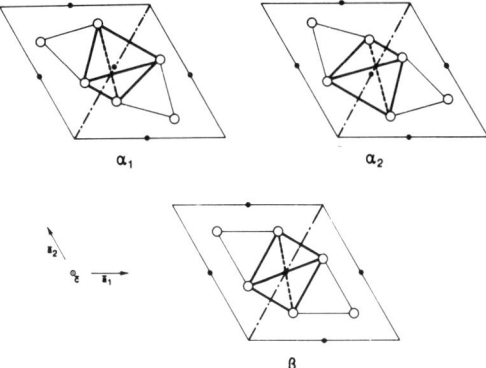

FIG. 2. Structure model for quartz as projected along the c-axis. a) α-quartz, one variant e.g. α_1. b) the other variant of α-quartz, α_2. c) projection of the β-quartz.

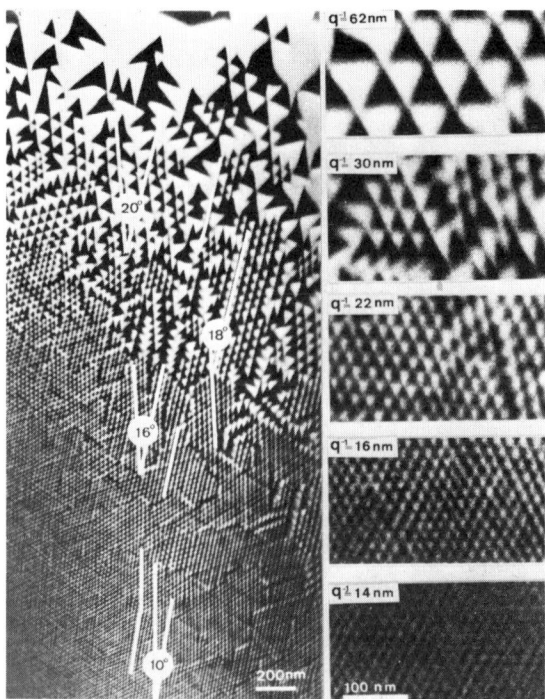

FIG. 3. Dark field images in the (301) reflection of the transition region from α to the incommensurate phase in a specimen cut perpendicular to the c-axis. Note the continuous transition from a clearly resolved lattice of triangular domains into a pattern of intersecting sinusoidal fringes: the corresponding q- vectors increase from 0.071 to 0.016 nm^{-1}. The variation as a function of temperature between the angle of the q-vectors is also indicated.

These observations inspired the theoretical considerations of Aslanyan and Levanyuk [10, 11], who used the Landau-Ginzburg theory to predict an incommensurate intermediate phase between α- and β-quartz, characterized by a triple q-vector with two possible orientations, symmetrical with respect to the Y-axis (Fig. 4). The q-vector and the angle 2ϕ should be temperature dependent. Based on their theory, Walker [12] conjectured that the Dauphiné twins boundaries should take two possible orientations, symmetrical with respect to the (100) orientation, and suggested the existence of macrodomains consisting of patches of one of the two above mentioned orientations for the triple q-vectors. These macrodomains are clearly visible in Fig. 5 : the presence of two orientations for the two macrodomains can be demonstrated by means of optical diffraction across the different macrodomains.

The detailed in situ observations of the phase transition in quartz are in very good agreement with the theoretical considerations concerning the occurrence of an incommensurate phase between α and β [13]. This phase is characterized by a double position triple q-vector and consists of periodic arrays of Dauphiné twin columns with vibrating boundaries (Fig. 6).

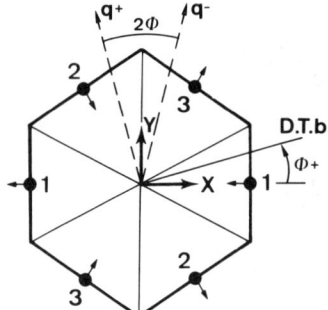

FIG. 4. Basal-plane projection of the silicon ions in the Wigner-Seitz cell of the quartz structures. The full circles indicate the silicon positions in the β phase, whereas the arrows give the directions of the silicon displacements in the transition to the α phase. The two symmetrical possible orientations for a triple q-vector are indicated, and the deviation of a Dauphiné twin boundary from the [100] direction.

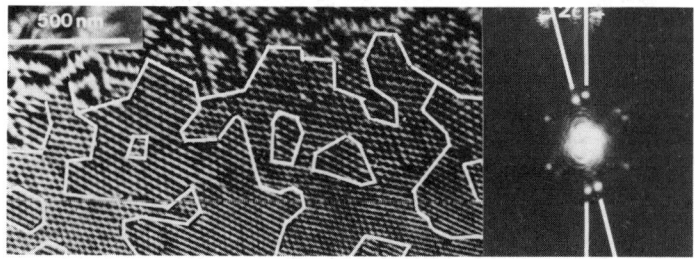

FIG. 5. Macrodomains of two orientations in quartz, marked by white lines. The inset corresponds to an optical diffraction pattern made across two different macrodomains.

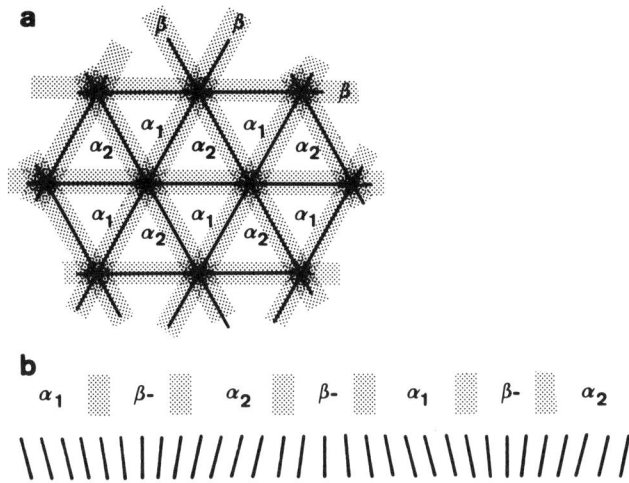

FIG. 6. Idealized model of the incommensurate modulated structure in quartz near the phase transition to the high temperature β quartz. a) View along the c-axis: the shaded area represents the vibrating boundaries between α_1 and α_2 which have on the average the β structure. b) Section through the Dauphiné twin domains : the line segments represent symbolically the tilt angle of the tetrahedra.

3. Two-phase microstructures in Co-Pt alloys

The Co-Pt system is based on a f.c.c. lattice and exhibits at low temperature the ordered phases $L1_2$, $L1_0$, $L1_2$ around the concentrations of 25, 50 and 75% of Pt. At high temperature the disordered phase is A1. This system is one of the rare ones that presents a two-phase ordered region $L1_0$-$L1_2$ around 60% of Pt [14, 15]. Despite of the tetragonality of the $L1_0$ phase, the $L1_0$ and the $L1_2$ structures essentially differ with the alternation of mixed planes and pure plane (A-B-A-B... for $L1_0$ and AB-B-AB-B... for $L1_2$). The $L1_0$ structure being tetragonal occurs in three orientation variants, depending on the direction of the c-axis, while for the cubic $L1_2$ phase no orientation variants are possible, but four different translation variants (ie. antiphase domains) can be formed. With the use of electron microscopy we showed that the antiphase boundaries in the $CoPt_3$ phase are wetted by the new phase (disorder or $L1_0$) in case of order-disorder or "order-order" transformations. More, the microstructure of the $L1_0$+ $L1_2$ alloys has been characterized in detail.

a) Starting from the disordered phase and by slow cooling , one can observe the formation of a peculiar microstructure, composed of a regular distribution of $L1_0$ and $L1_2$ germs (Fig. 7). Optical diffraction made on the two different variants present in the microstructure give the tetragonality of the $L1_0$ germs : it was found that the c/a ratio is .99, which means that this tetragonality is very weak. This explain why this microstructure can occur almost without elastic strain [14, 16].

b) Starting from the $L1_2$ phase, the antiphase boundaries are wetted by the $L1_0$ phase: they are transformed into small variants of the $L1_0$ phase (Fig. 8) and evolve to small plates after further annealing [14, 16].

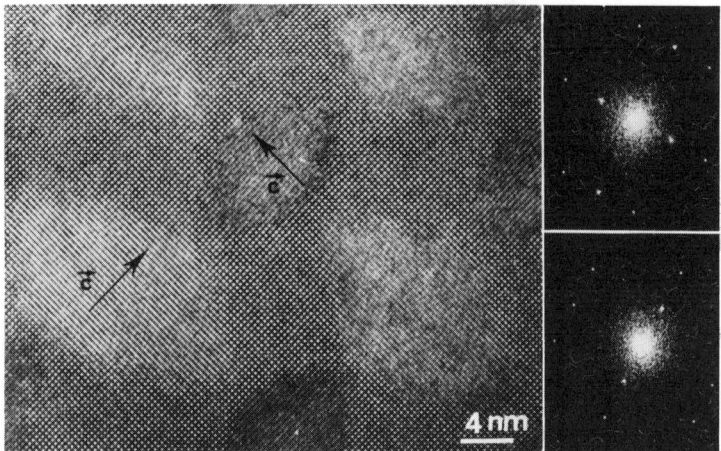

FIG. 7. HREM image of the "mosaïc" microstructure of a $Co_{40}Pt_{60}$ alloy, slowly cooled from the disordered phase. The insets correspond to optical diffraction performed on the two types of variants.

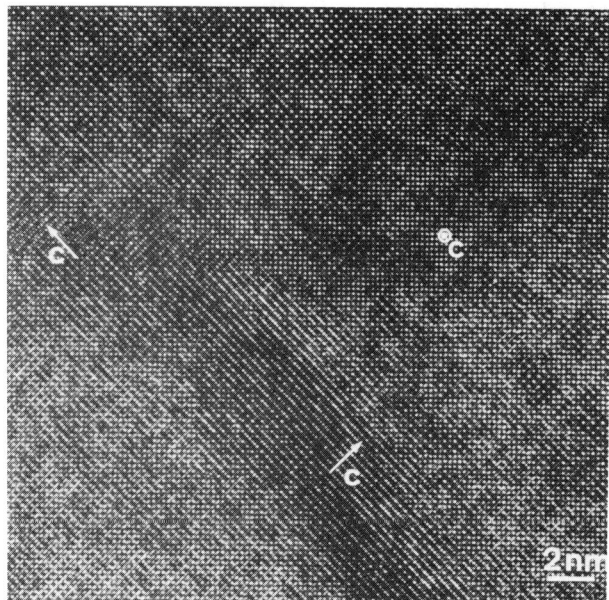

FIG. 8. HREM image of the microstructure of a $Co_{37}Pt_{63}$ alloy, slowly cooled from the disordered phase. Through the cooling, this alloy exhibits at first the $L1_2$ structure, and then the antiphase boundaries are transformed into the three types of $L1_0$ variants.

c) A more detailed study of the order-disorder transition $L1_2 \to A1$ in a $Co_{30}Pt_{70}$ alloy was performed by diffraction contrast and further by HREM. [14, 16, 17]. Fig. 9 illustrates the wetting of the APB by the disordered phase in the single phase $L1_2$. A characterization of these APB on an atomic scale was possible by HREM on quenched alloys, showing the disordered phase in the core of the APB and the diffuse profile of it (Fig. 10). Thus, electron microscopy provides evidence of various order-disorder related transformations in the Co-Pt system and furthermore allows to describe the way these transitions have occurred.

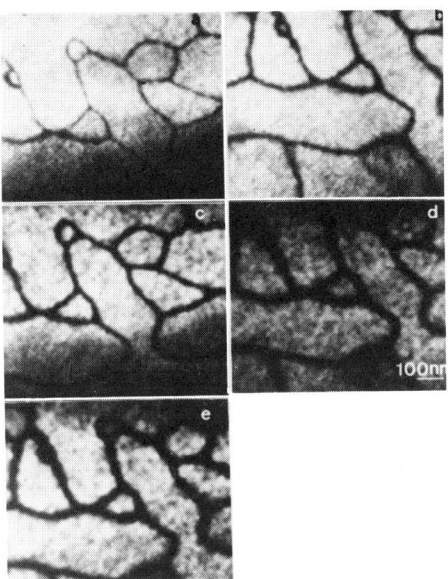

FIG. 9. In situ evolution of antiphase boundaries in $Co_{30}Pt_{70}$, 10° under the order-disorder transition temperature T_D. Dark field images taken 1 min. (a), 12 min. (b), 18 min. (c), 24 min. (d) and 28 min. (e) after reaching this temperature.

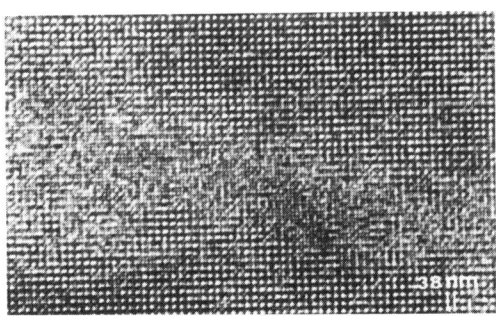

FIG. 10. HREM image on an antiphase boundary in a $Co_{30}Pt_{70}$ alloy annealed one day at 744 °C (T_D-10°) and quenched.

Acknowledgements

Thanks are due to S. Amelinckx, D. Broddin (University of Antwerp) and A. Loiseau (ONERA) for the use of results from joint studies. Parts of this work have been performed in the framework of the Institute for Materials Science under an IUAP contract with the Ministry of Science Policy (satellite project IUAP 9. ULB), and under an E.E.C. contract (SC1-0113).

References

[1] P.B. HIRSCH, A. HOWIE, R.B. NICHOLSON, D.W. PASHLEY and M.J. WHELAN, "Electronic Microscopy of Thin Crystals", 2nd Edition, Krieger, New-York, 1977
[2] S. AMELINCKX AND J. VAN LANDUYT, in "Diffraction and Imaging Techniques in Materials Science", eds. S. AMELINCKX, R. GEVERS, J. VAN LANDUYT, North Holland, Amsterdam, 1978, 107
[3] R. GEVERS, J. VAN LANDUYT and S. AMELINCKX, Phys. stat. sol. (a), 1965, **11**, 689
[4] J.C.M. SPENCE, "Experimental High Resolution Electron Microscopy" Clarendon Press, Oxford, 1981
[5] G. VAN TENDELOO, J. VAN LANDUYT and S. AMELINCKX, Phys. stat. sol. (a), 1976, **33**, 723
[6] G. DOLINO, J.P. BACHHEIMER, B. BERGE, C.M.E. ZEYEN, G. VAN TENDELOO, J. VAN LANDUYT AND S. AMELINCKX,
J. Phys. (Paris) 1984, **45**, 901
[7] B. BERGE, J.P. BACHHEIMER, G. DOLINO and M. VALLADE, Ferroelectrics, 1986, **66**, 73
[8] G. DOLINO et al., Europhysics Letters, 1987, **3**, 609
[9] K. GOUDHARA and N. KATO, J. Phys. Soc. Jap., 1984, 53, 2177
[10] T.A. ASLANYAN and A.P. LEVANYUK, Sov.Phys.LETP Lett., 1978, **28**, 70
[11] T.A. ASLANYAN, A.P. LEVANYUK, M. VALLADE and J. LAJZERROWICZ, J. Phys. C, Solid State Physics, 1983, **16**, 6705.
[12] M.B. WALKER, Phys. Rev. B, 1983, **28**, 6407
[13] J. VAN LANDUYT, G. VAN TENDELOO, S. AMELINCKX and M.B. WALKER, Phys. Rev. B, 1985, **31**, 2986
[14] CH. LEROUX, 1989, Thesis of the L. Pasteur University, Strasbourg, France
[15] CH. LEROUX, M.C. CADEVILLE, V. PIERRON-BOHNES, G. INDEN and F.HINZ, 1988, J.Phys. F : Met.Phys., **18**, 2033
[16] CH. LEROUX, A. LOISEAU, M.C. CADEVILLE, D. BRODDIN and G. VAN TENDELOO, 1989, to be published
[17] CH. LEROUX and A. LOISEAU, Film ONERA n° 1202, "In situ electron microscopy study of the $L1_2 \rightarrow A1$ transformation in $Co_{30}Pt_{70}$".
[18] CH. LEROUX, A. LOISEAU, M.C. CADEVILLE, D. BRODDIN and G. VAN TENDELOO, 1989, to be published

SELF-ORGANIZATION IN FAR-FROM-EQUILIBRIUM REACTIVE POROUS MEDIA SUBJECT TO REACTION FRONT FINGERING *

P. ORTOLEVA and W. CHEN
Indiana University *Geo-Chemical Research Assoc.*
Department of Chemistry *105 N. Jefferson*
Bloomington, IN 47405 *Bloomington, IN 47401*

* Research supported in part by a grant from the Basic Energy Sciences Program of the US Department of Energy,

ABSTRACT. When a porous medium constituted of grains that may dissolve or grow is subject to a flow-through of reactive fluids it may display a variety of self-organization phenomena. The system may support reaction fronts that can become morphologically unstable to the formation of scalloping, fingering and dendrites that themselves can become unstable to the generation of smaller spatial scale structure. The purpose of the present article is to introduce the reader to the physics and chemistry of these systems, the mathematical reaction-transport models that have been used to describe them and to the types of pattern formation that can arise. Our focus will be on results on the states of theses systems under very far-from-equilibrium conditions. We demonstrate the genesis of complex branching tree and dynamical dendritic states.

Flow Self-Focusing and Other Geochemical Patterning Phenomena

Geochemical systems exhibit a variety of self-organization phenomena on a wide range of space and time scales (Ortoleva et al., 1987a; Ortoleva, 1990). Effects include oscillatory intragranular compositional variations, stylolites, diagenetic bedding, metamorphic layering, banded skarns, and other mm to m scale differentiated mineral banding effects to km scale steady and oscillatory buoyancy driven flow cells.

As one concrete example of these phenomena, we focus here on the morphological instability of reaction fronts in porous media and the associated flow driven scalloping, fingering, branching, and dendrite phenomena. The reaction fronts in question arise when a fluid is injected into a rock or other porous medium such that the fluid reacts with and dissolves out at least one of the solid components composing the supporting matrix. The essential element of the system dynamics that leads to the morphological instability is that upon dissolution of part of the porous medium the Darcy permeability increases. The resulting flow self-focusing is suggested in Fig.1. There a bump in the reaction front is seen to attract flow to it due to the increased permeability on its upstream side within it; as a result dissolution of the soluble component at the reaction front at the tip of the original bump tends to elongate the initial bump. Ultimately if the bump becomes suficiently long, diffusion of the solutes that result from the dissolution of the soluble solid at the sides of the bump approach saturation with this solid as the reach the tip and the tip slows down. Thus the possibility for and the ultimate consequences of the morphological instability on the shape of the reaction fronts results from the balance between the saturation and flow self-focusing effects.

The reaction front fingering and related phenomena to be considered here are distinct from the classic Saffman-Taylor (Douglas et al., 1959; Glimm et al., 1985) problem. The S-T phenomena is the set of fingering effects that occur when one fluid is injected into a porous medium containing a second. If the fluids are immiscible and the original fluid is the more viscous then the interface separating the two fluids can lose stability to the development of fingers and other shapes. The S-T front moves with the fluid velocity while the reactionf ronts of interest here move slower than (typically by several orders of magnitude) the fluid speed. Furthermore, the stabilization effect for the S-T interface is surface tension while there is no surface tension in the reaction front problem. Finally, unlike for the reaction front problem of interest here, no change of porosity or permeability across the S-T interface is required to generalte structure.

There are some interesting relations between the problems. In both cases the reaction front instability occurs as what we have termed an "extrinsic" instability (Nitzan et al., 1974). For such an instability patterning first occurs at a length scale dictated by the size of the system as the system first becomes unstable upon being driven sufficiently far from equilibrium. In contrast we terme the case where a band of wavevectors becomes unstable that lies a finite interval about zero wavevector as extrinsic; hence the preferred wavelength is strongly dependent on system rate parameters and is essentially independent of system size (unless the system itself is of a size on the order of magnitude of the wavelength in the unstable interval). The two cases are illustrated in Fig.2.

Besideds the extrinsic nature of the symmetry breaking instabilities in the two problems, both phenomena are enhanced by a fluid viscosity contrast across the front. We have studied a mixed S-T-reaction front problem wherein the dissolution of the solid component across the front releases a molecular species into the fluid which increases its viscosity (Chadam et al., 1989). In this problem the S-T and dissolution mediated flow self-focusing are mutally reinforcing. This problem differs from the pure S-T case, however, in that the front moves slower than the fluid as it is tied to the dissolution front. In the present report we do not discuss the mixed case further.

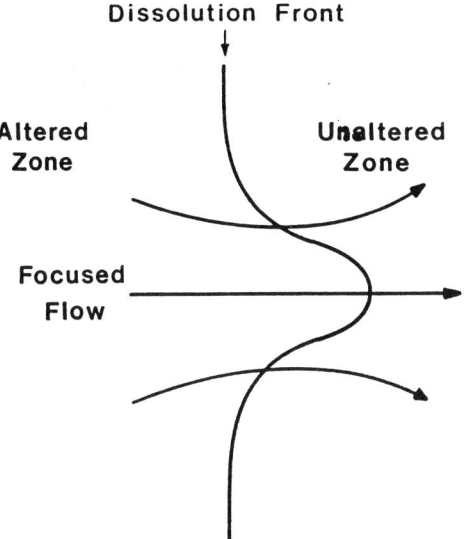

Figure 1: Schematic view of nonplanar reaction front illustrating flow self-focusing.

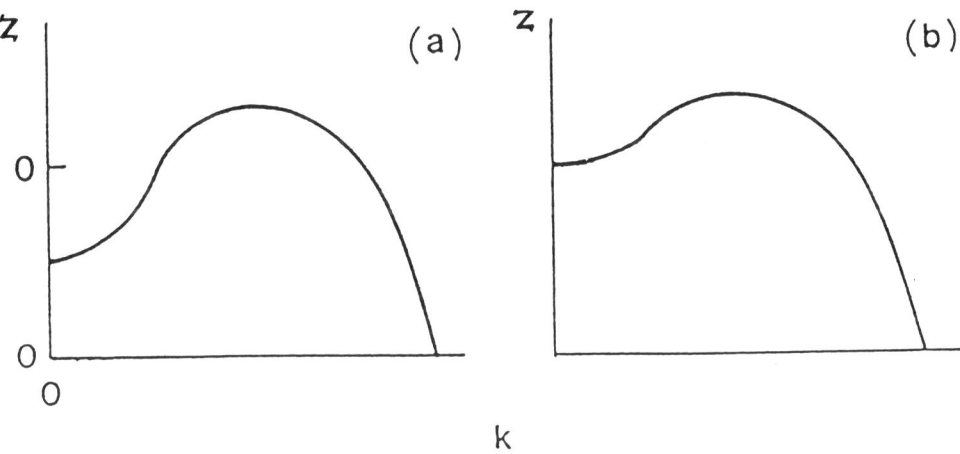

Figure 2: Stability eigenvalue z as a function of wavevector k showing intrinsic case (a) and extrinsic case (b).

Reaction-Transport Model

As a working example, we consider the alteration of a calcite cemented sandstone subject to an infiltration of fluids undersaturated with respect to the cements. The chemical model adopted is (Chen and Ortoleva, 1989)

$$\text{calcite} \leftrightarrow Ca^{2+} + CO_3^{2-} \tag{1}$$

$$HCO_3^- \leftrightarrow H^+ + CO_3^{2-} \tag{2}$$

$$H_2CO_3 \leftrightarrow H^+ + HCO_3^- \tag{3}$$

$$CaCO_3 \leftrightarrow Ca^{2+} + CO_3^{2-} \tag{4}$$

$$H_2O \leftrightarrow H^+ + OH^- \tag{5}$$

This model appears to capture the majority speciation when the imposed fluid is of pH ≤ 8 and low to moderate salinity (so that complexing of Ca^{2+} with Cl^- and OH^- do not play a significant role). Note that $CaCO_3^0$ is an aqueous phase complex.

The central feedback that allows for the destabilization of the planar calcite dissolution front is through the dependence of the Darcy permeability on the amount of soluble component (calcite cement) within the medium. In the present study we adapt the phenomenology described by the Fair-Hatch equation (Bear, 1972):

$$\kappa = \kappa_0 \, \phi^3 / F \tag{6}$$

where κ is the ratio of the Darcy permeability to fluid viscosity, ϕ is the porosity and κ_0 and F are multiple factors that depend on the shape and size of the mineral grains and mineral volume fractions (see the Appendix).

The temporal development of the flow focusing induced instability of a planar dissolution front is illustrated in Fig.3. There is a steady finger shaper dissolution front observed to result due to the influx of calcite undersaturated fluid into a calcite cemented sandstone. Fig.3 (as well as all the other figures to be shown) gives the location of a reaction front at a sequence of times. Each curve represents the locus of points at one instant of time wherein the porosity value is half-way between its original (chosen to be 10%) and that after the entire calcite cement has disappeared (chosen to be 20%). The inlet fluid, under a constant pressure, flows into the domain from the left-hand side wall. It flows out of the right-hand side wall with a constant imposed velocity (V_0) along the horizontal direction. Therefore, the dissolution front moves to the right with time. The left-most curve in the figure gives the location of the reaction front at starting time $t_0 = 0$; the first contour to its right represents the dissolution front at a later time $t_1 = t_0 + \Delta t$; the second one is at time $t_2 = t_1 + \Delta t$ and so on. Δt is the time interval between two adjacent contours plotted and is held to be a constant for each figure unless explicitly expressed in the figure captions. Reaction rate coefficients, diffusion and permeability data used are summarized in the appendix.

At time zero, the simulation domain in Fig.3 is uniformly cemented with calcite except for a small high porosity "bump" located just to the right of the inlet wall. There the permeability is higher according to (6). This small nonuniformity is amplified into a C-shaped reaction front during the dissolution process. The increase in permeability induced by calcite dissolution allows for the focusing of flow to the tip of the "dissolution finger" that sustains the nonplanarity. Diffusion of Ca^{2+} and CO_3^{2-} from the sides of the finger prevents the elongation from continuing indefinitely. These two effects tend to balance each other and the fingered reaction front advances down stream with a constant shape and velocity.

Flow self-focusing can only sustain a dissolution finger when the inlet flow speed exceeds a critical value. This value depends on inlet composition, initial calcite and inert quartz matrix volume fraction, and the channel width. A brief discussion of the mathematical model used and the simulation techniques adopted are given in the Appendix and further details are given in Chen and Ortoleva (1989). The purpose of this communication is to show that this fingering of the calcite dissolution front is just one of a rich class of morphological transformations that may develop when the inlet flow speed or the width of the flow channel increase.

The Stability of the Finger

The reaction front of Fig.3 developed for a channel with initial data symmetric with respect to the central line dividing its top and bottom halves. A question arises as to the stability of such a symmetric steady finger - i.e., what will happen if the initial porosity perturbation "bump" is displaced from the center of the inlet wall?

A set of simulations was carried out to answer this question, and the results are shown in Fig.4. The simulation system for Fig.4a is exactly the same as that of Fig.3 except the domain is twice as long for 4a. Simulations 4a-d only differ in the location of the initial perturbation at the left. We see that while the fingered reaction front develops and moves down stream, the finger tip gradually shifts toward the central line if the initial bump is off that line. The finger tips of 4.b and 4.c reach the central line after long enough time advancement and the shape of the reaction front becomes identical to that of 4.a. It is not clear from 4.d that the finger tip would reach the central line. However another simulation for a domain much longer than 4.d shows no further shift of the finger tip toward the central line before a steady state is reached.

We shall term the symmetric finger a "full finger". Fig.4 demonstrates that a full finger state is stable under a finite initial deviation from symmetry. In such a situation, a steady full finger may not reflect the initial state of the system. However, as seen in 4.d, the symmetry of the reaction front will be broken if the initial porosity bump is too far away from the symmetry position. This suggests that there exists an asymmetric state that is attained when the initial condition is sufficiently asymmetric. There can thus be multiple stable steady reaction front morphologies each with a finite basin of attraction. This is confirmed in the simulations of Figures 5 and 6.

The domains for Figure 5 and only half as wide as those for Figure 4, while the other parameters used are exactly the same. However, unlike 4.b and 4.c, 5.b and 5.c show the shift of finger tips toward the bottom boundary of the domain. After long enough times, the shape of the reaction fronts become identical to that of Figure 5.d in which the initial bump is put at the lower-left corner. We term this type of finger a "half finger". The half finger is located at the bottom boundary (the top boundary as well, because of the domain symmetry) is a stable solution with a finite "basin of attraction". Note the reaction front for 5.a is planar because the inlet fluid speed is lower than the bifurcation point for the full finger and the initial data was symmetric about the horizontal line dividing the system in half.

Figures 4 and 5 are good examples of the competition between tendencies to evolve distinct reaction front morphologies. As a result of this competition, the finger tips shift along the direction perpendicular to the overall fluid flow while the fingers develop and move down stream. For given initial data, the ultimate position of the finger tip depends on the basin of attraction in which the initial data resides. The finger amplitude varies with aquifer width as well as inlet fluid speed and composition.

Finger Tip Splitting

While flow focusing leads to the destabilization of a planar calcite dissolution front and the formation of a fingered front, it can destabilize a finger as well to form a pair or even a triplet of fingers. This phenomenon is termed "finger tip splitting".

Figure 4.d is in fact a good example of finger tip splitting. If we combine Figure 4.d with its mirror image obtained through a reflection across its bottom line, we obtain Figure 6.a. In other words, 6.a is a symmetric state for a system twice as wide as that of 4 with all the other system parameters the same. A comparison of the simulations 4.a, 5.a and 6.a, different only by their domain width, demonstrates that a reaction front is more unstable in a wider domain under the same inlet fluid speed and other parameters.

The channel has to be wide enough in order to support a fingered reaction front or else diffusion will drive the front to planarity. With increase of channel width, it will first become possible for the system to support a half-finger state. Then it will support a full-finger state as well as the half-finger state.

Figure 3: Temporal development of a steady finger-shaped dissolution front as described in section 2 for a system of 7.5x2 cm; obtained by using a grid of 76x21. The time interval between adjacent contours is 10^8 sec for the first 9 contours and 8.10^8 sec for the last 5 contours. The imposed flow along the horizontal direction (specified at the outlet) is 1.4×10^{-3} cm/sec.

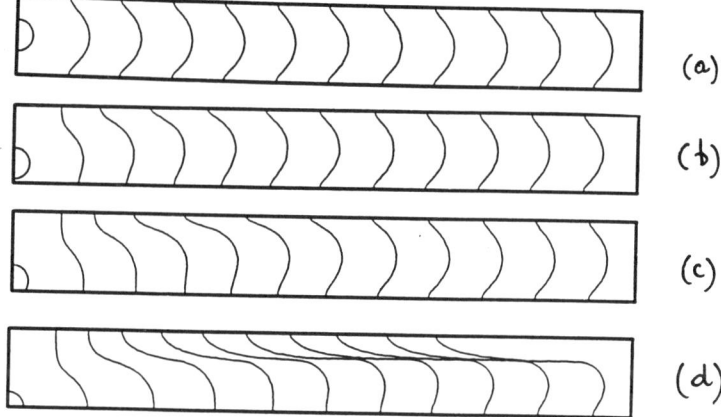

Figure 4: Stability of the finger due to symmetry breaking perturbation. The center of the initial porosity bump is placed on the inlet wall at : (4.a) $y_c = (1/2)\, y_L$; (4.b) $y_c = (1/4)\, y_L$; (4.c) $y_c = (1/8)\, y_L$; (4.d) $y_c = 0$ where $y_L = 1$ cm is the height of the domain. Note that the ultimate reaction fronts of 4.a through 4.c are identical although the initial perturbations are different, showing that the reaction front 4.a is stable with respect to symmetry breaking. 4.d illustrates the existence of the asymmetric state when the initial perturbation is sufficiently away from the symmetry point. The size of the domain is 15x2 cm; the grid used is 152x21. The time interval between adjacent contours is 10^9 sec. The imposed flow along the horizontal direction (specified at the outlet) is the same as that of fig.3.

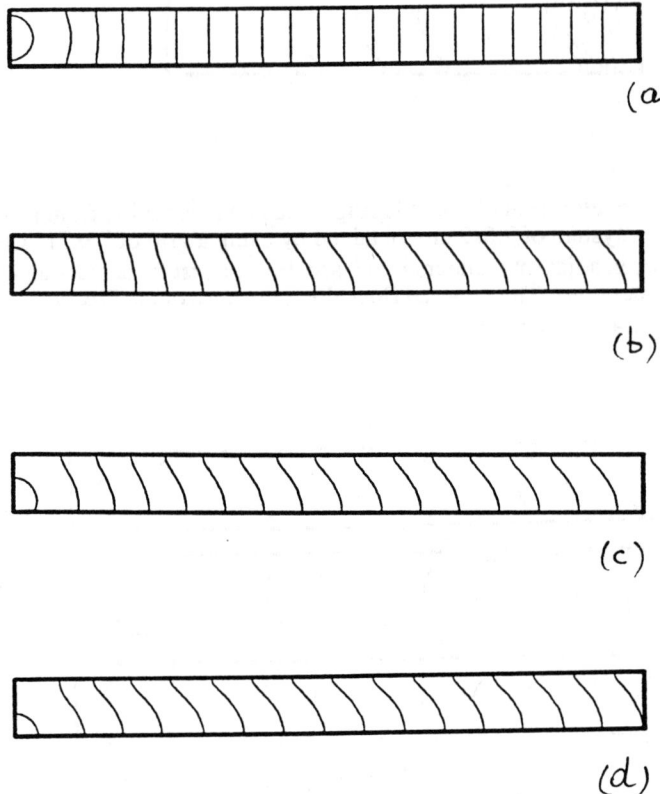

Figure 5: Evolution of a reaction front in a narrow domain showing that the asymmetry in the placement of the initial perturbation causes the front to evolve to a half-finger state, as in 4.d. The domain used here are half as wide as those of figure 4. The center of the initial porosity bump is placed on the inlet wall at : (5.a) yc = (1/2) yL ; (5.b) yc = (1/4) yL ; (5.c) yc = (1/8) yL ; (5.d) yc = 0 where y_L = 2 cm is the height of the domain. The size of the domain is 10x1 cm. The time between adjacent contours is $.5 \times 10^9$ sec. The imposed flow along the horizontal direction (specified at the outlet) is the same as in figures 3 and 4.

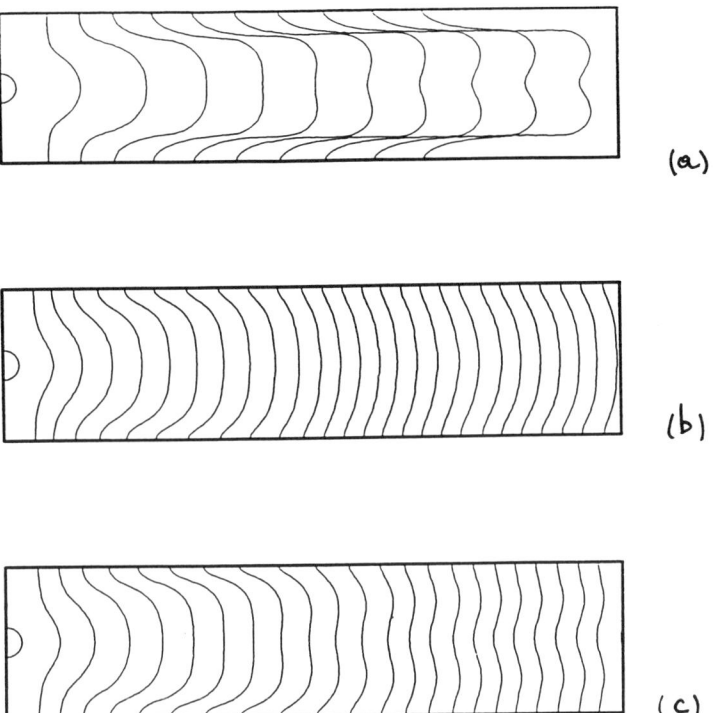

Figure 6: Finger tip splitting phenomenon whereby the reaction front divides into two branches. The imposed flow along the horizontal direction (specified at the outlet) has to be higher than the "tip splitting point " before which the phenomenon is observed. The imposed flow speeds are : (6.a) $v_0 = 1.4 \times 10^{-3}$ cm/sec; (6.b) $v_0 = 8.9 \times 10^{-4}$ cm/sec; (6.c) $v_0 = 1.0 \times 10^{-3}$ cm/sec, slightly higher than the splitting point. The domain is twice as wide as that of figure 4. Its size is 15x4 cm; the grid is 151x41. The time interval between adjacent contours is 10^9 sec.

The speed of the inlet flow is another factor that affects the stability of nonplanar front morphology. Figure 6.b shows a system exactly the same as that of 6.a except the inlet flow speed is reduced to half of that for 6.a. We observe that only a small amplitude finger develops whose tip does not split. Obviously, there exists another critical inlet fluid speed, denoted the "tip splitting point", below which tip splitting will not occur. It is rather interesting to observe that when the inlet flow speed exceeds the tip splitting point by just a small amount, the dissolution front finger will first elongate, then it will start to split. The amplitude of the steady front becomes even smaller than that of the same system under a slower inlet flow. This is shown in Figure 6.c for which the inlet flow is faster than that for 6.b, and just above the tip splitting point. The stable steady state observed here is termed "double finger" while the state observed in 6.b is termed "single finger".

Based on the above discussions, we expect that for a wider domain and/or faster inlet flow, the system will become even more unstable in that the finger tip may split again and with increasing complexity of the reaction front morphology. This is demonstrated in Figures 7 and 8.

In Figure 7 the temporal development of the half-height contour line shows the evolution of a reaction front starting from a small centered perturbation at the inlet. The initial disturbance elongates into a finger. The finger broadens until it is flat with two corners. The corners attract more fluid through them and develop into a pair of fingers. Note that the flatter central part of the finger in Fig.7 also develops a finger as well although this central finger is much weaker tan the other two that originated from the corners.

In Fig.8 the inlet fluid flow is faster than that of Fig.7, while all the other parameters are held the same. In this case the central finger is much stronger than putative side fingers. We term the development of small side fingers "budding". We see the buds arise periodically in time in a frame moving with the advancing finger. This suggests there is a critical value of inlet velocity at which the steady state finger undergoes a "Hopf bifurcation" (Marsden and McCracken, 1976) to a temporally oscillatory state.

The Stability of the Finger Doublet

Let us again address the stability of the fingers. A set of five simulations was carried out to probe this question for the finer doublet and the results are given in Fig.9. The simulations differ only in the location of the initial porosity bumps as shown, and will be identified later. Nevertheless, Fig.9 does show some obvious common features such as : 1) the fingered reaction fronts tend to split into a double finger state; and 2) the asymmetry allows one branch to dominate the other and the dominant one tends to stay away from the top and bottom boundary of the domain.

Fig.9a shows a symmetric state. The initial bump is centered vertically. The dissolution front is in a perfectly symmetric doublet form. In 9b, the initial bump is shifted toward the lower-left corner of the domain. The bump center on the inlet wall is located at $y_C=(5/12)y_L$ where y_L is the height of the domain. This down shift of the initial bump causes an uneven growth of two branches : the lower one being much stronger than the upper one. In 9c, the initial bump center is at $y_C=(1/4)y_L$. This time the upper branch is even weaker such that it only buds out periodically in a way similar to that observed in Figure 8 except the system is in an asymmetric state. In 9d, the initial bump center, located at $y_C=(1/12)y_L$, is even lower than in 9c. However, the upper branch is stronger now than the lower branch. Finally in Figure 9e, the initial bump center is at the lower-left corner of the domain. We again observed that the upper branch is stronger than the lower one. Note that the simulation domain for 9e is exactly the same as that of Figure 7 except the domain is twice as wide for Figure 7.

Another phenomenon demonstrated by Figure 9 is reinjection. Fig.9b suggests that the upper branch of the finger will reinject into the lower one. This is confirmed in Fig.10 which shows a system exactly the same as that of 9b except it is longer than 9b by 15%. For clarity, the first half-height contour line drawn in Fig.10 is at time step 65, which is the last contour shown in 9b. Also, the time interval between each half-height contour in Fig.10 is only 2/5 of that used for 9b. Apparently a side branch is generated but the faster advancing main shoot creates a pressure gradient that attracts the side shoot back into the main flow. After the upper branch is reinjected into the lower one, the overall profile of the dissolution front in Fig.10 is quite similar to that for Fig.9c. This implies that a budding state actually involves reinjection of a small branch into the main flow; the reinjection process of 9c may be masked by the level of grid resolution used in the calculation. This can be tested by repeating the simulation of 9c with better space resolution. We were unable to do this calculation beause it requires computing power much beyond that available for this study.

Fig.11 shows a simulation with an extended domain that is two times longer than that of 9d and the space resolution is accordingly reduced by a factor of two so as to cope with computing limitation. The location of the bump center is at $y_C=(1/6)y_L$. Under this reduced space resolution, we could not capture the lower branch of the finger as seen in 9.d. It is replaced by a bud instead. This again implies that a budding state is a representation of the reinjection process under a coarse space resolution.

In Fig.11, the dissolution front meanders within the flow channel. This meandering dissolution finger starts from a small local nonuniformity near the inlet while the flow channel is otherwise uniform. It is demonstrated that when the inlet flow is strong enough, a single nonuniformity may develop into two branches of dissolution fingers. If the nonuniformity is placed off-center, an oscillatory competition between two putative dissolution fingers may cause the dissolution front to meander within the aquifer. Interestingly, in a three dimensional channel with cylindrical symmetry, such a phenomenon may lead to a corkscrew structure.

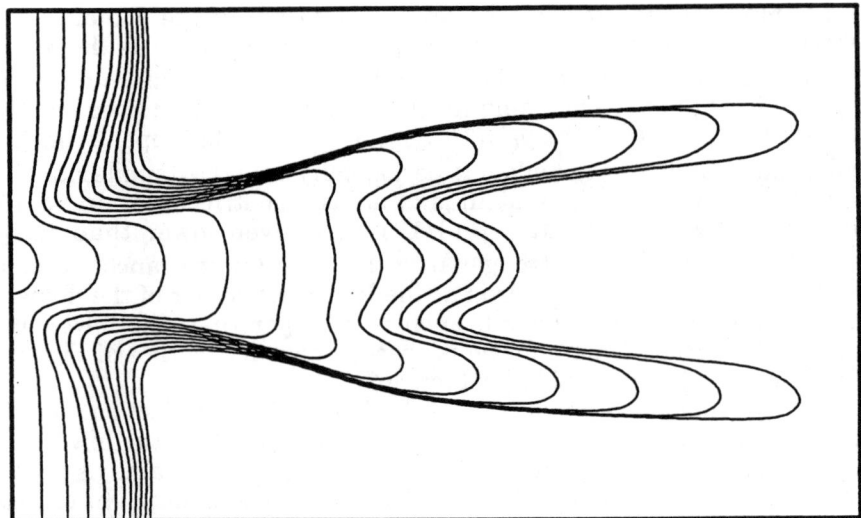

Figure 7: Triplet finger reaction front. The size of the domain is 12x120 cm. The time interval between adjacent contours is $.5 \times 10^7$ sec. The imposed flow along the horizontal direction (specified at the outlet) is 5.3×10^{-3} cm/sec which is much faster than that used for figure 6. The grid used is 201x121..

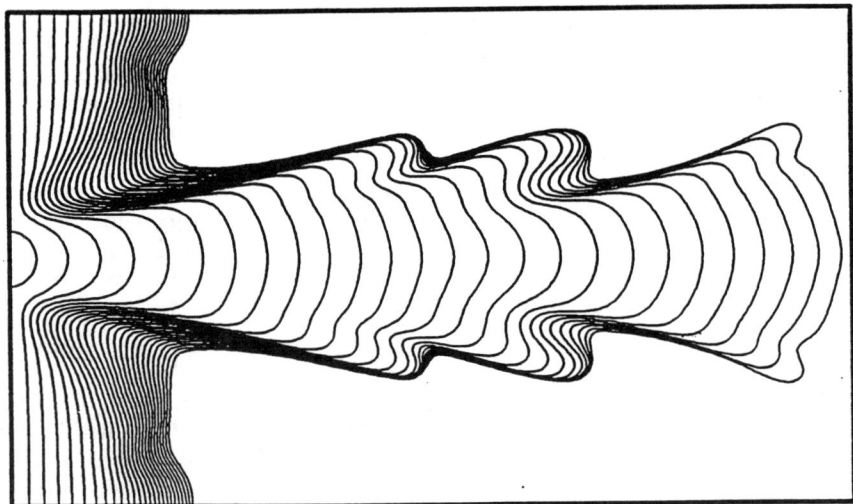

Figure 8: Small side fingers or "buds" emerge periodically in time in a frame moving with the advancung main shoot. The imposed flow is $v_0 = 7.2 \times 10^{-3}$ cm/sec, faster than that of figure 7. The time interval between adjacent contours is $.2 \times 10^7$ sec. All the other parameters are the same as that of figure 7.

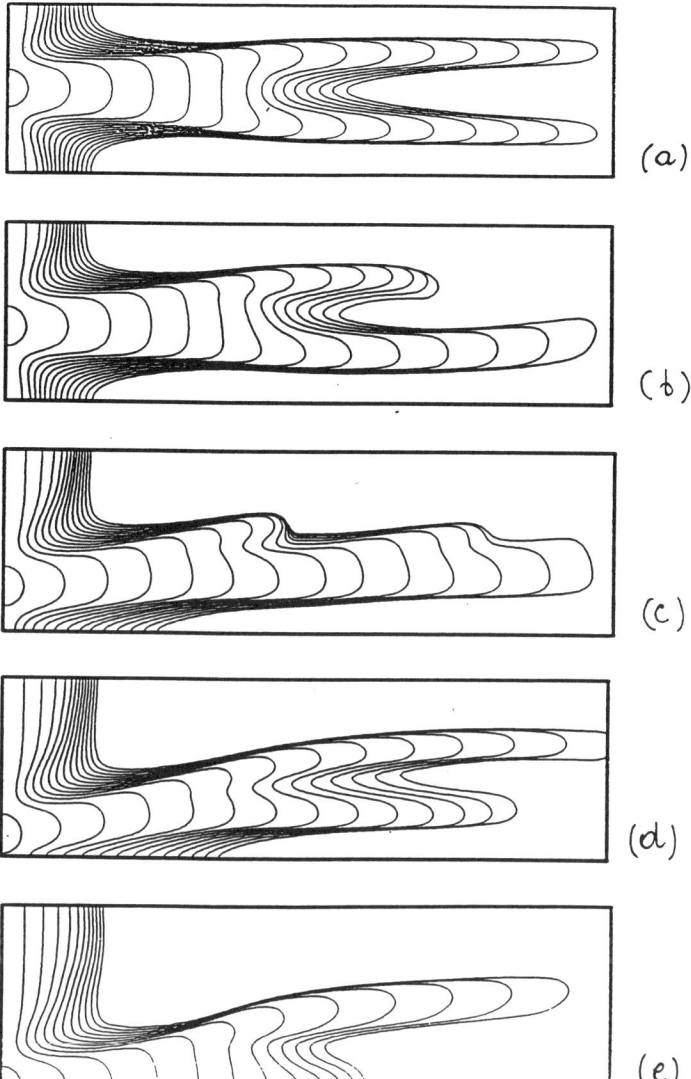

Figure 9: A host of multiple fingered front may be initiated with different initial data (here the position of the initial perturbation at the left) in a fast flow system with enough width. The center of the initial porosity bump is placed on the inlet wall at : (9.a) yc = (1/2) yL ; (9.b) yc = (1/4) yL ; (9.c) yc = (1/8) yL ; (9.d) yc = (1/12) yL (9.e) yc = 0 where y_L = 6 cm is the height of the domain. The profile of the five dissolution fronts are substantially different, showing that double fingers observed here are unstable with respect to the symmetry breaking of the system. The size of the domain is 20x6 cm. The time interval between adjacent contours is $.5 \times 10^7$ sec. The imposed flow along the horizontal direction (specified at the outlet) is $v_0 = 5.3 \times 10^{-3}$ cm/sec.

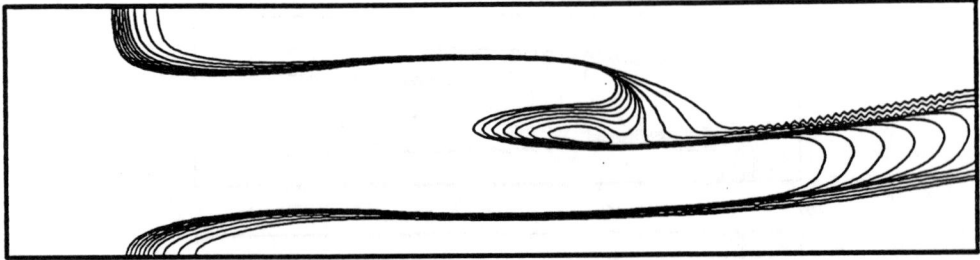

Figure 10: Reinjection of a small branch into the main branch of the dissolution firont finger. The system is exactly the same as in 7.b except that it is longer by 15%; the grid used is 231x61; The time interval between adjacent contours is 2×10^7 sec and the left most contour is at time 6.5×10^8 sec. The imposed flow along the horizontal direction (specified at the outlet) is the same as in figure 9.

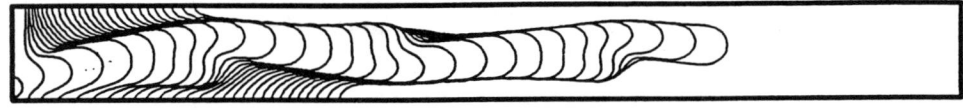

Figure 11: Meandering of the dissolution front finger. The center of the initial porosity ump is placed on the inlet wall at yc = (1/6) yL . The size of the domain is 60x6 cm, the same width as that of figure 9 but three times as long as that of figure 9. The grid used is 201x61. The time interval between adjacent contours is 15×10^8 sec and the imposed flow is the same as that used for figure 9.

The above calculations demonstrated that reactive infiltration fronts support a very rich class of static and dynamic structures. Each of these states exists in a well defined domain of system parameter space. At a given point in this space, the system may support multiple states. Which state is attained at long times depends on the state of the system in the remote past (i.e. on perturbations experienced by the front far upstream).

Summary

We have carried out numerical simulations of water-rock interaction in flow channels formed by calcite cemented sandstone. The two-dimensional reaction-transport equations have been applied to describe the interaction in which an inlet fluid undersaturated with calcite flows through the channel. It is demonstrated that a small local higher permeability nonuniformity located in the otherwise uniform channel can be strongly amplified to produce non-planar reaction fronts under the imposed flow. The resulting fronts may take on half, single or multiply fingered morphologies. Competition between multiple possible fingers may cause the front to meander within the channel. The important factors that determine the ultimate reaction front profile are the speed of the inlet flow and the width of the aquifer, both of which have to be higher than critical values in order for the non-planar morphologies to develop. The initial location of the nonuniformity may affect the ultimate front profile as well, especially in the cases of wide channels with large imposed flow where many basins of attraction may exist. A rich array of front morphologies can be generated by a small localized permeability nonuniformity. We believe that an even greater richness of phenomena will be observed when the simulations are done in multiple mineral or three dimensional systems. Preliminary results in multiple mineral simulations are demonstrated in another communication (Chen and Ortoleva 1989). Reaction-transport modeling as presented here will serve as an important guide in delineating the realm of the possible and aiding us in identifying these phenomena in natural systems or in association with enhanced oil recovery techniques such as acid flooding.

Appendix

The detailed discussion of the mathematical model and the numerical scheme used is given in Chen and Ortoleva (1989). The reaction-transport equations are used to simulate temporal development of the system starting from a given "texture" (i.e., the spatial distribution of average mineral grain size and number density and hence porosity) at initial time $t=t_0$. First, the fluid pressure (p) and hence the velocity (v) are calculated according to the water conservation equation (Eq.1) and Darcy's law (Eq.2)

$$\vec{\nabla}.\{\phi \kappa \vec{\nabla} p\} = 0 \qquad (1)$$

$$\vec{v} = -\kappa \vec{\nabla} p \qquad (2)$$

where k is the ratio of the Darcy permeability to fluid viscosity. The latter is then used to generate the spatial distribution of the composition of the aqueous fluid. This is carried out in terms of the variables h, b, and γ which satisfy the following coupled equations

$$\vec{\nabla}.[\phi D \vec{\nabla} \gamma - \phi \vec{v} \gamma] - 4\pi \rho \, nR^2 G = 0 \qquad (3)$$

$$\beta_1 = B - \gamma/(1+(b/K_3)) \qquad (4)$$

$$\beta_1 = h - K_{aq}/h + b(h/K_1 + 2h^2/(K_1 K_2)). \qquad (5)$$

These variables are defined to be :

$$h = [H^+]$$

$$b = [CO_3^{2-}]$$

$$B = [CO_3^{2-}] + [HCO_3^-] + [H_2CO_3]$$

$$= b(h/K_1 + 2h^2/(K_1 K_2))$$

$$\gamma = [Ca^{2+}] + [CaCO_3] = (1 + b/K_3)[Ca^{2+}]$$

The local average radius of the calcite grains is then advanced in time; we have

$$\frac{\partial R_{cal}}{\partial t} = G = k\left[\frac{\gamma b}{1 + b/K_3} - K_{cal}\right] \qquad (6)$$

The distribution of the texture is advanced to a later time t+Δt given its value at time t using a forward difference algorithm. Repeating the above sequence, we obtain the temporal advancement of the texture.

The symbols used in the above equations are defined as follows. The value used for each quantity is given in parenthesis if that quantity is a constant in the simulation.

D - Diffusion coefficient, defined by $D = D_0\phi^2$ where D_0 is a constant factor ($D_0 = 5 \times 10^{-5}$ cm2/sec).
k - Rate constant ($k = 10^{-1}$).
K - Equilibrium constant
$K_1 = [H^+][CO_3^{2-}]/[HC_3^-]$ ($K_1 = 4.13 \times 10^{-11}$).
$K_2 = [H^+][CO_3^-]\backslash[H_2CO_3]$ ($K_2 = 4.07 \times 10^{-7}$).
$K_3 = [Ca^{2+}][CO_3^{2-}]/[CaCO_3 (aq)]$ ($K_3 = 7.16 \times 10^{-4}$).
$K_{aq} = [H^+][OH^-]$ ($K_{aq} = 10^{-14}$).
K_{cal} = calcite equilibrium constant ($K_{cal} = 3 \times 10^{-9}$).
n - Calcite mineral number density ($n = 2.4 \times 10^7$ grains/cm^3).
p - Fluid pressure
R - Calcite mineral grain size
T - Time.
$\vec{v}$ - Fluid velocity.
V_0 - Imposed fluid velocity at the outlet wall of the domain. Its value for each simulation is given in the figure captions.
$\beta1, \beta2$ - sums of some key concentrations at the inlet wall of the domain.
ϕ - Porosity, defined by $\phi + \phi_{cal} + \phi_{qz} = 1$.

ϕ_{cal} = volume fraction of calcite.

ϕ_{qz} = volume fraction of quartz ($\phi_{Qz} = 80\%$).

K - Permeability divided by water viscosity. It is defined by (Bears, 1972)
$K = K_0\phi^3/F$ where $F = [(\phi_{Ca}/R_{Ca}) + (\phi_{Qz}/R_{Qz})]2$
$R_{qz} = 10^{-3}$ cm is the quartz grain radius,
$K0 = (J\mu_w \theta^2)^{-1}$, $J = 5$ is the packing factor,
$\mu_w = .01$ poise is the water viscosity,
$\theta = 6$ is the geometric factor.
ρ - Calcite molar density (= 27 mole/liter).

References

[1] Bear, J., *Dynamics of Fluids in Porous Media Elsevier*, Amsterdam (1972).
[2] Chadam, J., Hoff, D., Ortoleva, P., and Sen, A., "Reactive-Infiltration Instability" *SIAM Journal of Applied Mathematics* **36** (1986) 207-221.
[3] Chadam, J., Ortoleva, P., and Sen, A., "A Weakly Nonlinear Stability Analysis of the Reactive-Infiltration Interface", *SIAM Journal of Applied Mathematics* **48** (1988) 1362-1377.
[4] Chadam, J., Ortoleva, P., and Pierce, A., "Stability of Reactive Flows in Porous Media : Coupled Porosity and Viscosity Changes" (1989) submitted for publication.

[5] Chen, W., and Ortoleva, P., "Geochemical Self-Organization IV : Reaction Front Fingering in Carbonate Cemented Sandstones" in *Proceedings of the Workshop on Self-Organization in Geological Systems* (1989) to appear as a volume of Earth Science Reviews.
[6] Douglas, J., Peaceman, D. and Rachford, H., "A Method for Calculating Multi-Dimensional Immiscible Displacement", *AIME Trans.* **216** (1959) 297.
[7] Glimm, J., Lindquist, B. McBryan, O. and Tryggvason, G., "Sharp and Diffuse Fronts in Oil Reservoirs : Front Tracking and Capillarity" SIAM, Proceedings Math. and Comp. Methods in *Seismic Exploration and Reservoir Modeling*, Houston, January, (1985).
[8] Marsden, J.E., and McCracken, M., *The Hopf Bifurcation and its Applications* (1976), Springer-Verlag, N.Y.
[9] Nitzan, A., Ross, J. and Ortoleva, P., "Symmetry Breaking Instabilities in Illuminated Systems", *Journal of Chemical Physics* **60** (1974) 3134.
[10] Ortoleva, P., *Geochemical Sefl-Organization*, Oxford University Press (1990), in press.
[11]Ortoleva, P., Merino, E., Chadam, J. and Moore, C.H. Geochemical Self-Organization I : Feedback Mechanisms and Modeling Approach", *American Journal of Science* **287** (1987a) 979-1007.
[12] Ortoleva, P., Chadam, J., Merino, E., and Sen, A., "Geochemical Self-Organization II : The Reactive-Infiltration Instability", *American Journal of Science* **287** (1987b) 1008-1040.
[13] Ortoleva, P., Hallet, B., McBirney, A., Meshri, I., Reeder, R. and Williams, P., eds., *Proceedings of the Workshop on Self-Organization in Geological Systems*, Santa Barbara, June 1988, Earth Science Reviews, Elsevier, Amsterdam, in press.

NONLINEARITY AND SELFORGANIZATION IN PLASTICITY AND FRACTURE *

E.C. AIFANTIS
Dept. of Mechanical Engineering-Engineering Mechanics
Michigan Technological University
Houghton, MI 49931, USA

ABSTRACT. Even though there is strong evidence of multiplicity, nonlinearity and selforganization in plasticity, there is not a general consensus or systematic effort to utilize recent methods of nonlinear physics for understanding bifurcation and pattering of deformation phenomena. An approach recently advocated by the author and his co-workers is briefly outlined here. Emphasis is put on illustrating the origin of plastic instabilities at various levels of observation. Details are given on the problem of the persistent slip bands noted, for example, during cyclic deformation of Cu-monocrystals, and of the Portevin-Le Chatelier bands noted, for example, during monotonic deformation of Al-5% Mg polycrystals. Finally, the problem of viscoelastic peeling is outlined in the appendix and its relation to fracture is pointed out.

1. Pattern Forming Instabilities in Plasticity

It is well known that a dislocation may be viewed as a result of an instability inducing a local disturbance which persists, in the sense of metastability, over uniform lattice states. At somewhat larger or microscopic scales, such pattern forming instabilities result in static or travelling dislocation pile-ups. At even larger or mesoscopic scales, there is a plethora of phenomena related to the inhomogeneity of slip: slip lines, slip bands, slip bundles, and deformation bands. The cases of kink and coarse slip bands observed in monotonic deformation and of the persistent slip bands observed in cyclic deformation are characteristic examples of pattern forming instabilities at the mesoscopic scale. Finally, at macroscopic scales we

---

* An invited lecture on similar ideas was also delivered at the International Symposium "Plasticity 89" held in Tsu, Japan, July 31-August 4, 1989. A short version of this paper was printed in the proceedings of this symposium entitled Advances in Plasticity 1989, pp. 537-540, edited by A.Khan and M.Tokuda, and published by Pergamon Press 1989. The portion of the paper without the appendix on Stick-Slip Peeling is also to be printed in a NATO ASI series volume entitled Heterogeneous Materials : Interaction between Disorder and Behavior, edited by J-C Charmet

have the phenomena of shear band formation including the development of adiabatic shear bands. While many of these phenomena have been studied extensively in their own right, there has not been an underlying principle or framework to consider their description from a unifying point of view. In particular, no effort has been made to isolate their common characteristics and describe their occurrence and evolution with partial differential equations of nonlinear physics. It will be shown here that multiplicity or non-monotone equations of state is the basic feature in all these problems and that introduction of higher order gradients for capturing the evolution of the system in the post-instability regime is a feasible approach.

At the mesoscopic scale, in particular, a common framework for the analysis of dislocation patterning phenomena is obtained by employing the method and techniques of "selforganization". As in the case of typical physico-chemical systems which organize themselves into various forms and structures under the influence of an external driving force, the governing equations for the case of dislocation populations turn out to be nonlinear partial differential equations of the reaction-diffusion type. The physical origin of the reaction-diffusion terms is related to the production, annihilation, transport, and interaction of dislocation clusters and point defects. The corresponding mathematical analysis involves the application of stability and bifurcation theory together with the so-called method of "adiabatic elimination" or "slow mode dynamics" for describing the system in the nonlinear domain beyond bifurcation points. At macroscopic scales, partial differential equations for the plastic strain are derived exhibiting a higher order spatial dependence than the standard equations of classical continuum mechanics. The higher order gradients are introduced into the constitutive equations, e.g. in the yield condition or the expression for the flow stress. Their presence may be justified on the basis of the "adiabatic elimination" of internal variables with diffusive transport, such as point defects and dislocation ensembles, characterized by short life-times and evolving at scales smaller than the macroscopic scale. The explicit appearance of higher order gradients of the macroscopic variables is necessary in connection with non-monotone equations of state and the occurrence of softening in the stress-strain relationship.

Many of the implications of incorporating non-monotonicity and high-order gradients into the analysis of localization and patterning of deformation at the micro- and macro-level have been discussed by the author and his co-workers in recent publications (Aifantis and co-workers, 1983-1988). It was shown that the introduction of gradient terms in the equations of dislocation dynamics results in a predictive model for the wavelength of persistent slip bands and the symmetry breaking instabilities associated with these dislocation structures. The introduction of higher-order strain gradients into the constitutive expression for the flow stress results in a predictive model for the width of shear bands and the spacing and velocity of Portevin-Le Chatelier bands.

Some of this work is revisited here from a more general point of view. Moreover, other possible mathematical model equations describing pattern forming instabilities at various levels of plastic deformation are outlined.

2. Non-monotone Equations of State

In this section we give a brief account of non-monotone equations of state encountered in typical phenomena of plastic deformation.

We begin by considering an atomic lattice chain. Each atom of the chain is subjected to the influence of a body force field f due to the presence of the adjacent atomic layers, commonly known as substrate. In the simplest possible situation the effect of substrate is such that each atom of the chain is embedded in a periodic force potential of the form

$$f = sinu \qquad (1)$$

where f is normalized by a molecular constant and u a properly normalized (by $2\pi/b$, b = period of substrate) relative displacement. The form (1) is the most familiar non-monotone force-displacement relation at the atomic level. The atomic chain tries to maintain its integrity within the periodic potential imposed by the substrate through spring-like forces exerted among its atoms. While the most stable equilibrium configuration corresponds to each atom of the chain situated at each trough of the potential, other possible metastable equilibrium configurations can occur as a result of the competition between the "elastic" potential energy of the springs and the "long-range" potential of the substrate. Each atom may then progressively be displaced with respect to its nearest trough and the chain may eventually expand (contract) with the number of atoms being one less (more) than the number of troughs, thus giving rise to a negative (positive) dislocation. This is essentially the physical basis of the Frenkel-Kontorova model. The corresponding model equation is obtained by incorporating inertia and elastic strain gradients in (1) as it will be seen in the next section. It results to the well-known Sine-Gordon equation of soliton physics.

At the mesoscopic scale, non-monotonicity is inherent in the description of dislocation populations by nonlinear source terms of the form, for example,

$$g(\rho) = a\rho - b\rho^2, \qquad (2)$$

where ρ is the dislocation density, the coefficient a measures the production or generation of dislocations and the coefficient b measures annihilation or crowding effects. In fact, (2) is the simplest possible form of Verhulst dynamics and leads to the so-called "logistic equation" of animal populations. A cubic nonlinearity representing dislocation-dislocation dipole interactions is the next generalization of (2). Proper scaling of ρ and g in terms of the quantities u and f yields for (2) and its cubic generalization (by shifting variables and neglecting quadratic terms) the following expressions

$$f = u(1-u) \quad or \quad f = u(1-u^2). \qquad (3)$$

When different kinds or families of dislocations interact among themselves and point defects, (3) is replaced by a system of equations containing coupling terms determining species competition and coexistence.

Next, we discuss two examples of non-monotone equations of state related to the stress-strain and stress-strain rate responses. More details can be found in the work of Estrin and Kubin, (1988), where microscopic arguments pertaining to the concept of positive/negative strain hardening and strain rate sensitivity are outlined. The first example is concerned with the negative portion of the (microscopic) strain hardening ($h \simeq \partial\sigma/\partial\epsilon$) graph for small strains. This "unstable" regime is a result of the competition between forest hardening and dislocation production. It is associated with the formation of slip lines and slip bands. At large strains, another zero of the (macroscopic) hardening coefficient occurs and this is associated with the emergence of macroscopic shear bands. The second example is concerned with the non-monotone dependence of the stress on the strain rate. The negative slope regime (negative strain rate sensitivity $s \simeq \partial\sigma/\partial\dot\epsilon$) is consistent with microscopic theory based on the dynamical interaction of mobile dislocations and point defects. The aforementioned two cases are commonly known as strain hardening and strain rate sensitivity instabilities and they are respectively related to the regimes where the parameters $h = \partial\sigma/\partial\epsilon$ and $s = \partial\sigma/\partial\dot\epsilon$ are negative.

Finally, we mention the case of thermal softening as this is related to the occurrence of adiabatic shear bands at the macroscopic scale. This is explicitly seen upon the elimination of the temperature variable between the constitutive equation for the flow stress of the form

$$\sigma = k\epsilon^n \dot\epsilon^m (1 - d\theta), \qquad (4)$$

(for linear thermal softening) or

$$\sigma = k(1 - re^{-s\epsilon})\dot\epsilon^m \theta^\nu, \qquad (5)$$

(for power-law thermal softening) and the "homogeneous" part of the energy equation

$$\dot\theta = B\sigma\dot\epsilon, \qquad (6)$$

where σ, ϵ denote equivalent stress and strain and the rest of the quantities are material parameters.

3. Nonlinear Model Equations

By incorporating time- and space-dependent terms in non-monotone equations of state of the type discussed in the previous section, we derive classes of nonlinear partial differential equations which, in principle, can describe order, chaos, and pattern formation in plasticity. The time-dependent terms are second or first time derivatives for systems without or with friction respectively. The space-dependent terms are usually second or fourth order gradient terms for systems without or with diffusive instabilities. Due to space limitations we do not provide details of the derivation here but only give a few examples pertaining to the form of the relevant differential equation(s) and its basic properties.

In relation to the form (1) the following Sine-Gordon equation modelling the dynamics of a one-dimensional dislocation can be derived

$$u_{tt} - u_{xx} + sinu = 0 \ . \tag{7}$$

It is well-known that (7) possesses travelling wave and pulse-like solutions which may physically be identified with single dislocations and dislocation dipoles. It turns out that under certain conditions (for negligible damping and absence of external force) (7) also models the structure and motion of a dislocation viewed as a vibrating string in a Peierls potential.

In relation to the form (3) the following Fisher-Kolmogorov equation modelling the propagation of slip can be derived

$$u_t = u_{xx} + u(1-u) \ . \tag{8}$$

It is well known that this equation admits travelling wave solutions u(x-Vt) for $V > 2$. Analogous results but with different stability properties hold if the last term of (8) is replaced by $u(1-u^2)$. If, in addition, a fourth-order spatial term of the form $-\gamma u_{xxxx}$ is introduced in (8), then the fronts with oscillatory structure are possible.

In the case of interacting dislocation populations, as for example in the case with mobile and immobile dislocations during cyclic deformation, equations of the type (8) give their place to the system of differential equations.

$$\dot{\rho}_i = \nabla_i (D_{ij}^{(0)} - D_{ijk}^{(1)} \nabla_k^2) \nabla_j \rho_i + f(\rho_i) - b\rho_i + \sum_n c_n \rho_i^n (\rho_m^+ + \rho_m^-)$$

$$\dot{\rho}_m^+ = -v\nabla_x \rho_m^+ + \frac{b}{2}\rho_i - \sum_n c_n \rho_i^n \rho_m^+$$

$$\dot{\rho}_m^- = +v\nabla_x \rho_m^- + \frac{b}{2}\rho_i - \sum_n c_n \rho_i^n \rho_m^- \tag{9}$$

which is shown to lead to the prediction and symmetry breaking instabilities associated with persistent slip bands. The origin of this set of equations, the related physical considerations, and their mathematical implications are discussed in Section 4.

The early stages of monotonic deformation can also be modelled by a system of equations of the form

$$\dot{\rho}^+ = D\nabla^2 \rho^+ - E\nabla^4 \rho^+ - NL^+(\rho^-, \rho^+)$$
$$\dot{\rho}^- = D\nabla^2 \rho^- - E\nabla^4 \rho^- - NL^-(\rho^-, \rho^+) \tag{10}$$

where D may be negative due to attractive character of dislocations. It can be shown that the plastic flow in the form of a current or drift-like motion of dislocation u has a stabilizing effect on perturbations with wave vectors parallel to the glide direction, while it does not affect the development of patterns with wave vector orthogonal to the glide velocity with a preferred wavelength $\lambda_c = 2\pi\sqrt{2E/|D|}$.

The resulting patterns will then correspond to layered structures parallel to the slip planes.

Next, we discuss pattern forming instabilities at the macroscale by considering the solutions of appropriate differential equations for the plastic strain or strain rate. For the problem of stationary macroscopic shear bands it turns out that the "adiabatic elimination" of internal variables with diffusive transport leads to the following nonlinear differential equation for the shear strain distribution across the shear band

$$a(\gamma)\gamma_{xx} + b(\gamma)\gamma_x^2 = \kappa(\gamma) - \kappa_0 \quad , \tag{11}$$

where $a(\gamma) > 0$ and $b(\gamma)$ are phenomenological gradient coefficients and the flow stress $\kappa(\gamma)$ is non-monotone. It turns out that solution of (11) leads to the prediction of shear band widths in accordance with experiments.

For the case of viscoplastic materials with a gradient-dependent flow stress σ of the form $\sigma = h\epsilon + f(\dot\epsilon) + c\epsilon_{xx}$, with f being non-monotone and c constant, the differential equation of quasistatic equilibrium yields Lienard's equation

$$z_{\eta\eta} + \mu f'(z)z_\eta + (z - z_s) = 0 \quad , \tag{12}$$

for the strain rate $z = \dot\epsilon(x - Vt)$, with h being the travelling wave variable properly normalized. This equation possesses periodic solutions which can physically be identified with travelling Portevin-Le Chatelier bands. More details on the problem of Portevin-Le Chatelier effect are given below in Section 5.

We conclude by suggesting a model equation for a (unit density) viscous material with a higher order gradient dependent stress of the form $\sigma = \mu v_x - cv_{xx}$. Then, the one-dimensional momentum equation $\sigma_x = \dot v$ implies

$$v_t + vv_x + cv_{xxx} = \mu v_{xx} \tag{13}$$

which is recognized as the Korteweg-de Vries-Burgers equation.

4. A Dislocation Reaction-Transport Model for PSB's

Here we provide more details on a generalized dynamical system of coupled nonlinear differential equations describing the collective behavior of dislocation populations during fatigue experiments and discuss the related patterning instabilities leading to the formation of PSB structures. This dynamical model correctly predicts the wavelength of the ladder-like structure. Furthermore, the breaking of translational and rotational symmetries at the bifurcation naturally leads to layer-splitting and imperfection development in accordance with experimental observations. Below we discuss briefly the physical basis of the proposed model.

Consider a monocrystal oriented for single slip in a direction parallel to the x axis and submitted to cyclic loading. After an initial period where hardening occurs, the forest of immobile dislocations is already well developed and this motivates the distinction between two types of dislocations: "trapped" or nearly immobile ones of density ρ_i, and "free" or mobile ones gliding on the primary slip plane of density

ρ_m. In this regime, the dislocation density is sufficiently high ($10^{13} - 10^{14} m^{-2}$) and can be represented by a continuous concentration field on a space scale larger than a few lattice spacings. For each type of dislocation a balance equation is set up. The source and flux terms are determined by the following basic processes:

- The creation of dislocations under an applied stress is described by means of internal sources. In the case of PSB formation, a large number of dislocation clusters is already present in the crystal. Hence, the majority of the newly created dislocations will almost be immediately pinned by the forest of nearly immobile ones. Such mechanisms induce "source" terms $f(\rho_i)$ in the kinetic equation for the immobile dislocation density ρ_i.

- When the applied stress reaches certain thresholds such that thermal activation (along with internal stress- or secondary dislocation - induced activation) dominates local energy barriers, dislocations may break free and move rapidly with a stress-dependent velocity in their glide planes. The freeing processes of trapped dislocations lead to a linear source term $b\rho_m$ in the kinetic equation for the mobile dislocation density ρ_m and to a corresponding sink term in the immobile dislocation kinetics. The parameter b, specifying the freeing or mobilization rate, is vanishingly small below threshold and suddenly reaches a finite value beyond the critical stress level. This parameter will be taken as the bifurcation parameter of the system and its variation may depend on the strain rate, the temperature, and other mechanical, crystallographic, and internal structure parameters.

- The pinning of mobile dislocations by immobile dipoles or multipoles induces a sink term in their evolution equation and a corresponding source term in the equation describing the dynamics of the latter. This term is generally nonlinear and depends on the elementary interactions between the two types of dislocation populations. It is of the form $\sum_{n>1} c_n \rho_i^n \rho_m$ where the index n models the order of the cluster participating in the pinning process. The dipole formation corresponding to quadratic couplings between mobile dislocations is mainly efficient in the early stages of the deformation process. It will not be considered in the present discussion which is concerned with situations where the forest of multipoles is already well-developed.

- The motion of free or mobile dislocations is of a drift type giving rise to plastic flow. It is described by a "complete balance law" including both a material derivative and a divergence term. For the trapped or immobile dislocations, the motion in the slip and cross-slip directions is diffusive, their mobility being mostly due to thermal effects or to the coupling with vacancies and other defects.

By taking into account the aforementioned dynamical processes, the balance equations for the two dislocation populations may be written in the form given by (9)

$$\dot{\rho}_i = \nabla_i(D_{ij}^{(0)} - D_{ijk}^{(1)}\nabla_k^2)\nabla_j\rho_i + f(\rho_i) - b\rho_i + \sum_n c_n\rho_i^n(\rho_m^+ + \rho_m^-)$$

$$\dot{\rho}_m^+ = -v\nabla_x\rho_m^+ + \frac{b}{2}\rho_i - \sum_n c_n\rho_i^n\rho_m^+$$

$$\dot{\rho}_m^- = +v\nabla_x\rho_m^- + \frac{b}{2}\rho_i - \sum_n c_n\rho_i^n\rho_m^- \tag{14}$$

which are also listed here for convenience. We remark that a further distinction between positive and negative mobile dislocation densities ρ_m^+ and ρ_m^-, of a constant average glide velocity $v = v^+ = -v^- = \bar{v}sinwt$ (with $\bar{v}$ denoting the amplitude and w the frequency of the cyclic process), has been made. A mechanical justification for the origin of the various terms included in (14) on the basis of effective mass and momentum balances for dislocation populations has been given by Aifantis (1986-1987). By expressing the dynamical equations (14) in terms of the sum $\rho_m = \rho_m^+ + \rho_m^-$ and the difference $\Delta_m = \rho_m^+ - \rho_m^-$ we obtain

$$\dot{\rho}_i = \nabla_i(D_{ij}^{(0)} - D_{ijk}^{(1)}\nabla_k^2)\nabla_j\rho_i + f(\rho_i) - b\rho_i + \sum_n c_n\rho_i^n\rho_m$$

$$\dot{\rho}_m = -v\nabla_x\Delta_m + b\rho_i - \sum_n c_n\rho_i^n\rho_m$$

$$\dot{\Delta}_m = -v\nabla_x\rho_m - \sum_n c_n\rho_i^n\Delta_m \tag{15}$$

The terms $f(\rho_i)$ and $-b\rho_i$ in the above equations represent respectively creation/annihilation and mobilization of trapped dislocations. Due to the attractive character of the elastic interactions between dislocations, the diffusion coefficient tensor $D_{ij}^{(0)}$ may be negative definite in the high density regime while $D_{ij}^{(1)}$ remains positive definite.

Hence, a diffusive instability may occur for high creation rates or high stress levels, similar to Holt's theory of dislocation cell formation. The anisotropy of the underlying crystal structure is reflected here in the diffusion tensors.

On time scales larger than the period of the fatigue process and on space scales larger than the mean distance between obstacles, the effective motion of mobile dislocations is diffusive. It turns out (e.g. Aifantis [1987]) that this is a consequence of the combined effect of the back and forward motion of dislocations during each fatigue cycle and to the presence of a large number of pinning clusters. The effective diffusion coefficient D_M deduced from (15) is then proportional to the square of the mean velocity amplitude $\bar{v}$ [which may be related to the stress intensity t via phenomenological laws of the type, for example, $\bar{v} = \bar{v}_0 exp[-(\tau/\tau_0)^m]$ and inversely proportional to the total pinning rate. Specifically, it turns out that

$D_M \propto \bar{v}^2/2 \sum_n c_n \rho_i^{0n}$ and that its value is much larger than the values of the diffusion coefficients of the trapped dislocations. Our working version of the kinetic equation for ρ_m will then be given by

$$\dot{\rho}_m = D_M \nabla_x^2 \rho_m + b\rho_i - \sum_n c_n \rho_i^n \rho_m \qquad (16)$$

A less general version of (14) and (15) has been adopted and analyzed earlier by Walgraef and Aifantis (1985-1988). In that version the diffusivity D_{ij}^0, was assumed isotropic, the diffusivity $D_{ijk}^{(1)}$ was assumed to vanish and the index n was taken equal to 2 (thus accounting only for the trapping of mobile dislocations by immobile dipoles). It turns out that the generalizations introduced in this paper do not alter the basic features and properties of the elementary model discussed earlier.

To see this let us consider the pattern forming instabilities associated with the dynamical system (15) and (16). The linear stability analysis of the uniform steady state $[f(\rho_i^0) = 0, \rho_m = b\rho_i^0 / \sum_n c_n \rho_i^{0n}]$ gives, in Fourier space, the following linear evolution matrix

$$L = \begin{pmatrix} \beta - a - q_i q_j (D_{ij}^{(0)} + q_k^2 D_{ijk}^{(1)}) & \gamma \\ -\beta & -\gamma - q_x^2 D_M \end{pmatrix} \qquad (17)$$

where $a = -f'(\rho_i^0)$ is positive as a requirement of stability of the uniform steady state at low stress intensities. The coefficients γ and β are defined by the relations $\gamma = \sum_n c_n \rho_i^{0n}$, and $\beta = b(\rho_i^0 \frac{\partial \ln \gamma}{\partial \rho_i^0} - 1)$. The computation of the eigenvalues of this evolution matrix shows that, in the absence of diffusional instability ($D_{ij}^{(0)}$ is positive definite), by increasing the stress intensity (and since the motion of mobile dislocations in the primary slip direction is much faster than in any other), the uniform steady state becomes unstable versus density modulations first along the x direction at a bifurcation point given by

$$\beta = \beta_c = (\sqrt{a} + \sqrt{\gamma D_{xx}^{(0)}/D_M})^2,$$
$$\vec{q} = q_c \vec{1}_x, \quad q_c = \frac{2\pi}{\lambda_c} = (\frac{a\gamma}{D_{xx}^{(0)} D_M})^{1/4} \qquad (18)$$

where $\vec{1}_x$ is the unit vector in the x direction.

It follows that the expressions given in (18) are exactly the same with those obtained from the initial elementary model provided that the mobilization and pinning coefficients β and γ are properly defined.

Hence, beyond this pattern forming instability, ladder-like structures are expected to develop with a wavelength l which is a material property depending on the dislocation mobilities, as well as their multiplication and pinning rates. In fact, the wavelength selection results from a compromise between dislocation motion and interactions: the diffusion tries to increase it, while the interactions try to decrease it. By relating the parameters of the model to experimental quantities, it may be shown that the wavelength satisfies standard empirical or phenomenological relations such as $q_c \propto \sqrt{\rho_i^0} \propto \sqrt{\tau_s}$ where τ_s is the resolved shear stress.

When D_{ij}^0 is not positive definite, a diffusional instability occurs, leading first to the formation of cellular structures which may be associated with the vein structure of the matrix. By increasing the stress intensity, the freeing of trapped dislocations becomes more and more efficient. As a result of the highly anisotropic motion of free dislocations, this cellular structure is destabilized versus ladder-like structures with wave vectors parallel to the primary slip direction (Walgraef and Aifantis [1985-1988]).

Having found a pattern forming instability in the dynamical model, we must now consider the selection and stability properties of the pattern evolving beyond the instability. The nonlinear character of the dislocation dynamics does not allow, in general, analytic solutions for ρ_i and ρ_m. However, near the instability (or the weakly nonlinear regime around the bifurcation point) the dynamics may be reduced to much simpler forms, as a result of the time and space scale separations which occur between stable and unstable modes. This approximation procedure, justified by physical (slow mode dynamics) or mathematical (center manifold theorem) arguments leads in our problem to the following asymptotic dynamics for the dislocation microstructure, when attention is restricted to the primary x-y slip plane

$$\tau_0 \partial_t \sigma = [\epsilon - d_x(q_c^2 + \nabla^2)^2 + d_y \nabla_y^2]\sigma - v\sigma^2 - u\sigma^3 \tag{19}$$

where $\epsilon = (\beta - \beta_c)/\beta_c$, $d_x = D^{(0)}/\beta_c q_c^2$, $d_y = D^{(0)}/\beta_c(1 + \gamma/q_c^2 D_M)$, $\nabla^2 = \nabla_x^2 + \nabla_y^2$. The order parameter-like variable, σ, is a linear combination of the two dislocation densities while u and v may be obtained explicitely in terms of the parameters of the model. [The "characteristic time" like constant t is also directly related to the original parameters of the model and should not be confused with the reference yield-like or threshold stress τ_0 used in other places of the text.]

In isotropic cases ($d_y = 0$), the only stable steady states beyond threshold ($\epsilon > 0$) correspond to hexagonal or layered patterns, while in highly anisotropic media the cellular structures become unstable. It thus turns out again that layered structures with walls perpendicular to the primary slip direction will be the only possible ones. Their amplitude equation is given by

$$\tau_0 \partial_t A = \epsilon A + (\xi_x^2 \nabla_x^2 + \xi_y^2 \nabla_y^2)A - 3u|A|^2 A \tag{20}$$

where $\sigma(\vec{r}, t) = A(\vec{r}, t) exp\, iq_c x + A^*(\vec{r}, t) exp - iq_c x$ with * denoting complex conjugate, 1 1 magnitude, and $\xi_x^2 = 4q_c^2 d_x$, $\xi_y^2 = d_y$. This equation describes the slow amplitude modulations on the critical length scale of the pattern. The preferred solution (with respect to the associated Lyapounov function) for steady state modulations of uniform amplitude in a medium with natural boundary conditions is then given by

$$\sigma_0 = 2R_0 \cos(q_c x + \phi_0), \qquad R_0 = \sqrt{\frac{\epsilon}{3u}}, \qquad \phi_0 = \text{constant} \tag{21}$$

5. A Gradient Model for PLC Deformation Bands

Here we give a brief account of a higher-order strain gradient model pertaining to the PLC phenomenon and discuss its suitability in predicting shear band widths and spacings. First we show that higher order gradients in strain can be generated by "adiabatically eliminating" the concentration of defects obeying diffusing dynamics on scales much smaller than the macroscopic scale of interest. Having illustrated that second order strain gradients can indeed appear in the expression for the flow stress, we consider in detail the structure and spacing of propagating shear bands in a viscoplastic material with strain rate softening (PLC bands).

The underlying goal is to develop a nonlinear differential equation for the plastic strain rate by utilizing the equations of continuum mechanics. This is possible only if the standard constitutive equations for rigidly viscoplastic solids are properly modified to account for the strongly heterogeneous and strongly nonlinear character of deformation. In other words, the inhomogeneous (diffusive) evolution of the underlying microstructure (e.g. point defects or dislocations) enter implicitly into the problem in the form of higher order strain gradients which modify the standard constitutive equations of (homogeneous or nearly homogeneous) continuum mechanics and alter the qualitative character of the governing differential equations.

To render this argument more explicit we consider the generation of second order strain gradients into the one-dimensional constitutive equation describing a viscoplastic material $\sigma = h\epsilon + f(\dot\epsilon)$ [where σ is the stress, ϵ is the strain, and f a nonconvex function of the strain rate $\dot\epsilon$]. Under PLC conditions, the dislocations (which are accounted here for through Orowan's relation for the strain rate $\dot\epsilon \propto b\rho_m v$) interact strongly with the solute population ρ which, as usual, obeys a diffusive dynamics. We can thus expect that, in general, the system is described by the equations

$$\sigma = h\epsilon + f(\dot\epsilon) + g(\epsilon,\rho)$$
$$\dot\rho = D\rho_{xx} + r(\rho,\epsilon) \quad , \qquad (22)$$

where the subscript x denotes partial differentiation, the term $g(\epsilon,\rho)$ represents the solute-dislocation interaction, and $r(\rho,\epsilon)$ is a measure of the trapping/freeing mechanism of solute atmospheres in the dislocation cores. However, the solute concentration ρ is the fast variable of the system and can thus be "adiabatically" eliminated. Specifically, by assuming that g and r are linear forms, taking the Fourier transform of (22) and then solving for $\rho_q(\dot\rho_q \simeq 0)$, we can show that for "sufficiently large space scales" (22) is reduced to

$$\sigma = h\epsilon + f(\dot\epsilon) + c\epsilon_{xx} \quad , \qquad (23)$$

where c is a positive constant and the strain hardening modulus h is renormalized.

We note that (23) is a direct gradient-dependent generalization of the constitutive equation for viscoplastic materials discussed in detail by Kubin, Estrin and co-workers (see references quoted in Aifantis, [1987]). Specifically, they proposed the following constitutive equation, in one dimension, for materials exhibiting the PLC effect

$$\sigma = h\epsilon + f(\dot{\epsilon}) \quad , \qquad (24)$$

where σ and ϵ are the axial stress and strain and h is the strain hardening modulus. The function f is the viscous part of the flow stress and is assumed to be a single loop or nonconvex (negative slope regime i.e. $s = \partial\sigma/\partial\epsilon < 0$ for $\dot{\epsilon}_1 < \dot{\epsilon}_s < \dot{\epsilon}_2$). When the applied stress rate $\dot{\sigma}_0$ is such that the corresponding homogeneous steady state solution $\dot{\epsilon}_s (= \dot{\sigma}_0/h)$ lies in the negative slope region $\dot{\epsilon}_1 < \dot{\epsilon}_s < \dot{\epsilon}_2$, the homogeneous solution becomes unstable and a periodic succession of shear bands cross the specimen with constant velocity. To quantify the above discussion we assume a gradient dependent flow stress of the form given by (23).

With this simple modification to (23) and on assuming a travelling wave solution of the form $\epsilon = Z(x - Vt)$, (23) can be written as

$$Z_{\eta\eta} + \mu f'(Z)Z_\eta + (Z - Z_s) = 0 \quad , \qquad (25)$$

where $\eta = -\sqrt{h/c}(x-Vt)$, $\mu = V/\sqrt{ch}$. Equation (24) is the well-known Lienard's equation, a classical example of relaxation oscillations. According to LaSalle's theorem (see references quoted in Zbib and Aifantis [1988]) a stable periodic solution exists for $\dot{\epsilon}_1 < \dot{\epsilon}_s < \dot{\epsilon}_2$. Moreover, the natural speed of the travelling wave may approximately be obtained through the relation $V = 2\sqrt{ch}/|f'(Z_s)|$. More details on this problem together with appropriate numerical results illustrating the periodicity of the PLC bands, their velocity, as well as their direct influence on obtaining serrated and staircase stress-strain curves can be found in Zbib and Aifantis (1988) (See also Aifantis [1987]).

References

Aifantis, E.C. (1983). Dislocation kinetics and the formation of deformation bands, Defects, Fracture and Fatigue, (G.C. Sih and J.W. Provan, eds.), pp. 75-84, Martinus-Nijhoff.

Aifantis, E.C. (1984a). On the microstructural origin of certain inelastic models, J. Mat. Engng. Tech., 106, 326-330. Aifantis, E.C. (1984b). On the mechanics of modulated structures, Modulated Structure Materials, (T. Tsakalakos, ed.), pp. 357-385, Martinus-Nijhoff.

Aifantis, E.C. (1985). Continuum models for dislocated states and media with microstructures. Mechanics of Dislocations, (E.C. Aifantis and J.P. Hirth, eds.), pp. 127-146, Metals Park.

Aifantis, E.C. (1986a). Mechanics of microstructures I, II, III. Mechanical Properties and Behaviour of Solids: Plastic Instabilities, (V. Balakrishnan and C.E. Bottani, eds.), pp. 314-353, World Scientific, Singapore.

Aifantis, E.C. (1986b). On the dynamical origin of dislocation patterns. Mat. Sci. Eng., 81, 563-574.

Walgraef, D. and E.C. Aifantis (1985a). On the formation and stability of dislocation patterns - I, II, III. Int. J. Eng. Sci., 23, 1351-1358, 1359-1364, 1365-1372.

Walgraef, D. and E.C. Aifantis (1985b). Dislocation Patterning in fatigued metals as a result of dynamical instabilities. J. Appl. Phys., 58, 688-691.

Walgraef, D. and E.C. Aifantis (1986). Dislocation patterning in fatigued metals: Labyrinth structures and rotational effects. Int. J. Eng. Sci., pp. 1789-1798.

Walgraef, D. and E.C. Aifantis (1988). Plastic instabilities, dislocation patterns and nonequilibrium phenomena. Res Mechanica, 23, 161-195.

Triantafyllidis, N. and E.C. Aifantis (1986). A gradient approach to localization of deformation - I. Hyperelastic materials. J. of Elasticity, 16, 225-238.

Zbib, H.M. and E.C. Aifantis (1988a). On the localization and postlocalization behaviour of plastic deformation - I: On the initiation of shear bands; II: On the evolution and thickness of shear bands; III: On the structure and velocity of the Portevin-Le Chatelier Bands. Res Mechanica, 23, 261-277; 279-292; 293-305.

Zbib, H.M. and E.C. Aifantis (1988b). On the structure and width of shear bands. Scripta Met., 22, 703-708.

Zbib, H.M. and E.C. Aifantis (1988c). A gradient-dependent model for the Portevin-Le Chatelier effect. Scripta Met., 22, 1331-1336.

Aifantis, E.C. (1987). The physics of plastic deformation. Int. J. Plasticity, 3, 211-247.

Estrin, Y. and L.P. Kubin (1988). Plastic instabilities: Classification and physical mechanisms. Res Mechanica, 23, 197-221. [In: Material Instabilities, Special Issue of Res Mechanica, Vol. 23, E.C. Aifantis et al., eds.].

Appendix on Stick-Slip Peeling

In this appendix, we provide a brief description of stick-slip peeling on the basis of stability and bifurcation analysis. Numerical results are presented for pressure-sensitive adhesive tapes and an analogy between stick-slip peeling and dynamic fracture is made. The governing differential equation are analogous to the case of the PLC effect. A more detailed discussion of this problem will be given in a forthcoming paper by T. Webb and E.C. Aifantis.

Relaxation oscillations (stick-slip) are often observed in peeling of pressure-sensitive adhesive tapes. [See, for example, Aubrey, D.W., Welding, G.N. and Wong, T., J. Appl. Polym. Sci 13, 2193-2207, 1969. Also Maugis, D. and Barquins, M., in: Adhesion 12, ed. K.W. Allen pp. 205-222, Elsevier Appl. Sci. Publ, 1987.] Stick-slip peeling characterized by a saw-tooth peel force trace (Fig. 1) and by the appearance of periodic markings on the peeled surface of the adherand. The spacing between the periodic markings (length of the crack jumps) is related to the wavelength of the peel force trace.

Stick-slip peeling has been associated with a nonconvex crack resistance (G) vs velocity (v) curve shown in Fig. 2 in which dynamic losses decrease with increasing velocity over a certain velocity regime. This eventually gives rise to a limit cycle for the phenomenon as depicted in Fig. 2. As a result, we have an oscillatory

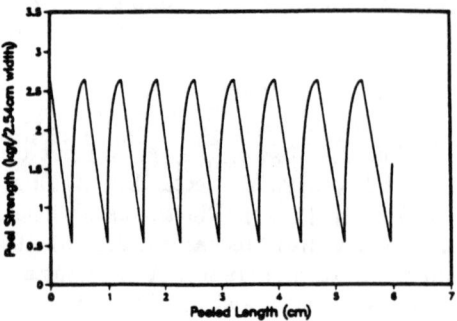

Figure 1 : Saw-tooth peel force trace.

velocity (Fig. 3) which, in turn, induces the periodic (jerky) behavior for the load depicted in Fig. 1.

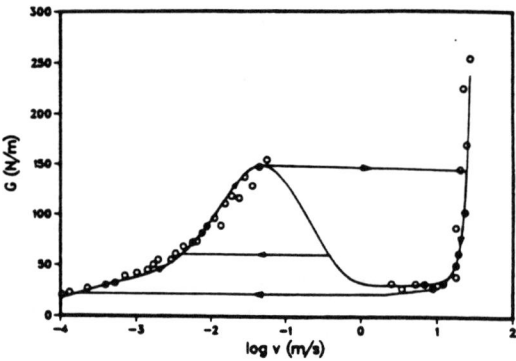

Figure 2 : G vs v curve fot scotsh tape (Maugis and Barquins, 1987) and limit cycle.

Below we describe a model by borrowing heavily from the work of Maugis and Barquins (1987). The energy release rate is given by the sum of three terms

$$G = \phi(v) + \psi(v) + I(\dot{v}) \quad . \tag{A.1}$$

The term $f(\dot{v})$ denotes the inelastic losses near the crack tip and is proportional to the surface energy. Roughly speaking, $f(\dot{v})$ represents the measured G vs v

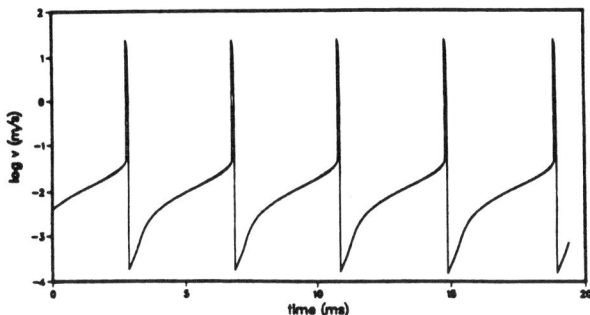

Figure 3 : v vs t profile for scotch tape peeled at a constant rate of 382 m/s.

curve shown in Fig. 2. The term $\psi(v)$ represents the inelastic losses away from the crack tip. When the length of the adherand is much greater than its thickness, the inelastic losses are very localized near the crack tip and $\psi(v)$ can be neglected. Finally, the term I(v) represents the effect of inertia which must be included due to accelerating (slip) and decelerating (stick) "crack" propagation. For the problem under consideration the energy release rate is easily computed as

$$G = \frac{P}{b} - (1 - cos\theta) + \frac{P^2}{2b^2 hE} \quad , \tag{A.2}$$

where θ is the peel angle, P represents the peel force, E is the elastic modulus and b and h denote the width and thickness of the adherand respectively (Fig. 4). For thin adherands, the second term in (A.2) is much smaller than the first term and can be neglected.

Taking the time derivative of (A.2) we have

$$m\dot{v} + b\frac{d\phi}{dv}\dot{v} - (1 - cos\theta)\dot{P} = 0 \quad , \tag{A.3}$$

where the peel angle was assumed constant ($\dot{\theta} = 0$) and $m = dI/d\dot{v}$ is a small parameter denoting an effective mass. From the geometry of the peel test (Fig. 4) it can be shown that the relationship between the constant peeling rate V and the peeling velocity v is given by

$$V = v(1 - cos\theta) + L\dot{\epsilon} \quad , \tag{A.4}$$

where the product $L\dot{\epsilon}$ denotes the stretching of an adherand of length L for an axial strain rate $\dot{\epsilon}$.

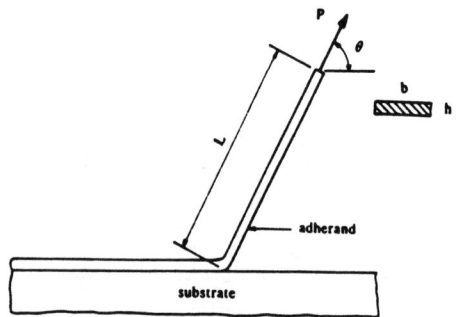

Figure 4 : Schematic of peel test.

Since pressure-sensitive adhesive tapes are viscoelastic materials and thin metallic films (e.g. copper) adhesively bonded to dielectric substrates are strain rate sensitive we adopt here, as a first approximation, a viscoelastic constitutive equation for the adherand of the form

$$P = bhE\epsilon + bh\eta\dot{\epsilon} \ . \tag{A.5}$$

By combining (A.3) - (A.5) with $\theta = 90^0$, we have

$$\ddot{v} + \mu\omega(\frac{d\phi}{dv} + c)\dot{v} + \omega^2(v - V) = 0$$
$$\dot{P} = K(v - V) - bd\dot{v} \tag{A.6}$$

where $K = bhE/L$, $\omega = \sqrt{K/m}$, $\mu = b/\sqrt{Km}$ and $c = h\eta/L$. When $\eta = 0$ (A.6) reduce to the equations first proposed by Maugis and Barquins (1987).

Letting $x = v - V$ and $F(x) = \phi(v) - \phi(V)$ and considering L constant (A.6) can be written in the form

$$\ddot{x} + \mu\omega(F'(x) + c)\dot{x} + \omega^2 x = 0 \ , \tag{A.7}$$

where μ, ω and c are constants. Equation (A.7) is Lienard's equation governing self-sustained oscillations and for m sufficiently large it is known to exhibit relaxation (jerky) oscillations.

In the following, we discuss the behavior of (A.6(1)) in the form

$$\dot{v} - \mu\omega[y - (\phi(v) + c)]$$
$$\dot{y} = -\frac{\omega}{\mu}(v - V) \tag{A.8}$$

implying that $(v = V, y = \phi(V) + cV)$ is a critical point. Upon linearization we obtain the corresponding eigenvalues as

$$\lambda_{1,2} = -\frac{\omega\mu(\phi'(V) + c)}{2} \pm \omega\sqrt{(\frac{\mu(\phi'(V) + c)}{2})^2 - 1} \ . \qquad (A.9)$$

Thus, when $\phi'(V) + c > 0$ the solutions are stable and when $\phi'(V) + c < 0$ the solutions are unstable and evolve into periodic solutions of a limit cycle type. This implies that when the constant peeling rate is in the negative slope region and $\phi'(V) < -c$ stick-slip peeling occurs. The viscous behavior of the adherand acts to stabilize peeling since an imposed rate within the negative slope regime does not necessarily imply stick-slip as in the elastic case.

In the special case when $\phi'(V) = -c$ Hopf bifurcation occurs in which small disturbances grow from uniform into periodic solutions. The bifurcation parameter σ in this problem is $-\phi'(V)$ and its critical value σ_c is c. By constructing a series solution about $v = V$ and $\sigma = \sigma_c = c$ it can be shown that the bifurcating periodic solution is of the form

$$v = V + \sqrt{\frac{-8(\phi'(V) + c)}{\phi'''(V)}} \sin\nu t$$
$$- (\frac{4\mu\phi''(V)(\phi'(V) + c)}{3\phi'''(V)})\sin 2\nu t + ...$$
$$\nu = \omega\frac{(\mu\phi''(V))^2(\phi'(V) + c)}{\phi'''(V)} + ... \qquad (A.10)$$

When $\phi'''(V) > 0$ near the maximum of the G vs v curve in the negative slope region, the bifurcation is supercritical (stable periodic solutions). When $\phi^{(2)}(V) < 0$ near the minimum of the G vs v curve in the negative slope region, the bifurcation is subcritical (unstable periodic solutions).

We now present some numerical results pertaining to pressure-sensitive adhesive tapes (Aubrey et al., 1969; Maugis and Barquins, 1987) for which G vs v curves have been measured. Fig. 2 shows the G vs v graph for scotch tape (Maugis and Barquins, 1987) and the phase plane with a limit cycle demonstrating an hysteresis between an accelerating "crack" and a decelerating "crack". Fig. 3 depicts the relaxation oscillations for the velocity when the constant peeling rate is in the negative slope region. Fig. 5 shows the G vs v graph for the data of Aubrey et al. (1969) where L is not fixed but is slowly varying. A similar analysis can be carried through for this case and analogous results are obtained for the peel force (Fig. 1) and velocity (Fig. 3). Fig. 6 shows the "discontinuous" nature of the crack growth which gives rise to the periodic markings on the adherand surface. We can see from Fig. 1 and Fig. 6 that the wavelength of stick-slip increases with the peeled length of the adherand demonstrating an increase in compliance, a fact in agreement with observations.

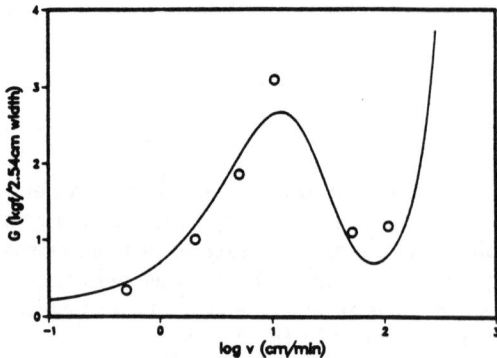

Figure 5 : G vs v curve for poly (n-butyl acrylate)-coated polyester film (Aubrey et al., 1969).

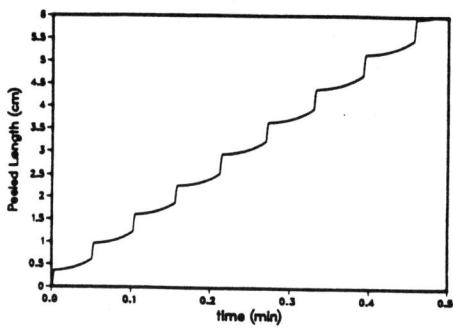

Figure 6 : Discontinuous (step-like) crack growth.

Finally, we briefly outline the analogy between the stick-slip behavior of peeling and dynamic fracture. For example during dynamic fracture of PMMA the G vs v graph is also of the form depicted in Fig. 2, i.e., it exhibits a negative slope regime. When the cross-head velocity (e.g., in a tapered double cantilever beam specimen) is such that the crack velocity is in the negative slope regime, discontinuous crack growth occurs. This, among other things, manifests itself as "periodically" spaced markings on the fracture surface. An analysis similar to that outlined earlier for the peel test, has also been employed for this problem and numerical results for the crack velocity and crack jumps are currently being obtained. The governing equations are

$$\ddot{v} + \frac{b}{m}\frac{d\phi}{dv}\dot{v} + \frac{8gP^2}{Ebma}\left(v - \frac{Eb\dot{\Delta}_0}{8gP}\right) = 0$$

$$\dot{P} = \frac{Eb}{8ga}\dot{\Delta}_0 - \frac{Pv}{a} \qquad (A.11)$$

where m is the effective mass, E is the elastic modulus, g is a geometric factor, b is the thickness, P is the load, $\dot{\Delta}_0$ is the constant cross-head velocity, $\phi(v)$ is given by the G vs v graph and v and a denote crack velocity and length respectively.

Acknowledgements

NSF support under grant CES-88 00459 is acknowledged. The long-term collaboration with Dr. D. Walgraef, and the joint work with my ex-student Professor H. Zbib and more recently with my graduate student Tom Webb were essential in putting together this review. Parts of the present article were based in earlier joint papers with these co-workers.

PLASTIC INSTABILITIES AND THE DEFORMATION OF METALS

H. Neuhäuser
Institute for Metal Physics and Nuclear Solid
State Physics
Technical University of Braunschweig
3300 Braunschweig
F.R.Germany

ABSTRACT. The paper reviews the main dislocation mechanisms relevant for plastic instabilities in metals and the patterns involved. The observability of plastic instabilities by various methods and the classification of micromechanisms (M-, h-, S-, T-type instabilities) as well as propagating modes of shear are summarized. Short excursions deal with instabilities by diffusionless structural transformations and with unstable localized deformation in amorphous alloys. Finally, the possibilities for describing the deformation kinetics on micro- and macroscale by the approaches of viscous glide and of generation rate of dislocations are discussed.

1. INTRODUCTION

One of the outstanding properties of metals is their good plasticity combined with considerable strength. Their deformation is commonly known to occur in most instances smoothly and continuously. However, there are a number of cases and deformation conditions where plastic instabilities occur on various levels of magnitude.

For instance, macroscopically visible serrations of load occur during deformation of certain alloys in an interval of relatively high temperatures, the socalled Portevin-LeChatelier effect, with abrupt bursts of strain in the order of $\delta\varepsilon \approx 10^{-4}$. Even for systems where no macroscopic instability can be resolved, the dislocations which produce the deformation in crystalline metals are often observed in the electron microscope to move in a jerky manner across the screen, with a strain of the order of $\delta\varepsilon < 10^{-9}$ produced by a typical jump. We will regard also such behaviour as a plastic instability which in favourite cases can be observed macroscopically as acoustic emission. On the other hand, we will not regard the very slightly jerky motion of dislocations between neighbouring pointlike obstacles (which would correspond to typical strain jumps of 10^{-14}) as an unstable motion, because in this case the average velocity of the dislocation is uniquely determined by the waiting times at obstacles according to the stress at the externally imposed strain rate. Although there is no sharp distinction, this classification seems to differentiate well enough the essential cases as will be shown below. In particular, plastic instabilities are considered to occur, if the (sometimes local) elongation rate of the specimen, $\dot{l}_{pl}$, exceeds the imposed elongation rate of the machine, $\dot{l}_m$, so that according to the equation

$$\dot{l}_m = \dot{P}/F + \dot{l}_{pl} \tag{1}$$

(P = load, F = effective spring constant of machine plus specimen) some resolvable temporal unloading of the system occurs.

In the following, we will briefly survey the basic mechanisms of plasticity by dislocations in various metal crystal lattices (sect. 2) in order to facilitate the understanding of the various mechanisms which may lead to plastic instabilities (sect. 4). These are often connected with pronounced patterning (sect. 3), e.g. a hierarchy of slip lines and dislocation cell structures, a consequence of the extreme nonlinearity of the laws governing plastic deformation, and the irreversibility of deformation which occurs far from thermal equilibrium (e.g. /1/). Short excursions will also deal with the instability and patterns in the deformation of amorphous alloys, and with the instabilities caused by diffusionless structural transformations, as both are quite prominent examples for instabilities in plasticity. We will conclude with a chapter on the problem to find appropriate averages in microscale for the description of deformation kinetics in the cases of continuous and discontinuous plastic flow.

2. ON DISLOCATION MECHANISMS OF CRYSTAL PLASTICITY

Plasticity, i.e. irreversible deformation, is produced on microscale in crystalline materials mainly by generation and movement of dislocations, at high temperatures also by a flux of vacancies. Below we will concentrate on dislocation mechanisms and very briefly review their strong dependence on the kind of crystal lattice (2.1.) and on the deformation temperature (2.2.) in order to provide a basis for understanding later the reasons for plastic instabilities in various instances.

Dislocations are line defects which may be considered as the border line of that part of the slip plane that has slipped relative to its neighbouring plane by the Burgers vector b (Fig. 1). By moving this line of atomic displacements across the slip plane, the shear is produced at stresses several orders of magnitude lower than would be neces-

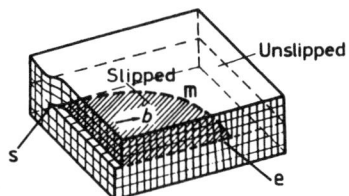

Fig. 1 Dislocation line as the border line of a slipped area (A). e = edge, m = mixed, s = screw dislocation, b = Burgers vector (after /2/)

sary for shearing the crystal planes as a whole. The sum of many of such slips produces the deformation of the crystal. Local shearing of the planes can be recognized by the steps arising at the surface (slip steps), and the dislocations can be made visible by their displacement field for instance in transmission electron microscopy.

2.1. Dislocation motion in various crystal lattices (/3/)

The crystal lattices of metals (fcc, hcp, bcc) are, according to the undirected nature of metal bonding, characterized by their dense packing of atoms, resulting in a relatively high mobility of disloctions. Nevertheless, definite differences exist, as shown for instance in Fig. 2, where the mea-

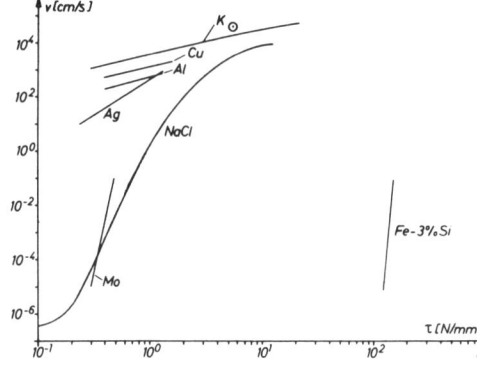

Fig. 2 Stress dependence of dislocation velocities, measured by pulse loading and etching for single dislocations in various materials at room temperature (from /4/, cf. refs. there)

sured dislocation velocities in dependence of the applied stress are shown for various materials. These differences can be traced back to the special configurations of the dislocations in the various lattices, expressed either by the Peierls stress (necessary to move the dislocation through the pure lattice at T = 0) or by the degree of spreading of the dislocation core. Widely spread dislocations (low Peierls stress) move more easily than those with narrow cores. Such spreading has been established by computer simulations of the atomic positions using appropriate interatomic potentials:

(a) pure fcc lattice: Here the displacement field of the dislocation is distributed mainly in a (111)-type "glide plane" (Fig. 3a) corresponding to a splitting of the perfect

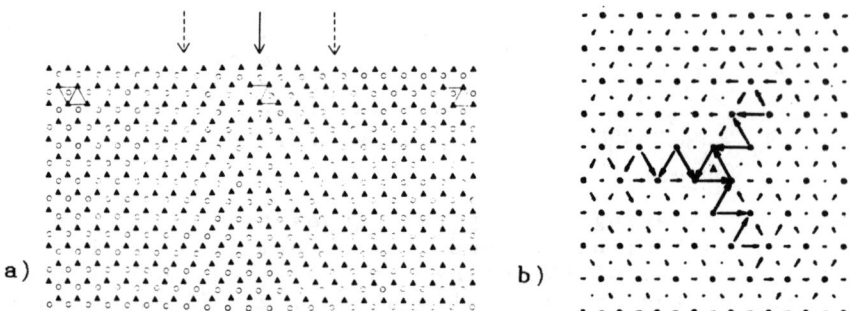

Fig. 3 Examples for atomic configurations around dislocations in model crystals: a) edge dislocation in fcc lattice split into two Shockley partial (dotted arrows) lying in the paper plane from /5/); b) three-fold splitting of a screw dislocation in the bcc lattice, dislocation line normal to paper plane (from /6/).

dislocation into two Shockley partials which are bound together by the stacking fault (with hcp stacking) in between. This unit may glide usually at very low stresses on its glide plane, and it may be transferred to another (111)-type plane ("cross slip") only, if the dislocation has screw character and if the splitting has been reduced locally. (We just mention that also rare slip on the less densely packed (100)-type planes with much less mobile dislocations has been discovered).

(b) pure hcp lattice: Large differences of dislocation mobility exist also for basal, pyramidal, and prismatic glide, with similar conditions as in the fcc case for the basal planes, but with complicated situations otherwise (which have been recently considered by Bacon et al. /7/).

(c) pure bcc lattice: The (110)-type planes are favoured as slip planes, and the screw dislocation (b = (1/2)<111>) is distinguished by its three-fold split core (Fig. 3b) which makes it difficult to move, while all other dislocation characters move more easily. Contrary to (a) and (b) with planar splitting, here the kink mechanism /8/ for propagating the screw dislocation is important (high Peierls stress).

(d) ordered alloys: In long-range ordered alloys dislocations tend to move in pairs, as the second dislocation removes the antiphase boundary produced by the first dislocation. The special core structure in some of these alloys with extension on two (111) planes explains their high strength at high temperatures /9/.

(e) solid solutions: In dilute solid solutions, the solute atoms, and in more concentrated ones small solute clusters act as local obstacles to moving dislocations, which can be overcome with the aid of thermal activation (e.g. /10, 11, 12/). The obstacle forces are due to size and modulus misfit of the solute atoms in the matrix. This simple picture has to be modified by the tendency of solutes to segregate to dislocations /13/ ("aging"), mainly due to their hydrostatic stress field, and, in case of fcc alloys, due to the stacking fault between partials ("chemical force" by Suzuki segregation). These effects cause resting or slowly moving dislocations in alloys to be additionally pinned, and after breakaway from their "cloud" (atmosphere) of impurities, the rapidly moving dislocation may feel less friction. This will be one of the reasons for instability and obviously it depends on the diffusivity of the solutes to the dislocation with large quantitative differences between substitutional and interstitial solid solutions. Another reason is the possible destruction of obstacles, for instance of short range ordered solute clusters in more concentrated alloys, or the shearing of precipitates by the dislocations cutting through them. In bcc solid solutions the effects of solute atoms may overwhelm the intrinsic lattice properties (c), so that such alloys largely behave similar to fcc ones.

2.2. Influence of temperature on dislocation mobility /14/.

The properties of dislocations indicated above in various situations cause different and characteristic behaviour with changing the deformation temperature.

A strong increase of the yield stress with decreasing temperature is observed always if the rate governing process is thermally activated, i.e. if it obeys an Arrhenius-type law. This occurs in pure metals when the double-kink production controls the movement of dislocations (e.g. in fcc crystals below few degrees K, in bcc roughly below room temperature), or when the dislocations have to cut through forest dislocations, or when cross-slip and jog dragging processes are involved; similarly in impure or alloy crystals, when the overcoming of local obstacles of solute atoms controls the dislocation movement. In such cases instabilities may occur if the temperature rises locally, or if the obstacles are destroyed partially by the dislocations (sect. 4.2.b).

An only negligible temperature dependence of the yield stress (corresponding to that of the elastic moduli) occurs if long-range stresses (e.g. passing stresses of dislocations), or obstacles which are too large for thermal activation (e.g. dislocation reaction products, large solute clusters or precipitates) control the dislocation movement. Long-range interactions may be involved in plastic instabilities, e.g. by stress concentrations in groups of dislocations moving on closely spaced or single slip planes, or in clusters of dipoles which may break up when their equilibrium is disturbed (sect. 4.2).

Another kind of instability occurs in dilute solid solutions at very low temperatures, when the dynamic friction (various phonon processes) drops to very low values. Then an inertial effect helps to overcome the local obstacles and exceedingly large path length of the dislocations are expected (sect. 4.2).

At high temperature the increasing mobility of solutes in solid solutions causes segregation in short times and prominent instability effects if the dislocations break away from their solute atmosphere, which will be considered in some detail in sect. 4 and 5. The instability vanishes again at higher temperatures, when the solute atmosphere is formed in times much shorter than the characteristic waiting time of the dislocations at obstacles, so that the complete atmosphere is "dragged along" with the moving dislocation. A typical dependence of the critical resolved shear stress crss, the yield stress at the onset of plastic deformation, on temperature for fcc alloy crystals, is demonstrated in Fig. 4, showing all the above-mentioned processes.

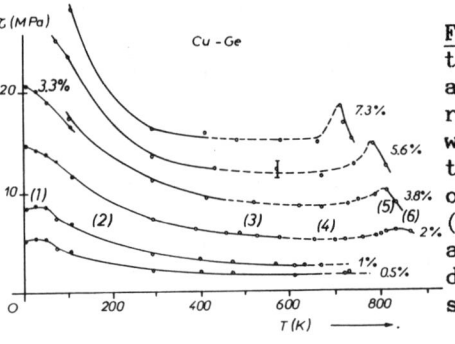

Fig. 4 Dependence of the crss on temperature for a typical fcc alloy showing the strongly temperature dependent part at low T (2) with anomaly at very low T (1), the plateau region (3) with stress oscillations due to the PLC effect (4, dotted line), the maximum (5) and following strong decrease (6) due to diffusive effects of obstacles at dislocations. /15,16/.

2.3. Survey of processes during work hardening /17, 18/.

While in the last section we mainly considered the onset of plastic deformation in a virgin crystal with friction forces on dislocation movement exerted by the lattice itself or by grown-in lattice defect, in the following we briefly deal with the mechanisms which increase the flow stress with increasing deformation (Fig. 5). This is not only of much practical importance but also gives rise to additional reasons for plastic instabilities.

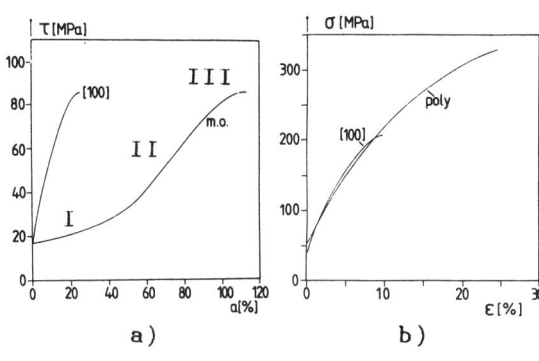

Fig. 5 Deformation curves for Cu-3.3at%Mn:
a) single crystals oriented for single glide (mo, with stages I, II and III) and for multiple glide [100], respectively;
b) [100]-crystal compared to a polycrystal /19/.
($\tau = \mu \cdot \sigma$, $a = \varepsilon/\mu$, μ = orientation factor).

a) crss and stage I.
The crss is determined by rapid multiplication of dislocations from the most favourable dislocation sources (e.g. /20/). In case of an upper and lower yield point, the grown-in sources are pinned by impurities or solutes with higher concentration than in the bulk (because of inevitable heat treatment during crystal growth), and after the first break-away at the upper yield point they can operate with high frequency and may produce more dislocations than necessary to accomplish the imposed deformation rate: the stress may drop to a lower level where a dynamical equilibrium is reached for multiplication and propagation of dislocations /21/. This is an example for a plastic instability based on rapid multiplication of overstressed dislocation sources, and it is usually connected with a pronounced inhomogeneity of slip along the crystal. In extreme cases, e.g. highly concentrated alloys or irradiated crystals, it may lead to the formation of a Lüders-band, where the deformation is concentrated to a small region of several 0.1 to 1 mm in width, propagating along the crystal from one grip to the other. Here macroscopic bending stresses are also involved keeping the externally measured load virtually constant.

A very low hardening rate is observed in stage I of pure single glide metals (Fig. 5a). It is connected with the gradual activation of sources in the whole crystal and with a rapid increase of the primary dislocation density to $>10^8$ cm^{-2} with important dipolar interactions and with a rotation of the lattice relative to the crystal axis by several degrees. This enables sources on the next favourable secondary slip system (i.e. the critical system) to be activated, which may also account for the hardening rate in stage I.

b) stage II.
With further change of the crystal orientation, the activation of secondary sources in the conjugate system, which is accompanied with the formation of strong dislocation reac-

tions of the Lomer-Cottrell-type (athermal obstacles), is considered to be the reason for the transition to the high and nearly constant work hardening rate in stage II. For symmetrically oriented crystals (Fig. 5) there is no distinction between primary and secondary sources, and after the same crss as in single glide the curve immediately continues with stage II work hardening. (Similar behaviour is observed in polycrystals (Fig. 5b)). Here the dislocation path length of all glide systems is similar, while in single glide crystals the path length of primary dislocations exceeds by far that of the secondary ones, implying somewhat different cell structures (sect. 3.2) evolving in stage II. With increasing deformation this characteristic dislocation pattern of cell walls with high dislocation density and nearly dislocation-free cell interior decreases in scale.

Plastic instabilities may be initiated in this (and later) regimes by a change in deformation path. By changing for instance the direction of loading or the deformation temperature, the newly activated sources find a type of forest structure which is alien to their orientation and thus can be easily destroyed by strong local slip (cf. sect.4.2).

c) stages III, IV, ...
When the dislocation density reaches critical values, recovery processes set in which may be of different nature: the onset of stage III is associated with massive cross-slip processes of screw dislocations, stage IV is supposed to be due to the onset of annihilation of edge dislocations by climb /22/. These stages of deformation are recognized in a plot of $\theta = d\tau/da$ versus τ shown for example in Fig. 6.

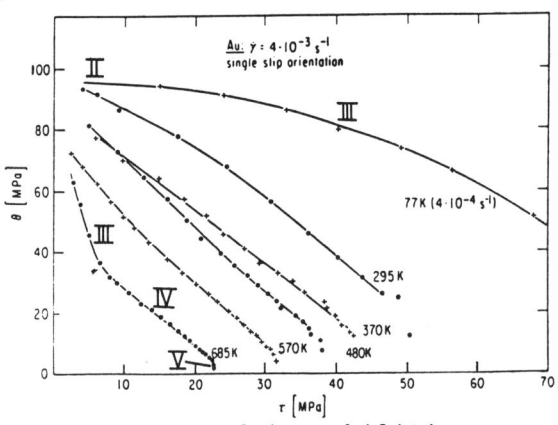

Fig. 6 Plot of the work hardening rate versus stress for Au at T = 77-865 K, to show in particulthe higher work hardening rates (II ... V). From /23/.

Various types of instabilities may occur in these structures of high dislocation density and correspondingly large stored energy, as indicated before. Further, shortly before

fracture, usually a geometrical instability sets in when due
to high local deformation and because of the limited work
hardening and high recovery rates from some random decrease
of the crystal cross section a neck may develop (Considere
instability) which can no longer carry the external load and
thus leads to catastrophic fracture (sect. 4.2.b).

3. PATTERNS DEVELOPED DURING DEFORMATION OF CRYSTALS

The processes discussed above usually favour localization of
slip: In particular the special nature of dislocation sources which may emit many correlated dislocations, and the
partial destruction of obstacles by cutting dislocations favour the development of groups of dislocations which move
under collective interactions. Apart from these micro-mecha-
nistic reasons, the formation of characteristic patterns of
slip may be considered from a more general point of view as
the consequence of the fact that plastic deformation occurs
far from thermodynamic equilibrium /24/, involving highly
non-linear (mostly exponential or power) laws and including
long-range interations between the microscopic constituents.
These patterns consist in (i) a characteristic sometimes
periodic arrangement of slip traces at the crystal surface
produced by dislocations in some strain interval, and (ii) a
characteristic, sometimes cellular dislocation structure
evolving in a shape which varies with stress or deformation.
Their main features will be briefly reviewed with emphasis
to the question of plastic instabilities.

3.1. Slip lines

Although some caution is necessary in conclusions from the
slip line features observed at the surface to the slip conditions in the bulk /17, 4/, valuable information can be
gathered from slip line studies about the distribution of
sources, the number of dislocations emitted by a source,
their nucleation rate, the propagation rate across the crystal, and the correlation between neighbouring sources. Slip
lines are visible because of the inherent inhomogeneity of
slip and local plastic instabilities (sect. 4.2); thus the
structure of slip lines gives first hints on the existence
of instabilities in particular in cases where these are not
directly observable as stress serrations. Slip traces often
show a hierarchy /25/ from slip lines (smalles units resolved by EM replica, produced by about 30 dislocation origina-
ting probably from single sources), to slip bands (clusters
of closely- spaced slip lines, probably produced by triggering or initiating new neighbouring dislocation sources, e.g.

by cross-slip), to slip band bundles (groups or clusters of slip bands with rather well-defined distances depending on crystal thickness and being separated by inactive crystal regions of similar or larger size), and sometimes to a Lüders band (consisting of one or several active slip band bundles). These units are exemplified in Fig. 7 and, within their limited range, can be shown to be fractal with an exponent of ≈ 0.7 /25/.

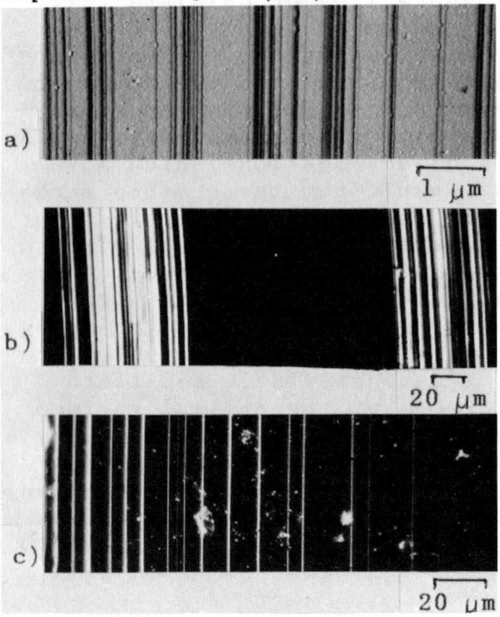

Fig. 7 Slip line structure after deformation of a single glide oriented Cu-30%Zn crystal (T = 300 K)
a) TEM replica of slip lines and slip bands (slip line cluster);
b) LM dark field micrograph of slip bands and slip band bundles (cluster of slip bands);
c) LM bright field micrograph of the Lüders-band front on a very thin plate-shaped crystal showing no slip band bundles.

The evolution of slip lines can be conveniently studied by microcinematography through a light microscope during deformation, which provides a direct check of the extent of plastic instabilities. Examples will be shown below. Further it gives insight into the processes which cause propagation of slip from line to line, from band to band, and from bundle to bundle. Double cross-slip of screw dislocations is a propable mechanism and it has often been observed, e.g. in neutron-irradiated Cu, as the important mechanism which governs the growth of slip band bundles /4/. We will show below (sect. 4.2) how this process during the PLC effect in alloys produces the plastic instabilities mentioned in the introduction. In the latter case, the termination of the instabilitiy may be due to the drop in load and time for aging of dislocations; further here and in other cases, the work hardening by the large increase of local dislocation density in the active crystal slab, with local activation of secondary slip as in sect. 2.2.b) may stop the slip band. It should be noted that the local strain may be orders of magnitude higher than the macroscopic average strain, and the active slip volume /26/ orders of magnitude smaller than the crystal volume, and that it is therefore not evident how to explain the macroscopically measured quantities of strain rate and strain hardening rate in microscopic terms.

3.2. Dislocation structures

In sect. 2.3 the most prominent dislocation structures found by TEM after deformation to various stresses have already been mentioned. As a rule, these structures can be made much more clean and distinct than in unidirectional tensile deformation, by cyclic deformation controlled to constant stress levels.

In single glide oriented crystals in the first stage of deformation dipolar dislocation arrangements dominate. These are especially unstable; on a small increase of stress parts of them break down and reform with smaller slip plane spacings of the dipoles. The corresponding catastrophic instabilities ("strain bursts") have been observed by Neumann /27/ in a fatigue test with increasing stress amplitude. The periodic arrangement of dipolar clusters found in fatigued specimen /28/ have been rationalized by simulations /29, 30/ and the transition between different structures could be explained in the reaction rate approach by Walgraef and Aifantis /31/.

With increasing lattice rotation in single slip, these dipolar clusters are stabilized by increasing secondary slip. The resulting dislocation structures (dislocation grids, multipolar braids) which are elongated in the slip plane, are subdivided and evolve towards an equiaxed cell structure at high deformations /32/.

In multiple slip (symmetrically oriented) crystals these cells develop in a nearly isotropic shape from the beginning, and with increasing strain become smaller and sharper dividing the volume into nearly dislocation-free cell interiors and high-density cell walls (Fig. 8). The latter are obviously mainly responsible for the increasing flow stress. The rapid crossing of the free cell interior by mobile dislocations is an example of plastic instabilities which cannot be recorded in the trace of load but is supposed to be the main reason for the well resolved acoustic emission /34, 35/ (cf. sect. 4.1.). Due to shorter path lengths of dislocations, plastic instabilities in symmetrically oriented crystals are less pronounced than in single glide. However, in case of ⟨111⟩-crystals, even slight (⟩

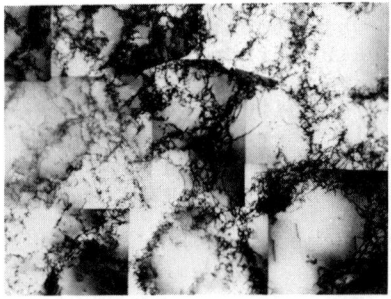

Fig. 8 Dislocation cell structure evolved in a ⟨111⟩ oriented Cu crystal after deformation to 28.5 MPa /33/.

1...2°) deviations from the symmetric axis cause local preference of a single glide system which then by a mechanism of cutting the "alien" dislocation structure (cf. sect. 2.3) may evolve in an unstable manner.

4. INSTABILITIES DURING PLASTIC DEFORMATION OF METALS

In the above sections we have seen the possibility of plastic instabilities of quite different origin in various deformation conditions. Below we will consider their observability and give a systematic classification along the lines proposed earlier by Estrin and Kubin /36/. Here several further mechanisms of instabilities will appear, such as structural transformations (e.g. mechanical twinning), and we will add a short account of localized unstable deformation of amorphous alloys, where the formation of shear bands has been observed.

4.1. Observability of plastic instabilities

As indicated in the introduction, we define plastic instabilities as those rapid plastic strain events which occur over a long enough time and with sufficient amount to be resolved by some macroscopic experimental method /37/. Such methods may consist in direct sensitive measurement of length changes during creep or stress relaxation in tension (e.g./38/), where a resolution of $\delta\varepsilon > 10^{-9}$ has been achieved, or more convenient, in monitoring the load drops occurring in the elastic system of the machine with specimen, when the latter suddenly elongates by some amount. Obviously, in such a setup the usually rather complex stress wave and vibration characteristics of the system have to be considered if quantitative interpretation of rapid load serrations is desired /39, 40/. Examples for records of load drops are shown in Fig. 9. In the first case (a) they are due to the rapid formation of single or several slip bands with high speed because of the sweeping of irradiation produced Frank loops by the dislocations which produce a soft cleared channel for local plastic flow (cf. sect. 4.2). In case (b) the alloy is

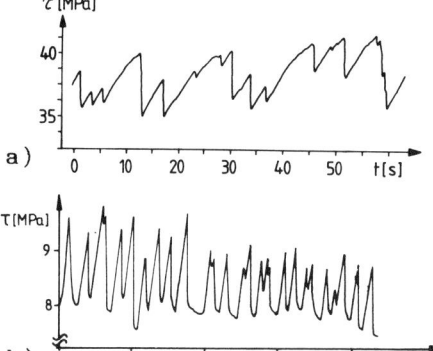

Fig. 9 Records of load during tensile deformation with plastic instabilities: a) thin flat n-irradiated Cu crystal, deformed at room temperature in the yield region; low inertia load cell /41/, b) 4 mm diameter Cu-8.7%Al crystal, T = 320 °C, yield region, conventional load cell /42/.

deformed in the PLC regime with dynamic strain aging of dislocations and breakaway from their solute atmosphere. Here a load drop could be assigned to the rapid successive formation of a cluster of slip bands /42/ (sect.4.2 and Fig. 14).

This assignment could be made by observing through the light microscope the formation of slip bands during deformation, which is a second, rather direct method to study plastic instabilities. Fig. 10 gives examples for the growth

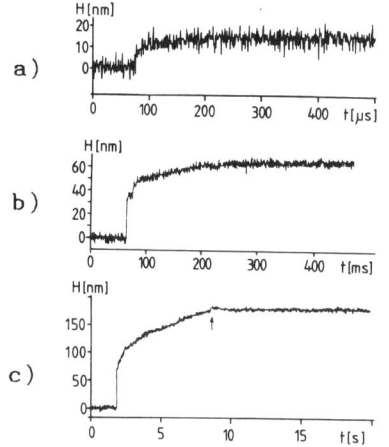

Fig. 10 Growth of slip step height (H) with time as recorded for two slip bands on the front surface of a thin flat Cu-30%Zn crystal by an optoelectronic technique /43/ using different resolutions in time.

of single slip bands. The evolution with time of the step height (corresponding to the number of dislocations $n = H/(b \sin \chi)$, χ = angle between slip plane and crystal axis) is measured from the scattered light intensity in dark field illumination; the value of the final height has been calibrated by interference microscopy. Fig. 10a gives a record with very high resolution in time, indicating that in this case the first 100 dislocations are emitted in about 10 µs; then the growth rate decreases continuously (sometimes interrupted by some break of inactivity) until the slip band is completed after a total time of several seconds. Most probably, this growth corresponds to the successive activation of neighbouring fine slip lines in the slip band (cf. Fig. 7), with the very first one being most rapid because its dislocation group is moving freely in the stress concentration present at the Lüders-band front, while the following lines feel opposing stresses from previous dislocations on neighbouring planes. The local shear rate in the slip band produces an elongation rate $\dot{l}_{pl} = dH/dt$ which in Fig. 10 exceeds the external deformation rate $\dot{l}_m = 1.15$ μm/s during the first 2 ms, i.e. during this time we have a plastic instability.

Another very sensitive method to reveal and to record plastic instabilities is called acoustic emission. It has been developed to high perfection /44/ and consists in the detection by piezoelectric transducers of the stress waves or vibrations produced by rapid local relaxations of load, for instance due to dislocation generation and movement. Recent investigations on the sources of acoustic emission have established the method as a quantitative tool for studying the glide processes /34, 35/ by recording the total acoustic emission (AE) power at various selected frequencies (or in a frequeny band) in relative units. Fig. 11 shows the

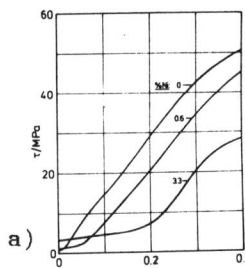 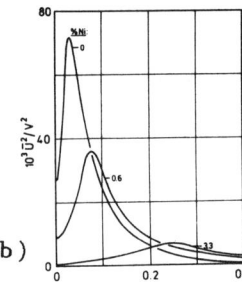

Fig. 11 Stress-strain curves (a) and acoustic emission power (b) in relative units for pure Cu, Cu-0.6 and -3.3at%Ni sin- single crystals. From /35/.

correspondence of stress-strain curves with acoustic emission curves taken for Cu-Ni crystals of various concentrations. The initial increase of the AE power with strain is supposed to be due to the rapid multiplication of dislocations, which in an activation event rush over relatively large areas of the slip plane (cf. sect. 2.3). With increasing strain hardening (transition to stage II) the plastic events become smaller and the AE power decreases. Increasing solute concentration also decreases the jumping distance of dislocations which decreases the AE power; the increase of the length of stage I with solute concentration is reflected in the shift of the maximum in the AE dependence on strain. The dependencies on temperature and on registration frequency have also been shown to be consistent with the picture that the AE power is generated by the rapid jumps of dislocation segments over small parts of the slip plane between two stationary positions at strong obstacles. While the burst-type acoustic emission can be quantitatively correlated to dislocation velocities /34/, it has not yet been possible to calculate the absolute value of the AE power as the integral over spatial and temporal distributions of those elementary events /35/. An application of this method to inhomogeneously deforming alloys /45/ will be discussed below (sect. 4.2.e).

Finally, we briefly mention the recording of current noise to monitor plastic instabilities, which was developed by Bertotti et al. /46/. Here the change of electrical resistivity by the rapid increase of dislocation density in the plastic burst is recorded. A special property of this method is its ability to differentiate between dislocation generation and annihilation through the sign of the resistivity change.

4.2. Survey of mechanisms for plastic instabilities in crystals

In an attempt to give an overview on the various mechanisms which may cause plastic instabilities in crystalline materials, we essentially follow the classification proposed by Estrin and Kubin /36/. They start out from a constitutive equation of the form

$$d\sigma = h \, d\varepsilon + S \, d(\log \dot{\varepsilon}) \qquad (2)$$

with h = strain hardening rate $(\partial\sigma/\partial\varepsilon)|_{\dot\varepsilon}$, S = strain rate sensitivity = $\partial\sigma/\partial(\log \dot{\varepsilon})|_\varepsilon$, and using the condition of constant volume during deformation ($d\sigma = \pm \sigma \, d\varepsilon$), find that according to a linear stability analysis the parameter

$$\lambda = [(\sigma-h)/S]\dot{\varepsilon} \qquad (3)$$

indicates stability ($\lambda < 0$) or instability ($\lambda > 0$). These conditions may be realized by various mechanisms.

a) M-type instabilities

M type instabilities are called those which arise from reasons of slip geometry: The resolved stress in a considered slip plane may increase at constant external load, if due to the shear of the active planes and simultaneous alignment of the specimen into the tensile (or compression) axis, it rotates into a more favourite position. This effect causes well-known "grip effects" (e.g. /47/) in single crystals, plays a role in the propagation of deformation fronts (sect. 4.2.e) and may also lead to texture softening in certain cases of polycrystals (for details cf. /36/). Instability occurs by this mechanism if, instead of the Considère criterion (h/σ < 1) /48/

$$h/\sigma + \partial(\log M)/\partial\varepsilon |_{\dot\varepsilon} < 1 \qquad (4)$$

is fulfilled with M = σ/τ (average Taylor factor), and it may be of practical importance in deformation processes with a change in strain path.

b) h-type instabilities

are those where h < σ with S > 0, i.e. when stress overcomes hardening.

(i) One famous example is necking in tensile deformation, where with increasing strain the hardening rate decreases while σ tends towards a saturation value. Often before it can be reached, necking sets in when the Considere criterion (h/σ < 1) is fulfilled. It may be noted that, if the stability analysis is carried through with variations of cross section δA rather than of strain $\delta\varepsilon$, then the Considère criterion is to be replaced by Hart's necking criterion /49/

$$h/\sigma + 1/m < 1 \qquad (5)$$

where m is the stress exponent of the strain rate ($\dot\varepsilon \propto \sigma^m$).

(ii) There are a number of dynamical effects which may cause h-type instabilities by structural softening in the locally slipped volume: For instance, moving dislocations may decrease the height of obstacles in the slip plane by cutting them (e.g., coherent precipitates /50/, irradiation-produced defect clusters /51/), or by sweeping them up (e.g. Frank loops /52/), or by destroying the previously built-up forest dislocation structure by reactions with dislocations of the new active system /53 - 55/. These processes clearly favour strong localization of slip: The following dislocations feel a reduced glide resistance in the previously activated slip plane, which favours the formation of dislocation groups infering collective effects by their mutual interaction. The high local stresses further enhance the strain localization in a synergistic way /56, 117 /. Examples for these prosesses are shown in Figs. 9 and 10 for slip band development in irradiated Cu and in Cu-30%Zn; in the latter the effective obstacles for thermally activated dislocation motion are supposed to be small shortrange ordered regions which are destroyed by the first few dislocations of the group /57/.

(iii) Even in pure crystals, the localization of slip in slip lines and slip bands /4/ indicates an h-type instability due to the rapid multiplication of dislocations (cf. sect. 2.2) in the activated crystal slab resulting in a local effective negative strain hardening rate /58/. This instability is restricted to the first stage of slip line formation because $\lambda > 0$, $d\lambda/dt < 0$, but $d\lambda/dt$ is not $\ll \lambda_o^2$ (the subsiduary condition /58/ for the growth of the instability, where $\lambda_o = \lambda(t=0)$), and it is damped within a small strain interval, contrary to the case of necking. Estrin and Kubin /58/ have shown that such instabilities can be described as the bifurcation of uniform to nonuniform slip, which is generally observed in the slip line patterns (cf. sect. 3.1), and that their development follows a course similar to that in Fig. 10. We may reverse the argument and evaluate from our measured curves of local strain rate (Fig. 12a) the local hardening rate, neclecting a possible stress

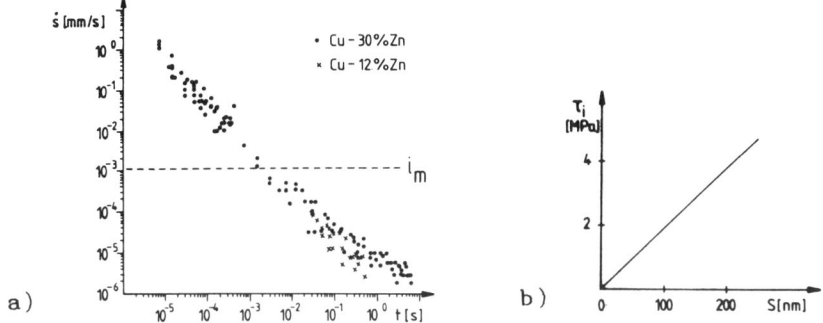

Fig. 12 a) Growth rate in height of slip bands $\dot S = d(H/\sin \chi)/dt$ versus time (a) from curves of development like Fig.10 for thin flat Cu-30%Zn crystals. The decrease approximates a logarithmic creep law corresponding to an increase of a internal opposing stress shown in (b).

stress concentration at the beginning and a drop of stress by rapid shear, i.e. assuming τ = const throughout the time for slip band formation. Fig. 12a suggests a variation of the local shear rate approximately according to

$$\dot S \approx S_o/t = \dot S_o \cdot \exp(-S/S_o) \tag{6}$$

with $S_o = 1.4 \cdot 10^{-5}$ mm which would arise if the dislocations move (or are produced, cf. sect.5) according to

$$v(\text{or } \dot N) = v_o(\text{or } \dot N_o) \cdot \exp[-(\Delta G_o - V(\tau - \tau_i))/kT] \tag{7}$$

with an opposing internal stress τ_i increasing according to

$$\tau_i = h^* S/B. \tag{8}$$

Then S_o is expressed as $S_o = (kT/V)(B/h^*)$; V is the activation volume, B is the width of the active slip band. Such a dependence is familiar from logarithmic creep curves ($S = S_o \ln(t/t_o)$). With $\tau_i = 0$ for $t = 0$, the increase of the opposing internal stress (local work hardening) is

$$\tau_i = (kT/V) \ln(\dot{S}(0)/\dot{S}(t)) = (kT/V) S(t)/S_o. \tag{9}$$

Taking the activation volume as determined from slip band growth in length /59/, we find τ_i as shown in Fig. 12b to reach about 1/3 of the flow stress for this example.

(iv) At very low temperatures pure metals and alloys show different types of instabilities, one will be discussed in sect. 4.2.d). Another one can be classified as h-type and is due to the fact that the dynamic resistance to dislocation motion exerted by various phonon mechanisms /60/ decreases to very low values. Then the dislocations may attain very high speeds and inertial effects may help with overcoming of obstacles /61, 62/. Schwarz and Mitchell /63/ observed prominent plastic instabilities during deformation of Cu-10%Al at 4.2 K which are supposed to be caused by inertial effects rather than by the local temperature rise (cf. sect. 4.2.d).

(v) At high temperatures dynamic recovery processes and dynamic recrystallization which occur in connection with dislocation activity may induce instabilities in the dislocation population. This case has been considered in detail recently by Boček /64/ providing another example for bifurcation in plastic flow, now in the range of high dislocation densities, where the nonlinear production and annihilation laws lead to oscillations in the dislocation density and in the macroscopic flow stress.

(vi) Finally, we mention static recovery processes, i.e. those occurring irrespective of dislocation motion, which by reduction of obstacle height (e.g. precipitation or grain coarsening, redistribution of phases) may result in plastic instabilities and localized flow.

c) S-type instabilities

are those where in the case of h > σ the instability parameter λ adopts positive values because of a negative strain rate sensitivity 0. Contrary to b) where λ crosses zero during the transition to instability, here λ crosses a singularity /36/.

(i) A very familiar mechanism with such a regime of negative slope in the relation between force and conjugate flux is the Portevin-LeChatelier effect, which is connected with the breakaway of dislocations from their solute atomsphere as originally proposed by Cottrell /13/. The picture has been repeatedly refined and it is now generally accepted that the formation of the "atmosphere" during the waiting times of the dislocations at obstacles is relevant /65, 66, 67, 68/.

Although it has been shown by Korbel /69/ that many of its characterstics can also be explained without diffusion just by collective effects of dislocation pile-ups and relaxation of their stress concentrations, the evidence for a contribution of dislocation pipe diffusion seems overwhelming and probably a theory combining both diffusion and collective effects would be most realistic /70/.

Usually the flow stress is considered to be composed of a dislocation interaction term σ_d and a friction term σ_f. The diffusion type theories result in a strongly nonlinear and nonmonotonous dependence of either σ_d /66/ or σ_f /65 - 68/ on strain rate as scetched in Fig. 13, where the domain

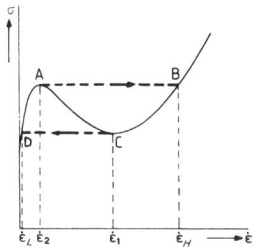

Fig. 13 Schematic dependence of the strain rate dependent part of the flow stress influenced by mobile solutes, as proposed by Penning /71/.

D-A corresponds to the case of the atmosphere being dragged along with the moving dislocation (i.e. aging during the waiting times can fully develop); A-B corresponds to breakaway (waiting times become too short for aging and B-C corresponds to the motion of solute-free dislocations escaped from their cloud (too short waiting times). C-D corresponds to the recapture of the dislocation by a cloud (i.e. its waiting times at obstacles become long enough for additional pinning). A-C is the range with negative strain rate dependence, i.e. if the external strain rate is selected in this range, serrations occur between the stresses $\sigma(A)$ and $\sigma(C)$ /71, 68/.

While in polycrystals often a critical strain is necessary for the onset of PLC serrations /72/ (which has been matter of much discussion, e.g. /65 - 69, 73/), in single crystals of, e.g. concentrated alloys of Cu-Al (> 4at%), the PLC effect is observed already in stage I at the onset of plastic flow in an appropriate temperature range depending on concentration. Simultaneous recording of stress and slip band development helps to reveal the microprocesses in some detail:

Starting at room temperature, the load curve is smooth and a Lüders-band (cf. sect. 3.1 and 4.2.e) propagates along the crystal, starting from one grip in tensile deformation. A small part of the crystal is active by initiating new slip band bundles at its front and adding new slip bands to several bundles in the Lüders-band, which are completed after some time while the front is propagating into the virgin crystal. The width of this active region decreases with increasing temperature, i.e. the number of simultaneously active slip band bundles decreases from several 100 to only a few in the temperature range where we already observe a fluctuating external load. These variations appear to reflect the fluctuations of activity of the few active bundles where only one or few successive slip bands are initiated. The true PLC serrations occur as soon as one single slip band bundle is active intermittently: As shown by the series of frames of cinematographic film (Fig. 14a) during the load drop (Fig. 14b) a rapid succession of new neighbouring slip bands is produced at the boundary of a slip band bundle. The evolution of the slip band number with time is shown in Fig. 14c, it closely reflects the course of load. As soon as the

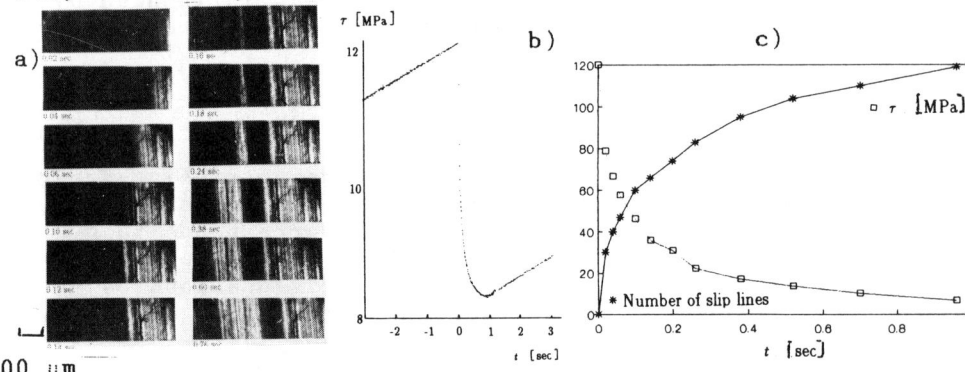

Fig. 14 PLC effect in Cu-10at%Al a) Series of cinematographic frames showing the development of slip bands at the boundary of bundle during the stress drop shown in b). This is drawn in extended scale in c) together with the increase of the number of slip bands evaluated from a). /74/

waiting time between successive slip bands exceeds a critical value, additional pinning by segregation of solute atoms sets in and abruptly stops the propagation of slip. A new avalanche can start only after an (elastic) increase of stress up to the level of breakaway for dislocations aged during that period. We recognize that, although the principal features of Fig. 13 are realized, the processes are more intricate in detail than in the simple picture dealing with single dislocations (cf. simulations in /75, 76/), and that the processes of slip transfer from the activated to fresh crystal region (probably cross-slip processes, cf. sect.

5.2) including local stress concentrations are involved in the course of the serrations and in the propagation of the active zone. The knowledge about possible source processes is, however, still poor at present. A few remarks on PLC bands and their propagation will follow below (sect. 4.2.e).

(ii) A second family of mechanisms which provoke S-type instabilities are stress-induced diffusionless structural lattice transformations such as mechanical twinning or martensitic transformations. In the first case macroscopic elongation is produced by formation of lamellae parallel to the twin plane with the same but mirror-reflected crystal structure; in the latter case the crystal structure is changed from fcc (austenite) to tetragonally distorted bcc (martensite) in spreading needles, which are often internally twinnned to reduce the strain energy /77/.

The stress to nucleate twins is usually very high, therefore the onset of twinning in metals is mostly observed in the highly workhardened condition and preferably at low temperatures. If the stacking fault energy is not too low, the growth of twins (and martensites) is more easy than nucleation; then extremely rapid load drops occur, often connected with audible "clicks" in the tensile machine, as a substantial elongation of the specimen is accompanied with this diffusionless and therefore very rapid lattice transformation, maybe only limited by the velocity of sound. These quasi-periodic instabilities occur with constant stress maxima, corresponding to the nucleation of a twin which extends without work hardening until during relaxation of stress locally the limit of propagation stress is reached. Examples for twinning in Cu-Al alloys can be found in /78 - 81/. We add in parenthesis that there are cases, i.e. low stacking-fault alloys, where propagation of the twins is more difficult than nucleation; then a smooth stress-strain curve results /79/ with a smaller work hardening rate than expected for pure dislocation glide.

In the twinning mode, an increase of strain rate is observed to cause a decrease of the maximum (nucleation) stress (Fig. 15), while the minimum stress of the drops be piled-up dislocation loops with extrinsic stacking faults. At higher strain rates with higher dislocation velocity the splitting of partials is expected to be larger than at lower strain rates /82/ which facilitates nucleation similar to a lower stacking fault energy.

It should be noted here again that - owing to the very remains the same. We have, therefore, a negative strain rate sensitivity, $S < 0$, which can be understood in terms of the critical size of the twin nuclei /78/. These are supposed to rapid athermal cooperative character of twinning - a correct

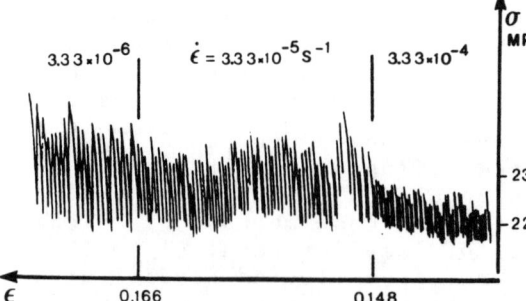

Fig. 15 Load serrations produced by twinning in a ⟨111⟩-oriented Cu-7.3%Al single crystal, as observed for different strain rates at room temperature. Note the negative strain rate sensitivity. From /79/.

registration of the size and shapes of load drops requires special attention to the dynamics of the mechanical testing system used. Such investigations have been performed by Demirski and Komnik /40/ also on CuAl crystals.

An example for plastic instabilities by martensitic transformations is given in Fig. 16 showing the pseudoelasticity in an Ag-Cd

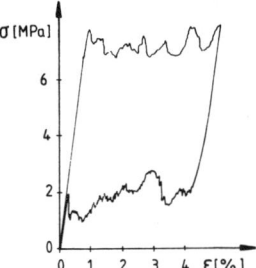

Fig. 16 Pseudoelasticity in Ag-Cd arising from martensitic transformation during tension (upper part) and reverse transformation (lower part of the curve) during unloading. From /83/.

alloy /83/. A pronounced hysteresis is observed indicating that a considerable chemical driving energy is necessary to overcome the resistive energy (mainly stored elastic energy), and its spatial variation and the interaction of needles cause the serrations in external load. Related effects are thermoelasticity and the shape memory effect /84/.

d) T type instabilities

The last type of instabilities which is to be dealt with is connected with the fact that > 85% of the work expended by the tensile machine during deformation is dissipated as heat in the specimen /85, 86/. The corresponding increase of temperature may facilitate thermal activation of slip processes which often follow an Arrhenius-type relationship, e.g. eq. (7). The temperature rise may be substantial if the specific heat is low and if the heat conduction is small, so that the locally produced heat will not dissipate too rapidly to the surroundings. In metals these conditions are usually (i.e. at ordinary strain rates) fulfilled only at very low temperatures, where plastic instabilities have often been observed (e.g. /87 - 91/). (Another type of low temperature plastic instabilities, connected with inertial effects due to vanishing phonon damping, has been considered in sect. 4.2.b).

The possible situations have been carefully analysed by Estrin and Kubin /92/ who give a classification for the various kinds of load serrations. The governing parameter is the heat exchange with the surroundings. Serrations arise in a certain range of strain rates (Fig. 17), below which the heat production rate by dislocation avalanches is too low to rise the temperature enough because of rapid heat exchange with the surroundings. Above the upper limit of the instability range, the periods between plastic jumps are too short for the released heat to relax by heat exchange with the surroundings, i.e. the specimen remains warm and flow gets smooth again.

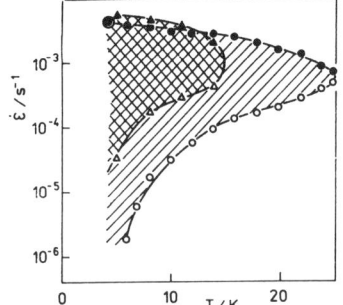

Fig.17 Range of jump-like deformation (shaded) for various temperatures and strain rates for single (o) and polycrystals (△) of Cu-14at%Al. From /91/.

Within the instability range, a negative strain rate sensitivity is observed /91/. The increase of thermal activation may also be interpreted as a "thermal softening" and incorporated formally into an extended Considere-type criterion /93, 92/

$$h/\sigma + (\partial(\log M)/\partial\varepsilon)|_{\dot\varepsilon} + (\partial(\log \sigma)/\partial T)|_{\dot\varepsilon} \cdot (dT/d\varepsilon) < 1 \quad (10)$$

where the last term for an Arrhenius-type relation between $\dot\varepsilon$, σ, and (eq. (7)) becomes $-(1/T)(\Delta G_o/V\sigma)(\dot T/\dot\varepsilon)$, and due to its negative sign plays a destabilizing role. Its magnitude depends on the competition between the rates of heat generation and heat conduction

$$c\dot T = K \cdot \Delta T + \sigma\dot\varepsilon(1 - \dot W_s/\dot W_p) \quad (11)$$

(c = specific heat, K = heat conduction coefficient, $\dot W_p$ = rate of plastic work, $\dot W_s$ = storage rate of defects), which together with the boundary condition at the specimen surface and the machine equation (eq. (1)) can be shown to yield a Hopf bifurcation /92/. In metals heat conduction is high enough to spread the heat evolved in each slip band rapidly so far that at ordinary strain rates the temperature of the sample can be considered as nearly uniform, so that the heat exchange with the surroundings is essential for the instabilities.

In high-speed deformation, on the other hand, the adiabatic approximation is approached resulting even at room temperature in the strongly localized formation of so-called adiabatic shear bands (e.g. /94, 95/) which also can be consistently explained by the thermomechanical coupling (e.g. /96, 97/).

e) Propagating modes of shear

As we could see above each of the mechanisms which results in plastic instabilities strongly favours localization of shear. With few exceptions (e.g. necking in crystalline, shear banding in amorphous materrial), this localization is limited by rapid hardening in the deforming part of the specimen, and slip has to continue elsewhere. In many cases this process occurs in a quite well organized way by continuing slip in the vicinity of the previously deformed slab, resulting in a propagation of a deforming region across the specimen. There are different types of these bands of deformation, depending on material, prestrain and deformation conditions. We indicate a few frequently observed cases. The first three are processes occurring in random distribution along the crystal length, the remaining ones are those with orderly propagation along the crystal.

(i) The formation of slip bands as clusters of fine slip lines in the first stages of deformation (cf. sect. 3.1 and 4.1.b) may be considered as a propagation of slip in a limited region of the crystal. It may be followed up to high stages of deformation by electrolytic polishing of the crystal surface before each successive straining interval /98/. Cross-slip of dislocations or/and dislocation source activation by local stress concentrations may be relevant micromechanisms.

(ii) In higher deformation stages by activation of secondary systems socalled kink bands and deformation bands occur in pure crystals, as investigated intensively by Takamura and coworkers /99/. Here also long-range internal stresses seem to be involved to activate those slip systems in order to compensate for strong crystal plane rotations which would occur in single slip.

(iii) In crystalline materials at very high strains, and in particular in situations of change in the deformation path, strongly correlated heavily localized slip occurs in socalled shear bands which sometimes take noncrystallographic orientations and may cross grain boundaries sometimes without change of their direction /100/, often with an extremely high speed due to the destruction of the "alien" dislocation structure (cf. sect. 4.2.h). They are built up of "microbands" which, propably again by cross-slip and stress activation, trigger slip in the near neighbourhood /101/. Shear banding in high-speed deformation by "adiabatic" heating has been mentioned already in sect. 4.2.b, cf. also /102/.

(iv) In irradiated metals or concentrated alloys the obstacle destruction mechanisms (sect. 4.2.b) causes coarse slip in the primary system with correspondingly higher local stresses which appear to be essential in formation of slip band bundles and the Lüders-band /103, 43/. Cross-slip processes and stress concentrations again seem to propagate slip from the deformed into the virgin crystal part. These are the "G-type" Lüders bands with a front parallel to the slip plane. Another type, "K-band" with their front normal to the primary slip plane, usually shows much stronger instabilities connected with the sudden collective activation of dislocations forming a kink band /104/. These Lüdersbands in single crystals have to be distinguished from those in polycrystals; in the latter the front is usually oriented along the direction of maximum shear stress, i.e. about $45°$ to the tensile axis, and the incompatibility stresses in the grains play an important role in the propagation process.

There is an extensive literature on Lüders-bands (e.g. /105/) and their role in stable and unstable modes of deformation has been considered recently by Schlipf /106/. While the Lüders-band usually occurs with h-type instabilities, often in the first stages of deformation, filling up the crystal with dislocations to a high density and being followed by quasi-homogeneous slip all along the specimen length in the next deformation stage, the following

(v) Portevin-LeChatelier-bands (PLC-bands) occur with S-type instabilities in particular in the processes of dislocation breakaway and recapture in alloys at higher strains, and by their propagation do not change the microstructure of the sample (i.e. this is the same in front and in the rear of the PLC-band /36/). For the case of loading with constant stress rate $\dot{\sigma}$, Kubin and Estrin /107/ were able to describe the propagation of PLC-bands and to explain the observations on Al-Mg-alloys /108/ which show different behaviour (e.g. repeated bands, several bands running simultaneously) depending on strain rate at a fixed temperature.

An essential point is that a propagation mode is not yet contained in the formal description indicated in sect. 4.2.c, but requires as a further ingredient either an appropriate length scale by assuming or supplying the experimentally determined propagation rate /107/ or gradients in stress or dislocation density /109/. These are still phenomenological formal quantities and there is a lack in understanding the details of the propagating process. First attempts to incorporate microprocesses in simulations of the formation and propagation of dislocation structures have been done by Kubin and Lepinoux /76/ who demonstrated by a cellular automata simulation the growth of slip bands in width, when a cross-slip process was allowed for. More experimental information may be provided by slip line observations (cf. the results indicated in sect. 4.2.c and in Fig. 14) or by acoustic emission (AE) studies as done recently by Càceres and coworkers /45/. They also find, from the strain rate dependence of the AE power during Lüders-band and repeated PLC-band propagation in Cu-30%Zn and Al-2.5%Mg alloys, that the stress concentrations at the band front are essential for propagation: At low strain rate, due to a further degree of relaxation at the band front, reactivation of sources produces the observed higher level AE than at higher strain rate, where relaxation of stresses and static aging at the band front are less pronounced (cf. /69/).

4.3. Localized unstable deformation of amorphous alloys

In the following short excursion we wish to show that plastic instabilities are not bound to crystals with their various possible dislocation mechanisms and structural transformations. In amorphous alloys also unstable propagation of localized shear has been observed as one possible deformation mode:

Homogeneous deformation of the amorphous structure occurs at low strain rates, at low stresses, and prevails at high temperatures (if embrittlement by structural relaxation or phase separation does not intervene). It is supposed to be produced by the local shear of small volumes containing in the order of 100 atoms which are less densely packed than the average ("free volume") and from their surroundings feel appropriate local stresses for easy shear /110, 111/. Each such local shear transformation produces some dilatation of its surroundings which rapidly relaxes by diffusive processes at high enough temperature and low enough strain rates. If, however, the imposed strain rate is too high, new shear processes will occur preferentially in the dilated region near prior shear, if the dilatation cannot relax fast enough. This causes an avalanche of shear in form of a shear band which follows the pattern of the external stress and is influenced by the boundary conditions in these extremely thin (≈ 40 μm) flat specimens. As no dislocations and correspondingly no work-hardening mechanism exists in these amor-

phous alloys, the resulting instability leads to rupture of
the specimen in a tensile experiment. This may be avoided by
performing a bending experiment, where the shear bands stop
near the neutral plane and many of them can be observed in
succession (Fig. 18): We find that a shear band is completely developed within a few 10 μs. In an analysis in terms of

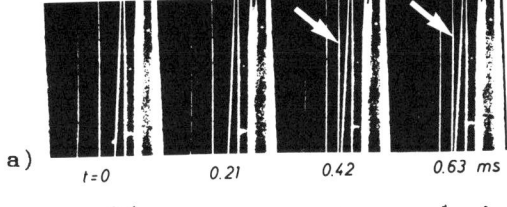

Fig. 18 Shear bands deve-
loping during bending of a
metallic glass ribbon:
a) series of neighbouring
frames of a high-speed film
taken at a frame rate of
4800 s^{-1};
b) optoelectronic record of
the growth of shear band
step height S with time.

of the coupled kinetic equations for the activation of local
shear transformations and their relaxation Argon /112/ and
Spaepen et al. /113/ have explained the transition from
homogeneous deformation to the unstable mode of shear band
formation in amorphous alloys.

5. CONNECTION OF MICRO- TO MACRO-KINETICS OF PLASTIC FLOW

5.1. Viscous flow versus generation rate approach

Orowan /114/, one of the discoverers of the dislocation, already gave the relations which connect the dislocation generation and motion with the kinetic of macroscopic plastic
flow. According to Fig. 1 the macroscopic strain rate is given by the rate of the slip plane area F swept by all mobile
dislocations (lotal length L_a) with Burgers vector b in the
deforming volume V_o in a general form /115/

$$\dot{a}(t) = b \dot{F}(t)/V_o = (b/V_o) \int_{L_a} v(t) \cdot n(t) \, dl \quad (12)$$

where each segment dl may be characterized by its normal n
and moves with velocity v. Orowan /116/ pointed out two possible approaches using average values which are commonly
used:

The first approach considers the movement of dislocations as a flow of mobile entities (density ϱ_m) moving with
average speed v, i.e. $\dot{F}$ is represented as $\dot{F} \approx L_a \cdot v = \varrho_m \cdot V_o \cdot v$, giving the famous Orowan relation

$$\dot{a} = b \, \varrho_m \cdot v. \quad (13)$$

It should be noted here that this "viscous glide approach" equals $\overline{\rho_m \cdot v} = \overline{\rho_m} \cdot \overline{v}$ which is quite questionable if one looks into the details of dislocation loop generation and expansion /117/. However, this approach serves quite well with ρ_m and v as operational quantities in cases of (quasi)continuous dislocation motion, when velocity and generation rate of dislocation are closely connected.

We also note that, as slip is confined to only few slip lines evolving simultaneously (cf. sect. 3.1), the average strain rate (13) may differ by many orders of magnitude from the true strain rate in the "active slip volume" V_a /26,15/

$$\dot{a}_{true} = \dot{a} \cdot (V_o/V_a). \tag{14}$$

This would not matter much, if V_a would be constant in various deformation conditions. Unfortunately, however, this is not the case /26, 15, 117/: V_a increases slightly with increasing strain rate and decreasing temperature, as measured by microcinematograpphy of slip line formation /4, 15/.

These difficulties may be avoided by using the second, the "generation rate approach", where the details of sweeping the slip plane are not considered. If a fixed average total area F_t can be assumed which is swept after each generation event of a disloction loop, $\dot{F}$ can be represented by the generation rate $\dot{N}$ of these events per unit volume as $\dot{F} = \dot{N} \cdot F_t \cdot V_o$ and the strain rate becomes

$$\dot{a} = b \dot{N} F_t. \tag{15}$$

Clearly, this approach is in particular appropriate in all situations of serrated deformation where $F_t b/V_o$ gives the total strain produced by each serration. It may also be applied in steady-state conditions /118/, where the generation rate of dislocation density $\dot{\rho}^+$ equals the immobilization rate $\dot{\rho}^-$ and if this can assumed to be constant over a time $\gg$ life time t_L = L/v, with the path length L of dislocation segments.

We recognize that the external strain rate is quite generally a measure for the generation rate of dislocations or dislocation avalanches, however, the description of the development of the avalanche itself requires additional information. If the strain rate is represented by the averages of mobile density and velocity of dislocations, these quantities and their relation has to be specified. H.Suzuki /119/ has favoured the generation rate eq. (15) since many years for explaining the yield stress of solid solutions, while Haasen /10/ and many others prefer the viscous glide eq. (13). We will take the following position according to our observations on micro- and macroscale.

5.2. Example: Deformation of Cu-Al in stage I

In the following, we take again the example of concentrated Cu-Al single crystals in single glide in stage I to discuss the application of these concepts to experiments of strain-rate change and stress relaxation which are performed to obtain information on the rate-determining processes.

(i) At low (but not too low) temperatures up to intermediate values, the smooth course of the load-elongation curve suggests a dynamical equilibrium between the generation of dislocations and their motion away from the sources through the lattice. This would correspond to the simple picture of a Frank-Read source where a new dislocation is emitted whenever the back-stress exerted by the previous ones has sufficiently decreased by expansion of the loop. Then the generation rate of dislocations by the source $\dot{n}_{FR} \propto v$ (cf. /120, 121/), and the two relations (13) and (15) coincide. Experiments by strain rate changes (relevant for generation rate) and stress relaxation (approximately relevant for velocity) should yield the same activation parameters, which is often observed. Differences found in some cases (e.g. /122, 15/) indicate differences in the kinetics of generation and of motion, which may be traced back to the details of source operation. They cause the changes of the active slip volume as well as flow stress transients after rapid changes of strain rate /122, 4/.

(ii) At intermediate to high temperatures the solute atoms become mobile and cause additional pinning of dislocations during their waiting times at obstacles. This may be expressed according to /67/ as an increase of the friction component τ_f of the applied stress τ ($= \tau_f + \tau_d$, τ_d = dislocation component):

$$\tau_f = \tau_{fo} + \Delta\tau_f(1 - \exp(-t_w/\theta)) \tag{16}$$

with the characteristic time $\theta = (1/\nu_s)\exp(U/kT)$ for the strongly temperature dependent segregation (pipe diffusion) of solutes at the dislocation during the waiting time t_w at an obstacle. Schwarz /67/ has applied this to internal friction /123/, where the mobile segments (density ρ_m, average length l, average path length L produce the strain rate (13)

$$\dot{a} = \rho_m bv = \rho_m bL/t_w = \rho_m bL\nu_d \exp(-(\Delta G_o - V\tau)/kT) \tag{17}$$

with the activation volume $V \approx b^2 l$. With (16), $\nu_s \approx \nu_d$, and $\tau = \tau_f + \tau_d$ (τ_d = const) the strain rate sensitivity becomes /67/

$$S = \partial\tau/\partial(\ln\dot{a}) = (kT/V) - \Delta\tau_f \cdot \alpha \cdot \exp(-\alpha) \tag{18}$$

with $\alpha = \rho_m bL/\theta\dot{a}$. S takes a minimum value at $\alpha = 1$, corresponding to $t_w = \theta$.

Applying a similar idea to the case of the PLC with big plastic avalanche after a breakaway of an aged dislocation source, the term ϱ_m in eq. (13) or in α becomes unclear and we rather use the generation rate approach (15) with the frequency of avalanches $\dot{N}$ and the total strain $\Delta a = bF_t/V_o$ in one serration. As the former is again influenced by the waiting time during which aging of a mobile dislocation takes place, it may be assumed to follow an analogous Arrhenius type equation as before:

$$\dot{N} = \dot{N}_o \cdot \exp(-(\Delta G_o - V\tau)/kT), \text{ with } \tau = \tau_f + \tau_d, \quad (19)$$

leading again to the strain rate sensitivity S of eq.(18) now with $\alpha = \Delta a/\theta \mathring{a}$. For large avalanches ($\Delta a$) the term $\Delta \tau_f \alpha \cdot \exp(-\alpha)$ may be so large that a negative strain rate sensitivity for the breakaway stresses results (cf. sect. 4.2.c), and this is the case in the PLC region. In Fig. 19a recent strain rate change measurements on Cu-15%Al /74/ are

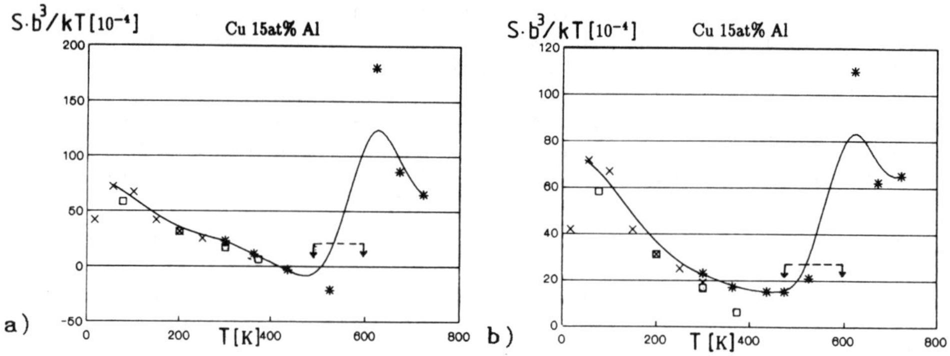

Fig. 19 Normalized strain rate sensitivity $S \cdot b^3/kT$ for Cu-15%Al single crystals measured in stage I for different temperatures, a) from strain rate changes, b) from stress relaxations (outside PLC range, indicated by arrow) or from stress drop Fig. 14b in the PLC range) /74/. Data (x) from /124/, (□) from /125/.

plotted together with data from literature /124, 125/ as the normalized strain rate sensitivity Sb^3/kT (which is approximately equal to b/l if the diffusive effect is negligible) in dependence of temperature. The gradual decrease of S with increasing T is supposed to reflect a spectrum in the obstacle distances (l) along the dislocations in the low temperature region. The steeper decrease to a minimum corresponds to the PLC region (arrows in Fig. 19). According to the observations of slip band generation in the PLC region in this case (Fig. 14), the very first dislocation breaking away at the onset of a serration has been aged during $\Delta t = \Delta a/\dot{a}$, i.e. essentially during the elastic loading time between serrations which is so long that $\Delta \tau_f \alpha \cdot \exp(-\alpha)$ is large enough to produce S < 0 (Fig.19a). On the other hand, the following slip bands growing <u>during</u> the serration are initiated by fresh or only slightly aged dislocations; here α remains small and S > 0. These processes may be measured by stress relaxation experiments (e.g. /126/), during the PLC region from the trace in load (S = $\partial \tau/\partial \ln(-\dot{\tau})$) or the increase of number of slip bands (S = $\partial \tau/\partial \ln \dot{N}$) during one serration as indicated in Fig. 14c. The resulting S values are shown in Fig. 19b (*) where some low temperature values from literature for strain rate changes are included as they do not differ from stress relaxation. We emphasize again that the relevant waiting times which determine S at higher temperatures are not simply the average waiting times of all dislocations at solute obstacles but, depending on the type of experiment, the waiting times either for breakaway of single more or less aged dislocations initiating the avalanche (strain rate changes) or for propagating the deformation front (stress relaxation), probably by cross-slip assisted by local stress concentrations.

Near the high temperature end of the PLC region, $\Delta \tau_f \alpha \cdot e^{-\alpha}$ approaches zero, S increases (Fig. 19) to values even higher than expected from (16) propably because of the very small obstacle distance l along the dislocations which are then practically saturated with solute atoms.

(iii) At still higher temperatures, when diffusion is rapid enough to fully age the dislocations even during the shortest waiting times, the serrations vanish. The motion of the dislocations is smooth /127/ and the viscous glide approach is clearly applicable: the solute atoms simply act as a smooth viscous friction force to the moving dislocations. In fact we do no longer observe slip bands but just fine slip /74/ and the strain rate sensitivity is similar from strain rate changes and stress relaxation. Its decrease with increasing temperature may be due to the increasing effect of diffusion which tends to destroy the solute atmosphere and to equlilize the solute distribution. Possibly, short range order effects are also involved.

In conclusion, we have shown that there are various different reasons and mechanisms which may produce unstable plastic flow in various levels of magnitude and which are usually connected with localization of shear, sometimes in pronounced hierarchical structures. Although most of the mechanisms are quite well understood qualitatively, open questions still remain in particular concerning the slip transfer process from active to fresh crystal parts which according to observations controls the overall deformation rate in many cases. Contrary to many simulations of dislocation motion across obstacle fields /3/ and many pulse loading experiments on dislocation velocities /3, 4/, these processes of slip transfer have only scarcely been investigated in detail and are still not well understood. They may involve cross-slip /128, 129/ and probably are sensitive to local stress concentrations which make the interpretation of externally measured stresses somewhat difficult.

Acknowledgements

Thanks are due to Prof. Dr. Ch. Schwink for continuous support and discussions, and to Mr. J. Plessing for supplying recent results. The financial support of our work by the Deutsche Forschungsgemeinschaft, now in SFB 319, is gratefully acknowledged.

REFERENCES

1 A.Seeger, W.Frank, in: Non Linear Phenomena in Materials Sience (eds. L.P.Kubin, G.Martin), Trans.Tech.Publ., Aedermannsdorf 1988, p. 125
2 R.E.Smallman, Modern Physical Metallurgy, Butterworths, London 1970
3 E.Nadgornyi, Dislocation Dynamics and Mechanical Properties of Crystals, Progr.Mat.Sci. $\underline{31}$ (1988) 1
4 H.Neuhäuser, in: Dislocations in Solids Vol.5 (ed. F.R.N. Nabarro), North Holland P.C., Amsterdam 1983, p. 321
5 R.M.J.Cotterill, M.Doyama, Phys.Rev. $\underline{145}$ (1966) 465
6 M.S.Duesbery, V.Vitek, D.K.Bowen, Proc.Roy.Soc. A $\underline{332}$ (1973) 85
7 D.J.Bacon, in: Strength of Metals and Alloys (ICSMA 1988) (eds. P.O.Kettunen, T.K.Lepistö, M.H.Lehtunen), Pergamon Press, Oxford 1988, p. 3
8 A.Seeger, in: Dislocations 1984 (eds. P.Veyssiere, L.Kubin, J.Castaing), Edition du CNRS, Paris 1984, p. 141
9 M.Yamaguchi, V.Paidar, D.P.Pope, V.Vitek, Phil.Mag. A $\underline{45}$ (1982) 867
10 P.Haasen, in: Physical Metallurgy (eds. R.W.Cahn, P.Haasen), North Holland P.C., Amsterdam 1983, p. 1341; in: Dislocations in Solids Vol.4 (ed. F.R.N.Nabarro), North Holland P.C., Amsterdam 1979, p. 156
11 H.Suzuki, in: Strength of Metals and Alloys (ICSMA 7) (eds. H.J.McQueen et al.), Pergamon Press, Toronto 1986, p. 1727

12 Ch.Schwink, in: Mechanisms and Mechanics of Plasticity
   (eds. J.Castaing, J.L.Strudel, A.Zaoui), Revue Phys.
   Appl. 23 (1988) 395
13 A.H.Cottrell, Phil.Mag. 44 (1953) 829
14 U.F.Kocks, A.S.Argon, M.F.Ashby, Thermodynamics and Kinetics of Slip, Progr.Mat.Sci. 19(1975) 1;
   G.Schoeck, in: Dislocations in Solids Vol.3 (ed. F.R.N.
   Nabarro), North Holland P.C., Amsterdam 1980, p. 63
15 H.Traub, H.Neuhäuser, Ch.Schwink, Acta Met. 25 (1975)
   437, 1289
16 Th.Wille, W.Giesecke, Ch. Schwink, Acta Met. 35 (1987)
   2679
17 Z.Basinski, S.Basinski, in: Dislocations in Solids Vol.4
   (ed. F.R.N.Nabarro), North-Holland P.C., 1979, p. 261
18 P.B.Hirsch, in: The Physics of Metals, Vol.2 Defects
   (ed. P.B.Hirsch), Cambridge Univ.Press 1975, p. 189
19 Ch.Schwink, F.Springer, to be publ.
20 O.B.Arkan, H.Neuhäuser, phys.stat.sol.(a) 99 (1987) 385;
   100 (1987) 441
21 H.Alexander, P.Haasen, Solid State Phys. 22 (1968) 28
22 M.Zehetbauer, V.Seumer, W.Witzel, in: Strength of Metals
   and Alloys (ICSMA 8) (eds. P.O.Kettunen et al.) Pergamon Press Oxford 1988, p. 451
23 H.Mecking, in: Dislocation Modelling of Physical Systems
   (eds. M.F.Ashby, R.Bullough, C.S.Hartley, J.P.Hirth)
   Pergamon Press, Oxford 1981, p. 197
24 M.Bocek, Physica Scripta T29 (1989) 213
25 H.Neuhäuser, in: Non Linear Phenomena in Materials Sci.,
   (eds. L.P.Kubin, G.Martin), Trans.Tech.Publ. Aedermannsdorf 1988, p. 349
26 Ch.Schwink, phys.stat.sol. 18 (1966) 557
27 P.Neumann, Z.Metallk. 58 (1967) 780; 59 (1968)927;
   Acta Met. 17 (1969) 1219
28 H.Mughrabi, in: Strength of Metals and Alloys (ICSMA 5)
   (eds. P.Haasen, V.Gerold, G.Kostorz), Pergamon Press,
   Oxford 1980, p. 1615
29 J.Kratochvil, Revue Phys. Appl. 23 (1988) 419
30 J.Lepinoux, L.P.Kubin, Scripta Met. 21 (1987) 833
31 D.Walgraef, E.C.Aifantis, J.Appl.Phys. 58 (1985) 688;
   Int.J.Eng.Sci. 23 (1985) 1351, 1359, 1365
32 T.Ungar, H.Mughrabi, D.Rönnpagel, M.Wilkens, Acta Met.
   32 (1984) 333
33 E.Göttler, Phil.Mag. 28 (1973) 1057;
   Ch.Schwink, E.Göttler, Acta Met. 24 (1976) 173
34 D.Rouby, P.Fleischmann, C.Duvergier, Phil.Mag. A 47
   (1983) 671, 689
35 W.Schaarwächter, H.Ebener (1989) to be publ.
36 Y.Estrin, L.P.Kubin, Res Mechanica 23 (1988) 197

37 H.Neuhäuser, in: Mechan. Prop. and Behav. of Solids: Plastic Instabilities (eds. V.Balakrishnan, C.E. Bottani), World Scientific, Singapore 1986, p. 209
38 R.F.Tinder, J.P.Trzil, Acta Met. 21 (1973) 975
39 P.P.Gillis, D.J.Shippy, Appl.Polym.Symp. 12 (1969) 165
40 V.V.Demirski, S.N.Komnik, Acta Met. 30 (1982) 2227
41 H.H.Potthoff, phys.stat.sol.(a) 77 (1983) 215
42 A.Hampel, M.Schülke, H.Neuhäuser, in: Strength of Metals and Alloys (ICSMA 8) (eds. P.O.Kettunen et al.), Pergamon Press Oxford 1988, p. 349
43 A.Hampel, H.Neuhäuser, phys.stat.sol.(a) 104 (1987) 171
44 T.P.Drouillard, F.J.Lauer, Acoustic Emission (Bibliography), Plenum Press, New York 1979
45 C.H.Caceres, A.H.Rodriguez, Acta Met. 35 (1987) 2851
46 C.Bertotti, M.Celasco, P.Fiorillo, P.Mazetti, Scripta Met. 12 (1978) 943; J.Appl.Phys. 50 (1979) 6948
47 R.K.MacCrone, J.Appl.Phys. 38 (1967) 705
48 A.Considere, Ann. ponts et chausses 9 (1885) Ser.6, 574
49 E.W.Hart, Acta Met. 15 (1967) 351; J.Appl.Phys. 38 (1967) 705
50 E.Hornbogen, H.Gleiter, phys.stat.sol. 12 (1965) 235, 251
51 U.Essmann, A.Seeger, phys.stat.sol. 4 (1964) 177
52 A.J.E.Foreman, J.V.Sharp, Phil.Mag. 19 (1969) 931
53 P.J.Jackson, Z.S.Basinski, phys.stat.sol. 9 (1965) 805; 10 (1965) 45
54 A.Luft, Ch.Ritschel, phys.stat.sol.(a) 72 (1982) 225
55 M.Szczerba, A.Korbel, Acta Met. 35 (1987) 1129
56 V.Z.Bengus, Cryst.Res.Techn. 19 (1984) 757
57 J.Olfe, H.Neuhäuser, phys.stat.sol.(a) 109 (1988) 149
58 Y.Estrin, L.P.Kubin, Acta Met. 34 (1986) 2455
59 H.Flor, H.Neuhäuser, Res Mech. 5 (1982) 101
60 V.A.Alshits, V.L.Indenbom, in: Dislocations in Solids Vol.7 (ed. F.R.N.Nabarro), North-Holland P.C., Amsterdam 1986, p. 43
61 A.V.Granato, Phys.Rev. B 4 (1971) 2196
62 V.D.Natsik, Sov.J.Low Temp.Phys. 5 (1979) 191
63 R.B.Schwarz, J.W.Mitchell, Phys.Rev. B 9 (1974) 3292
64 M.Bocek, Z.Metallk. 79 (1988) 132
65 S.H.van den Brink, A.van den Beukel, P.G.McCormick, phys.stat.sol.(a) 30 (1975) 469; P.G.McCormick, Acta Met. 36 (1988) 3061
66 A.van den Beukel, U.F.Kocks, Acta Met. 30 (1982) 1027
67 R.B.Schwarz, Scripta Met. 16 (1982) 385; R.B.Schwarz, L.L.Funk, Acta Met. 33 (1985) 295; R.B.Schwarz, in: Strength of Metals and Alloys (ICSMA 7) (eds. H.J.McQueen et al.) Pergamon 1986, p. 343
68 L.P.Kubin, Y.Estrin, Acta Met. 33 (1985) 397

69 A.Korbel, J.Zasadzinski, Z.Sieklucka, Acta Met. 24 (1976) 919; A.Korbel, H.Dybiec, Acta Met. 29 (1981) 89
70 G.Schoeck, Acta Met. 32 (1984) 1229
71 P.Penning, Acta Met. 20 (1971) 1169
72 M.Mayer, O.Vöhringer, E.Macherauch, phys.stat.sol.(a) 49 (1978) 473
73 L.P.Kubin, Y.Estrin, Acta Met. (1989) in press
74 J.Plessing, H.Neuhäuser, to be publ.
75 G.Ananthakrishna, D.Sahoo, J.Phys.D Appl.Phys. 14 (1981) 2081; G.Ananthakrishna, M.C.Vasakumar, J.Phys.D Appl. Phys. 15 (1982) 2171
76 J.Lepinoux, in: Non Linear Phenomena in Materials Science (ed. L.P.Kubin, G.Martin), Trans.Tech.Publ. 1988, p.389
77 J.W.Christian, Transformations in Metals and Alloys, Pergamon Press, Oxford 1965
78 J.Vergnol, J.Grilhe, J.de Phys. 45 (1984) 1479
79 F.Tranchant, J.Vergnol, J.Grilhe, in: Strength of Metals and Alloys (ICSMA 8) (eds. P.O.Kettunen et al.) Pergamon Press, Oxford 1988, p. 167
80 V.V.Demirski, S.N.Komnik, Phys.Met.Metall. 47 (1980) 162
81 A.Korbel, M.Szczerba, Acta Met. 30 (1982) 1961
82 S.M.Copley, B.H.Kear, Acta Met. 16 (1968) 227
83 L.Delaey, R.V.Krishnan, H.Tas, H.Warlimont, J.Mat.Sci. 9 (1974) 1521
84 J.Perkins, Shape Memory Effects in Alloys, Plenum Press, New York 1975
85 M.Bever, D.L.Holt, A.L.Titchener, Progr.Mat.Sci. 17 (1973) 1
86 D.Rönnpagel, G.Schulz, Thermochim. Acta 22 (1978) 289; D.Rönnpagel, Ch.Schwink, Acta Met. 26 (1978) 319
87 Z.S.Basinski, Proc.Roy.Soc.A 240 (1957) 229; Austral.J. Phys. 13 (1960) 354
88 L.P.Kubin, B.Jouffrey, Phil.Mag. 24 (1971) 437
89 E.Kuramoto, S.Takeuchi, T.Suzuki, J.Phys.Soc.Japan 34 (1973) 1217
90 R.Zürchner, V.Gröger, F.Stangler, phys.stat.sol.(a) 84 (1984) 475
91 S.N.Komnik, V.V.Demirski, V.Z.Startsev, Cryst.Res.Techn. 19 (1984) 863; Czech.J.Phys.B 35 (1985) 230
92 Y.Estrin, Scripta Met. 14 (1980) 1359; L.P.Kubin, Ph. Spiesser, Y.Estrin, Acta Met. 30 (1982) 385
93 A.S.Argon, in: The Inhomogeneity of Plastic Deformation (ed. R.E.Reed-Hill), ASM Metals Park 1973, p. 161
94 H.C.Rogers, Ann.Rev.Mat.Sci. 9 (1979) 283
95 D.Pierce, R.J.Asaro, A.Needleman, Acta met. 30 (1982) 1087
96 A.Molinari, J.theor.and appl.mech. 4 (1985) 659
97 A.Needleman, Revue Phys.Appl. 23 (1988) 585

98 S.Mader, Z.Phys. 149 (1957) 73
99 J.Takamura, Y.Ohtani, K.Higashida, N.Narita, in: Strength of Metals and Alloys (ICSMA 5) (eds. P.Haasen et al.) Pergamon, Oxford 1979, p. 17
100 A.Korbel, J.D.Embury, M.Hatherly, P.L.Martin, H.W.Erbsloh, Acta Met. 34 (1986) 1999
101 A.Korbel, P.Martin 34 (1986) 1905
102 J.W.Hutchinson et al. (viewpoint on shear bands), Scripta met. 18 (1984) 421
103 K.Shinohara, S.Kitajima, M.Kutsuwada, Acta Met. 34 (1986) 2335
104 R.Kuromoto, H.Suzuki, Trans.JIM 17 (1976) 683
105 D.W.Moon, Mat.Sci.Eng. 8 (1971) 235
106 J.Schlipf, Z.Metallk. 75 (1984) 517; Mat.Sci.Eng. 77 (1986) 19
107 L.P.Kubin, Y.Estrin, Acta Met. 33 (1985) 397
108 L.P.Kubin, K.Chihab, Y.Estrin, Acta Met. 36 (1988) 2707
109 H.M.Zbib, E.C.Aifantis, Res Mech. 23 (1988) 261, 279 293
110 A.S.Argon, J.Phys.Chem.Sol. 43 (1982) 945
111 D.Srolovitz, V.Vitek, T.Egami, Acta Met. 31 (1983) 335
112 A.S.Argon, Acta Met. 27 (1979) 47
113 P.S.Steif, F.Spaepen, J.W.Hutchinson, Acta Met. 30 (1982) 447
114 E.Orowan, Z.Phys. 89 (1934) 605, 614, 634
115 J.J.Gilman, Microdynamics of Flow in Solids, McGraw Hill, New York 1969
116 E.Orowan, Proc.Phys.Soc.Lond. 52 (1940) 8
117 H.Neuhäuser, in: Strength of Metals and Alloys (ICSMA 5) (eds. P.Haasen et al.) Pergamon 1980, p. 1531
118 H.Mecking, K.Lücke, Scripta Met. 4 (1970) 427
119 H.Suzuki, in: Strength of Metals and Alloys (ICSMA 5) (eds. P.Haasen et al.) Pergamon Oxford 1980, p. 1595
120 P.S.Steif, R.J.Clifton, Mat.Sci.Eng. 41 (1979) 251
121 V.D.Natsik, K.A.Chishko, Sov.Phys.Sol.St. 17 (1975) 214; 20 (1978) 1117
122 N.Himstedt, Acta Met. 26 (1978) 351
123 R.B.Schwarz, L.L.Funk, Acta Met. 31 (1983) 299
124 T.J.Koppenaal, M.E.Fine, Trans.AIME 224 (1962) 347
125 S.N.Komnik, V.V.Demirskii, Czech.J.Phys.B 31 (1981) 187
126 H.Flor, H.Neuhäuser, Acta Met. 28 (1980) 939
127 F.Monchoux, H.Neuhäuser, J.Mat.Sci. 22 (1987) 1443
128 B.Escaig, in: Dislocation Dynamics (eds. A.R.Rosenfield et al.), McGraw Hill, New York 1968, p. 655
129 P.J.Jackson, Progr.Mat.Sci. 29 (1985) 139

DISLOCATION PATTERNS AND PLASTIC INSTABILITIES

L.P. KUBIN[*], Y. ESTRIN[**] and G. CANOVA[***]
[*]CNRS-ONERA, OM
29, Av. de la Division Leclerc, BP 72
92322 Châtillon Cedex
France
[**] T.U. Hamburg-Harburg
Eissendorfer Strasse 42, Postfach 90 14 03
2100 Hamburg 90
F.R.G..
[***] L.P.M.M., Université de Metz
Ile du Saulcy
57045 Metz Cedex
France

ABSTRACT. Two important types of plastic instabilities occurring in crystalline materials are discussed within a phenomenologic frame. They involve anomalies in the behavior of the strain hardening rate (h) or of the strain rate sensitivity (S) of the flow stress. These two quantities are calculated in terms of the evolution of the densities of mobile and forest dislocations during plastic flow. This approach is exemplified by a study of the critical conditions for slip line patterning and for necking (type h) and by an investigation of the critical strains for the occurrence of jerky flow (type S). Experimental features associated with dislocation patterning are reviewed and analytical approaches through reaction-transport or reaction-diffusion are discussed. Because of the nonlocal character of internal stresses, these models are, for the moment, confined to the study of fatigue patterning. For this reason numerical simulations have been developed. The preliminary results presented here were obtained with a 3 D simulation of collective dislocation dynamics, based on a discretization of time and space.

1. Introduction

The plastic flow of crystalline materials is controlled by the motion of dislocations. Carriers of plastic deformation, the mobile dislocations have to cross over various types of obstacles, in particular those associated with the lattice periodicity (Peierls relief), solute atoms or precipitates, subgrain boundaries and other dislocations. When studying the dynamics of plastic deformation, one usually distinguishes between two types of interactions between dislocations and obstacles: localized interactions with obstacles of nearly the atomic size such as the lattice periodicity or impurity clusters, and long range interactions, e.g. with large precipitates or subgrain boundaries. Localized barriers can be overcome with the help of thermal activation and this establishes a relation between the effective local stress, σ^*, the strain rate and temperature. Localized obstacles are thus responsible for most of the strain rate and temperature dependence of the flow stress.

Extended obstacles or obstacles characterized by long range stress fields cannot be overcome by thermal activation and lead to a weakly temperature dependent contribution to the flow stress, often referred to as the internal or athermal stress, σ_i. The temperature dependence is associated with that of the shear modulus which enters σ_i as a scaling factor. It is usually assumed that the effective stress σ^* and the internal stress σ_i are additive, their sum being equal to the externally applied stress, σ:

$$\sigma = \sigma^* + \sigma_i \tag{1}$$

Mutual interactions between dislocations are quite complex. Dislocations interact at short range, e.g. during cutting processes, but they also interact over large distances through their long range ($\propto 1/r$) elastic fields. One has to distinguish between at least two dislocation populations: those which are mobile and the "forest" dislocations which are trapped at various fixed obstacles or entangled in particular configurations like dipoles or reaction segments. Because of the occurrence of various multiplication, annihilation, and trapping mechanisms, the densities of these two populations evolve during plastic flow while remaining coupled. The increase in forest density is responsible for the increase of internal stress during plastic flow and therefore for most of the strain dependence of the flow stress.

It follows from these considerations that the mechanical properties of crystalline materials should in principle be predictable in terms of individual and collective dislocation behaviour by expressing the right-hand side of eq. (1) in terms of dislocation densities and of the appropriate interaction mechanisms. Such a microscopic theory of plasticity does not exist, particularly because dislocation populations exhibit self-organization, a feature which is not clearly understood to date.

In macroscopic as well as in microscopic terms, plasticity is inherently a dissipative and irreversible process far from equilibrium. Self organization manifests itself at several scales of observation: the dislocations of the forest tend to form more or less well defined patterns as soon as their mutual interactions prevail over their interactions with other defects. Such patterns are commonly observed in transmission electron microscopy of thin foils prepared from deformed specimens. Mobile dislocations are also spatially organized, which manifests itself by the occurrence of slip markings which appear to be periodic at optical scale but seem to bear more complex forms of organization at smaller scale (see Neuhäuser, this volume). The problem of dislocation patterning is examined in Section 3. It is perhaps the most fundamental problem still untouched by dislocation theory. While individual dislocation properties are now reasonably well understood, at least with regard to their relation to macroscopic plasticity, the study of collective properties is still in its infancy.

At meso or macroscales one observes what metallurgists call plastic instabilities, i.e. anomalies on the deformation curves or more precisely situations where a material is able to flow with an increased strain rate under a decreasing applied stress. In such conditions, plastic flow cannot be homogeneous, and each "instability" is associated with a particular type of strain localization which may or may not lead to the rupture of the specimen or to the initiation of microcracks. These instabilities are obviously detrimental to material properties and this has been the main motivation for their study. To investigate them, one has to proceed from uniform dislocation densities, which leads to the phenomenological formulation discussed in Section 2.

The first detailed observations of plastic instabilities date back to the beginning of this century, see e.g. [1] for Lüders bands and [2, 3] for the Portevin-Le Châtelier effect. Studies in this field primarily concentrated on the determination of criteria for the onset of instabilities and on a qualitative description of unstable flow, leaving many features unexplained. A large amount of data on forest patterning has been accumulated since the early fifties, following from the expansion of T.E.M. studies. In the phenomenological models which have been proposed to account for it, the length scales are usually input parameters whose values are fixed in conformity with experiment. Sometimes criteria derived from nonequilibrium or even from equilibrium thermodynamics are also invoked to explain the spontaneous emergence of forest patterning.

Except for a few isolated studies [4], the investigation of dislocation patterning and of plastic instabilities in terms of the bifurcation from uniform to nonuniform solutions dates back to the beginning of the eighties.This approach has yielded realistic criteria for the onset of instabilities and some insight, in particular through numerical simulations, into the post-bifurcation behavior and into the definition of length scales in terms of the physical properties of the material considered. A summary of recent advances and of the experimental situation can be found in the references [5-7].

2. Plastic Instabilities

2.1. THE STRAIN DEPENDENCE OF THE DISLOCATION DENSITIES

In this phenomenological approach, dislocations are assumed to be uniformly distributed in space. Their densities evolve with time as a result of various creation, annihilation or trapping events and this can be described by the so-called kinematic balance equations [8]. The simplest kinematic equations involve only one density, for instance the classical description of stage II hardening assumes that the mobile density does not evolve [9]. As will be shown below, many interesting features cannot be accounted for by such models. On the other hand, to take into account the full complexity of plastic flow it would be necessary to distinguish between dislocations of both signs, of various characters (e.g. at least edge and screw), and to express the number and nature of active slip systems which can interact. A reasonable compromise consists in lumping the dislocations of all characters, signs and slip systems into two species: the mobile dislocations (with density ρ_m) and the forest dislocations (with density ρ_f) which act as obstacles to mobile dislocations.

As an example of a basic annihilation process, we consider the mutual annihilation of two mobile dislocations of opposite sign. The rate of annihilation, $d\rho_m/dt$, is obtained by considering two populations of density $\rho_m/2$ moving with a relative velocity $2v$ with respect to each other. Annihilation takes place when two dislocations of opposite sign move past each other on slip planes whose distance is smaller than a critical value y. y is the critical distance for annihilation, a quantity which is usually assumed to vary as the inverse of the applied stress. During a time interval dt, the volume within which annihilation takes place is $dV = 2\rho_m$ v y dt. The total density annihilated is ρ_m dV, so that the loss term can be written as $(d\rho_m/dt)^- = -2yv\rho_m^2$ and has, as expected, a quadratic form. We transform this expression with the help of Orowan's law, which expresses that the plastic strain rate, $\varepsilon = d\varepsilon/dt$, is proportional to the dislocation flux: $\dot\varepsilon = \rho_m bv$, b being the magnitude of the Burgers vector. As a result, we obtain: $(d\rho_m/d\varepsilon)^- = -(2y/b)\rho_m$. Other short range interactions can be treated in the same manner, leading to the following system of two coupled differential equations for the evolution with strain of the densities of mobile and forest dislocations [10, 11]:

$$d\rho_m/d\varepsilon = C_1/b^2 - C_2\rho_m - (C_3/b)\rho_f^{1/2}, \qquad (2)$$

$$d\rho_f/d\varepsilon = C_2\rho_m + (C_3/b)\rho_f^{1/2} - C_4\rho_f \qquad (3)$$

The first term at the right-hand side of eq. (2) is related to the multiplication of mobile dislocations at fixed sources, the second term describes the mutual annihilation and trapping of two dislocations while the third one accounts for the trapping of mobile dislocations by randomly distributed or regularly arranged forest dislocations. In both cases, the mean free path of the mobile dislocations between obstacles is proportional to $\rho_f^{1/2}$. These last two terms in eq. (2) appear as generation terms in the balance equation for the forest density, eq. (3), while the negative loss term in this equation accounts for recovery mechanisms.

Several distinct material parameters are lumped into the material constants C_i. For instance C_2 contains both the critical annihilation distances for edge and for screw dislocations, which may have quite different values (see Section 3.). For this reason, these constants were calculated numerically using heuristic arguments based on experiment, rather than on theoretical estimates. In addition to the initial values of the dislocation densities, four conditions are required to determine a reference set of the material constants.

An average ideal material is then defined, which has the following properties. The initial densities are fixed to the values $\rho_{mi} = 10^{10}$ m^{-2} and $\rho_{fi} = 10^{11}$ m^{-2}. By setting $d\rho_m/d\varepsilon = 0$ and $d\rho_f/d\varepsilon = 0$ in eqs. (2) and (3), one can calculate the values of the densities at saturation which are taken to be typically $\rho_{ms} = 5.10^{13}$ m^{-2} and $\rho_{fs} = 10^{14}$ m^{-2}. The density of the forest saturates with a characteristic strain ε_f of the order of 0.3 and, finally, the relative weight of the two mechanisms, trapping and annihilation, which contribute to the saturation of the mobile density is set to a reasonable value, 0.1, meaning that trapping by forest obstacles is more efficient than mutual annihilation or trapping. Fig. 1 shows the strain dependence of the mobile and forest densities as obtained by numerical integration of the systems of equations (2) and (3). Beyond $\varepsilon = 1$, the density of mobile dislocations slightly decreases before reaching its saturation value.

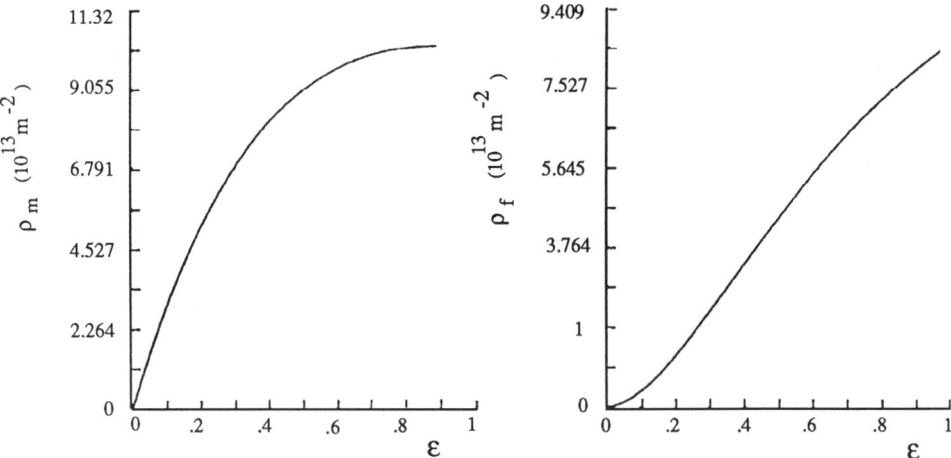

Figure 1. The strain dependence of the mobile (ρ_m) and forest (ρ_f) dislocation densities. The parameter values used in this calculation are defined in the text.

This system of two coupled kinematic balance equations has many possible applications and can be developed to include other obstacles present in the microstructure which may interact with dislocations. It may also be used to simulate deformation curves. In what follows, we include it into a constitutive description of plastic deformation and examine two typical cases of instabilities.

2.2. CONSTITUTIVE EQUATIONS AND THE STABILITY OF PLASTIC FLOW

The average internal stress is of the form $\sigma_i = \alpha G b \rho_f^{1/2}$, where G is the shear modulus and α is a dimensionless coefficient whose dependence on the arrangement of the forest appears to be quite weak [12]. Both experiment and phenomenological theories seem to yield values of α of the order of 0.3 although no consistent theoretical derivation of this value is available to date. The effective stress σ^* can now be expressed in terms of the applied stress and of the forest density through eq.(1).

The thermally activated nature of dislocation mobility is accounted for by expressing the velocity of mobile dislocations in Orowan's law by an Arrhenius-type relation:

$$\dot\varepsilon = A\Omega(\varepsilon)\nu \exp[(\sigma - \alpha G b \rho_f(\varepsilon)^{1/2})/S] \tag{4}$$

Here A is a constant, $\Omega(\varepsilon) = b\rho_m \rho_f^{-1/2}$ is the elementary incremental strain obtained when all dislocations accomplish a successful activation event (this quantity is discussed in more detail in Section 2.4.1.) and ν is a constant attempt frequency. The effective stress has been expressed according to eq. (1) and $S = kT/V^*$ is the strain rate sensitivity (k = Boltzmann constant; T = absolute temperature; V^* = the activation volume representing the spatial extension of the obstacle to be overcome with the help of thermal activation as well as the shape of the interaction force-distance profile). Equation (4) provides a constitutive relation between stress, strain, strain rate and temperature and it is of quite general use. Together with the evolution equations (2) and (3) it enables a description of plastic deformation in a closed form. To investigate stability of plastic flow, we rewrite eq.(4) in differential form, assuming isothermal conditions

$$d\sigma = h d\varepsilon + S d\ln\dot\varepsilon, \tag{5}$$

$$h = \alpha G b [\partial \rho_f(\varepsilon)^{1/2}/\partial \varepsilon]_{\dot\varepsilon} - S[\partial \ln\Omega/\partial \varepsilon]_{\dot\varepsilon} \tag{6}$$

Eq.(5) is a well-known incremental constitutive form which expresses that the flow stress depends on strain as well as on strain rate increments through two coefficients which may be considered as partial derivatives: $h = (\partial\sigma/\partial\varepsilon)_{\dot\varepsilon}$ and $S = (\partial\sigma/\partial\ln\dot\varepsilon)_\varepsilon$. In the present elaboration the first term of the strain hardening rate, h, depends only on strain, which is an approximation. The strain rate sensitivity depends on temperature, but it may also depend on other quantities, e.g. on the strain, via the activation volume. The constitutive forms (4) or (5) and (6) are entirely defined through the knowledge of the strain dependence of the dislocation densities and of the nature of the thermally surmountable obstacles.

To examine stability of the uniform uniaxial deformation, we introduce a small local fluctuation in strain $\delta\varepsilon$. Assuming that the constitutive equation (5) is fulfilled throughout the material, i.e. in the localized region under consideration, we can replace the differentials $d\sigma$ and $d\varepsilon$ in eq. (5) by the local deviations $\delta\sigma$ and $\delta\varepsilon$. This suggests that the localized region is simply ahead of the uniformly deforming bulk in its development. The fluctuation is expressed in the usual form $\delta\varepsilon = \delta\varepsilon_0 \exp\lambda t$, where $\delta\varepsilon_0$ is a constant and λ is the growth parameter. Assuming incompressibility of the specimen and the constancy of the applied load along the specimen axis, we obtain the condition [13, 14] $\delta\sigma = \pm \sigma \delta\varepsilon$ where the upper sign refers to tension and the lower one to compression. The growth parameter can, then, be extracted from eq. (5):

$$\lambda = \dot{\varepsilon}(-h \pm \sigma)/S \qquad (7)$$

In compression, plastic flow becomes unstable if either (h+σ) or S become negative. At small strains, σ is usually small compared to h and in such conditions, the instability criterion is h < 0, S > 0 or h > 0, S < 0. In tension, the same inequalities express the instability conditions. At large strains, there is an additional instability when (h-σ) < 0, with h and S > 0. This last instability is associated with necking (Considère instability) of the specimen. Necking occurs because the reduction in cross-sectional area in tension cannot be compensated for by the increase in flow resistance stemming from strain hardening.

For instabilities of type h, λ changes sign by passing through zero and various types of bifurcations may occur depending on the microstructural mechanism involved. These mechanisms are listed in [15] (see also Neuhäuser, this volume), but the conditions for their occurrence as well as the spatial patterns which further develop have not, up to now, been theoretically investigated in much detail. One particular case will be treated in the next paragraph. It is concerned with the negative strain hardening rate being produced by the fast creation of mobile dislocations.

Instabilities of type S are quite particular since in that case λ changes sign by going through infinite values. Because of continuity requirements, stresses and strains must remain continuous, so that these instabilities are necessarily associated with strain rate jumps. In Section 2.4. we examine the Portevin-Le Châtelier effect, the most well understood example of instabilities of type S to date.

2.3. INSTABILITIES OF TYPE h AND SLIP LINE FORMATION

Inserting into eq.(6) the strain dependence of the dislocation densities calculated above, we obtain the strain dependence of the strain hardening rate which is reproduced on Fig.2. On this figure two bifurcations appear: one at large tensile strains (the Considère instability) and one at small strains. The latter stems from the fact that at the beginning of plastic flow the contribution of mobile dislocations (through the elementary strain Ω) to the strain hardening rate predominates over that of the forest dislocations. The result is softening, as can be seen from eq. (6). Virtually all crystalline materials are prone to this instability, as has been proved numerically. Therefore, as a rule, plastic flow starts in a nonuniform manner which explains the universal observation of slip lines and slip bands at the surface of deformed crystals. The first value of strain ε_1 at which the growth coefficient λ passes through zero is

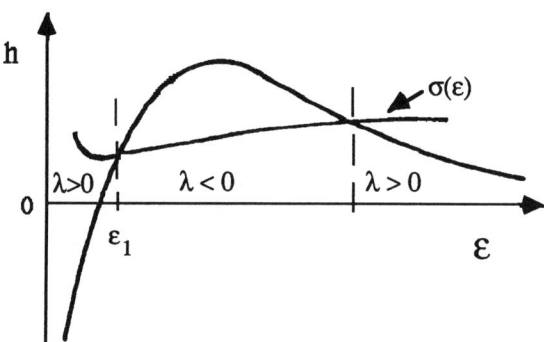

Figure 2. The strain dependence of the strain hardening rate, h. Uniform plastic flow is unstable at small strains in tension and in compression, and at large strains in tension.

representative of the strain produced during the fast initial stage of the development of a slip line. Its value is typically of the order of a fraction of 10^{-2} but in some particular cases, it can be quite large, up to the order of 0.1. Beyond ε_1, slip tends to spread out along the gauge length of the specimen, either in an orderly or in a disordered manner and fresh slip lines or bands are initiated while the strain rate decreases in the old ones. The plastic strain, then, tends to become more uniform.

The spatial aspects of slip line formation have not been yet investigated. Such an investigation would require modelling the rate of growth of slip lamellae as well as of their mutual interactions. In particular there exists no criterion defining the conditions under which slip line initiation takes place in an ordered manner or randomly along the specimen gauge length. In addition there is a well-known type of instability during which slip lines or bands are produced in succession, being initiated at one end of the specimen and propagating to the other end with a well-defined velocity. This type of behavior (Lüders-like behavior) is certainly a topic of interest for the near future.

2.4. INSTABILITIES OF TYPE S: THE PORTEVIN-LE CHATELIER EFFECT

2.4.1. *Dynamic Strain Ageing and the Strain Rate Sensitivity.*
An improved version of earlier elaborations is presented here, the focus being on the critical strains at which jerky flow starts and ceases.

Dynamic strain ageing and its macroscopic manifestation, the Portevin-Le Châtelier effect arise from the dynamic interaction between mobile dislocations and solute atoms which are also mobile via diffusion mechanisms. Fig. 3 schematically depicts the situation. A mobile dislocation moves in its slip plane and is temporarily arrested in front of a forest obstacle. The average waiting time t_w is the time needed for the dislocations to overcome the obstacle with the aid of thermal activation. During this waiting time, solute atoms diffuse towards the arrested dislocation, the driving force stemming mainly from the attractive elastic interaction between the two types of defects. This mechanism is referred to as dynamic strain ageing (DSA).

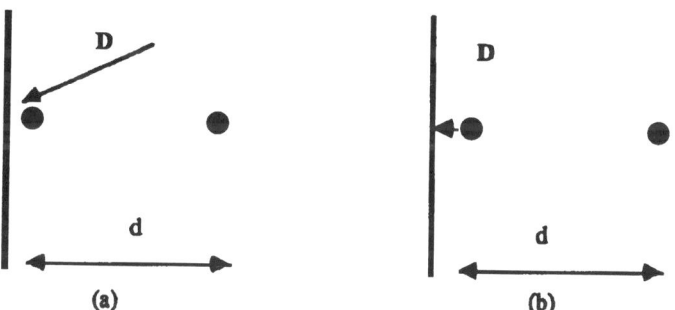

Figure 3. Two possible mechanisms for dynamic strain ageing. (a) Bulk diffusion of solute atoms towards the arrested mobile dislocation (straight line). (b) Solute transport to the arrested mobile dislocation via pipe diffusion through forest dislocations. (Full circles indicate the interceipts of the forest dislocations with the glide plane of the mobile dislocations.)

Assuming that the free-flight time of mobile dislocations between forest obstacles is negligible compared to the waiting time, the dislocation velocity is proportional to (d/t_w), where $d \approx \rho_f^{-1/2}$ is the obstacle spacing. Substitution into Orowan's relation yields

$$\dot{\varepsilon} = (b \, \rho_m \, \rho_f^{-1/2})/t_w = \Omega(\varepsilon)/t_w \qquad (8)$$

(a constant geometric factor has been omitted for simplicity). This defines Ω, the elementary incremental strain obtained when all mobile dislocations accomplish a successfull activation event. Since the dislocation densities evolve with strain, so does Ω. In deformation tests the strain rate can be externally controlled and usually a constant strain rate is imposed. It follows that the waiting time also evolves with strain and this is the basis for all strain dependent effects.

In a normal thermally activated process, the strain rate sensitivity, S, is always positive, meaning that the obstacle can be overcome at a smaller stress if more time is left for thermal activation. In the presence of DSA, the "forest" obstacles become harder with increasing time because of the increased hardening contribution from solute atoms forming a cloud round the mobile dislocation line. As a consequence the strain rate sensitivity, S, decreases in the presence of diffusing solute atoms and may even become negative. The solute contribution to stress, f, is proportional to their concentration and is conveniently represented by

$$f = f_0\{1 - \exp[-(t_w/\tau)^{2/3}]\} \qquad (9)$$

The exponential form ensures that the concentration saturates for long waiting times, while for short waiting times it increases as a power (2/3) of time as expected from an interaction through the size effect [16]. τ is the characteristic time associated with diffusion, which contains the solute diffusivity. Depending on the type of alloy considered and on the temperature, the activation energy for diffusion can represent bulk diffusion, which may be assisted by vacancies in the case of substitutional solutes, or pipe diffusion if solute atoms diffuse into the obstacle region through forest dislocations [17] (cf. Fig. 3). f_0 is the maximum hardening reached at saturation, which is proportional to the solute concentration.

Dislocation mobility is then determined by a competition between two adverse mechanisms [17]. The normal process of activation tends to reduce the flow stress as waiting times increase and it is associated with a strain rate sensitivity, S_0, which we will consider as a constant during a deformation test. Ageing by solute atoms on the contrary makes the obstacle harder with increasing waiting time and, by definition, its contribution to the SRS is

$$S_{ageing} = (\partial f/\partial \ln\dot{\varepsilon})_\varepsilon = - df/d\ln t_w \qquad (10)$$

Theory and experiment indicate [18] that the resulting total strain rate sensitivity, S is simply the sum of these two contributions:

$$S = S_0 - df/d\ln[\Omega(\varepsilon)/\dot{\varepsilon}] \qquad (11)$$

where use has been made of eq. (8). During DSA, the strain rate sensitivity is therefore a function of strain, strain rate, temperature and solute concentration. It may become negative if the contribution from ageing is large enough. Then, uniform plastic flow ceases to be stable and jerky flow occurs.

2.4.2. *The Portevin-Le Châtelier Effect.* Inserting eqs. (8) and (9) into the expression for the SRS, eq.(11), and substituting the result into the constitutive equations (4) or (5), one obtains a complete description of the macroscopic mechanical properties associated with DSA and the PLC effect. Here, we only list the main experimental features (see Neuhäuser, this volume) which can then be explained. More detail can be found in [14, 19-21].

Given a value of strain and provided that the concentration of solute atoms is larger than a critical value (of the order of 1 at.% for substitutional alloys), the criterion $S = 0$ for the onset of the PLC effect is fulfilled within a well-defined range of strain rates and temperatures.

- Inside the instability domain, the stress as a function of the dislocation velocity is bivalued and the specimen exhibits a type of bistable behavior. For the same value of the local stress, dislocations can either move slowly, dragging along a significant concentration of solute atoms, or they can move at high speed, being almost free from solute clouds. The evolution of the local strain rate with time is governed by a differential equation which is a first-order degenerate form of the Liénard equation. Jerky flow can be, then, interpreted as a type of relaxation oscillations. The deformation curve consists of one long sub-period during which deformation is uniform, dislocations being pinned by solute atoms and of a short sub-period, the jerk, during which a band of fastly moving dislocations is initiated and traverses the specimen with a well-defined velocity.

- Some characteristic features of the jerks can also be explained. For instance, at a fixed temperature, the effect of the strain rate on the size of serrations is as follows. Serrations appear in a limited range of strain rates, their amplitude being a maximum at the lower boundary of this strain rate range. It decreases with increasing strain rate and at the upper boundary, the magnitude of the serrations vanishes and they disappear altogether. The transition to uniform deformation is then smooth at the upper boundary of the PLC range. This behaviour is explained within the model [21].The periodicity of the PLC bands as well as the average strain and strain rate inside the bands can also be calculated in terms of the characteristic properties of the defects involved and of material properties.

- Finally, the width and velocity of PLC bands are related by a simple expression but, to

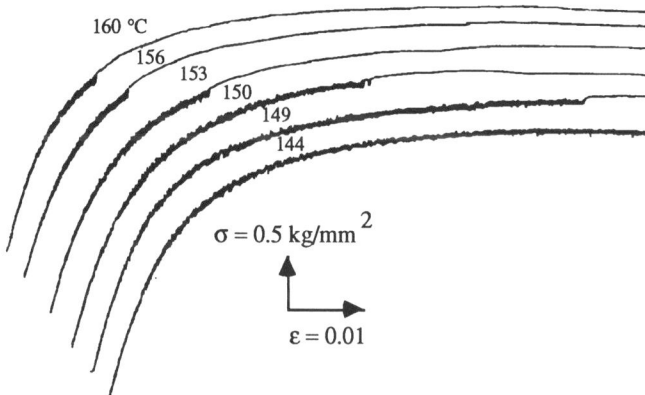

Figure 4. Serrated stress-strain curves for an Al- 5% Mg alloy. The strain rate is 3.33×10^{-4} s^{-1} (After Guillot and Grilhé [22]).

date, they have not been consistently determined. Resolving this indeterminacy is the major problem for further studies on the PLC effect.

2.5. THE CRITICAL STRAINS

2.5.1 *Experimental Evidence.* Very often, plastic flow starts in a uniform manner and jerky flow appears beyond a critical strain value of the order of one to about ten percent. PLC serrations then occur throughout the deformation curve or sometimes disappear at large strains, as shown in Fig. 4 which is reproduced from the work of Guillot and Grilhé [22]. There are, however, experimental observations of more than two critical strains along the same deformation curve. Fig. 5 reproduces the stress-strain curves of a Cu-3.3 at.% Sn alloy strained at 255 °C with various imposed strain rates and with an imposed strain rate of 6.5×10^{-5} s^{-1} at various temperatures, after Räuchle, Vöhringer and Macherauch [23]. Such observations of the simultaneous occurrence of several critical strains on the same deformation curve are quite rare, probably because this requires investigations within a very narrow range of temperatures or strain rates, as can be seen from Figs. 4 and 5.

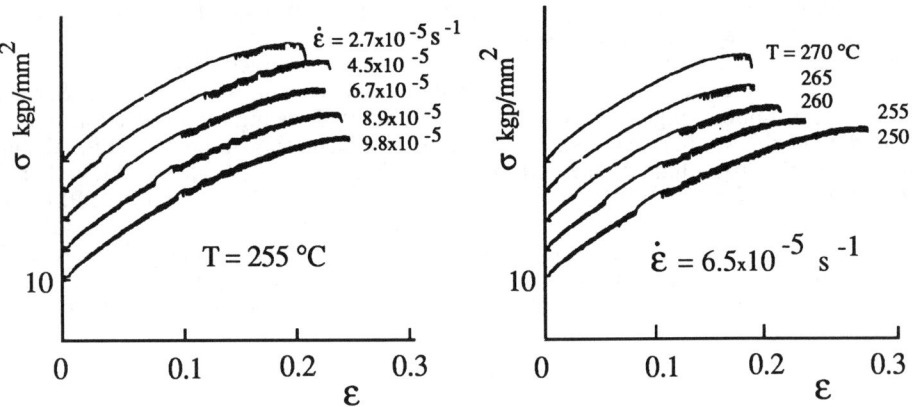

Figure 5. The stress-strain curves of a Cu-3.3 at.% Sn alloy at a fixed temperature for various imposed strain rates and at a fixed strain rate for various temperatures. After Räuchle, Vöhringer and Macherauch [23].

2.5.2. *The Model.* Expressing the critical condition for the occurrence of the PLC effect, S = 0, in eq. (11) and inserting f from eq.(9), we obtain the relation

$$X \exp(-X) = \alpha , \qquad (12)$$

where $\alpha = (3/2)S_0/f_0$ and $X = t_w/\tau = (\Omega/Z)^{2/3}$, with $Z = \dot{\varepsilon}\tau$. Therefore, if we consider a deformation test performed at a given temperature with a prescribed strain rate, all the strain dependence is incorporated in the dimensionless quantity X through the elementary incremental strain Ω. $X\exp(-X)$ is a bell-shaped function which has a maximum value of 1/e at $X = 1$, i.e. at $t_w = \tau$. Depending on the value of α, i.e. in particular on the volume concentration of solute atoms, eq. (12) has two solutions, denoted X_1 and X_2 ($X_2 > X_1$), for $\alpha < 1/e$, which degenerate into a single solution when $\alpha = 1/e$. There is no solution when $\alpha > 1/e$ and plastic flow, then, remains uniform throughout the deformation curve. Given the strain rate and temperature, this defines two critical values of the elementary incremental strain,

$$(\Omega_{1,2}) = Z(X_{1,2})^{3/2}, \qquad (13)$$

the SRS being negative when Ω is within the interval (Ω_1, Ω_2).

The strain dependence of $\Omega = b \, \rho_m \rho_f^{-1/2}$ can be calculated from the strain dependences of ρ_m and ρ_f given in Fig. 1. This strain dependence of Ω, represented on Fig. 6, can be understood as follows. At low strains, the increase in mobile density is predominant and Ω increases with increasing strain. Once the mobile density has saturated, the strain dependence of Ω is determined by the increase in forest density. Hence Ω undergoes a maximum, typically around $\varepsilon = 0.1$-0.15 and saturates at large strains, asymptotically approaching a saturation value Ω_s. The strain dependence of the SRS following from eq.(11) is represented on Fig. 7. It has been calculated using numerical values ($S_o = 5$ MPa, $\tau = 1$ s, $f_o = 30$ MPa) which derive from a study on Al-Mg alloys around room temperature [21]. In this case, which is the most usual one, there is a finite domain of strain within which jerky flow occurs. This domain is bounded by two critical strains, ε_c and ε_c'; they are associated with the ascending and descending branches of the elementary strain profile (cf. Fig. 6), respectively. In some other cases, one of these critical strains may be too small or too large to be recorded, or the situation is more complicated with up to four critical strains on the same deformation curve.

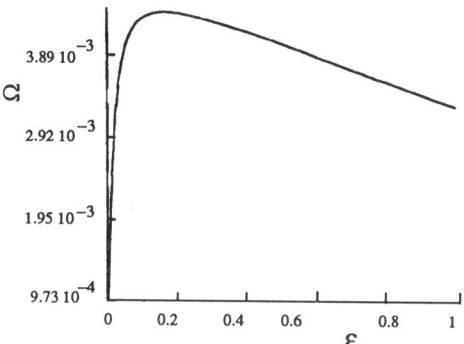

Figure 6. The strain dependence of the elementary incremental strain Ω, as calculated from Fig. 1.

To have at least one critical strain the following two conditions must be simultaneously fulfilled:
(i) Eq.(12) must have at least one solution, which is realized for $\alpha \le 1/e$, i.e. the maximum strengthening f_o produced by DSA must be larger than a threshold value, typically a few times S_o.
(ii) In the graph of Fig. 6, the band bounded by the horizontal lines $\Omega = \Omega_1$, $\Omega = \Omega_2$, within which the SRS is negative, must intersect the Ω vs. ε profile. The number of intersections depends on the position and width of this band, i.e. on strain rate and temperature. On Fig. 8 (a), there are two intersections, which corresponds to the

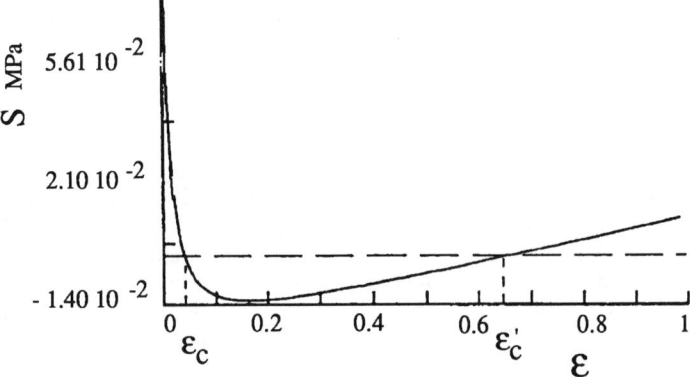

Figure 7. A typical strain dependence of the strain rate sensitivity, S. Jerky flow occurs in the strain interval between the two critical values ε_c and ε_c' where S is negative.

experimental situation of Fig. 4, but one sees that it is possible to obtain up to four critical strains, two on the ascending branch of the profile and two others on the descending branch.

Since the critical values Ω_1 and Ω_2 are proportional to the strain rate and inversely proportional to the temperature, it is possible to derive various possible scenarios for the

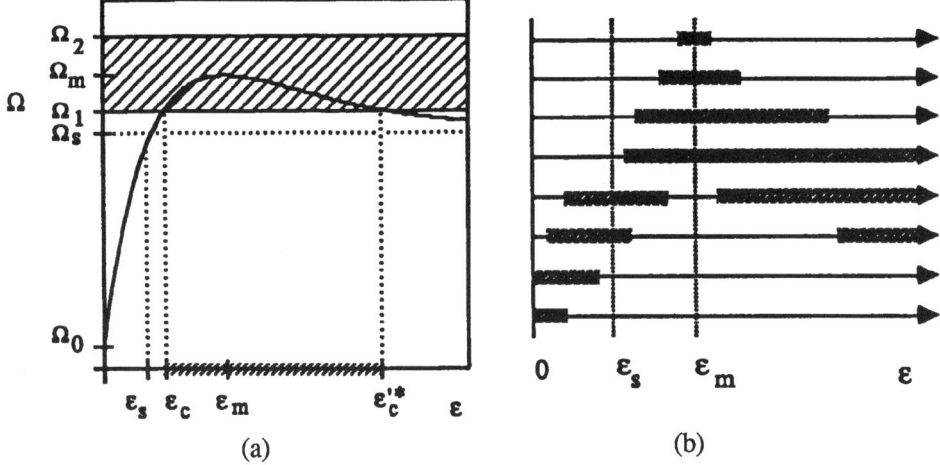

Figure 8. (a) The determination of the critical strains at given strain rate and temperature. ε_m is the strain corresponding to the maximum value of the elementary incremental strain profile and ε_s is the strain at which the ascending branch intersects the asymptotic saturation value $\Omega = \Omega_s$. (b) The evolution of the critical strains with strain rate (increasing from bottom to top) at a fixed temperature or, alternately, with temperature (decreasing from bottom to top) at a fixed strain rate.

occurrence and evolution of the critical strains. Actually, four scenarios are possible [11], but a detailed investigation shows that they do not differ much from each other. The most representative one is reproduced on Fig.8 (b) where the hatched parts indicate the regions of negative SRS, for various strain rates at a given temperature (or various temperatures at a given strain rate). Fig. 8 (b) reproduces quite well the experimental observations of Fig. 5, and other data from the literature can be understood as particular cases of the possible scenarios as well.

The temperature and strain rate dependence of the first critical strain has often been used to estimate the diffusion coefficient of the solute atoms entering through the characteristic time τ in the ageing stress f. This is still possible within the present model, which, in addition, provides an understanding of the other critical strains.

3. Dislocation patterns

3.1. QUALITATIVE ASPECTS

3.1.1. *Forest Patterning*. Because of its effect on the strain hardening, the formation of dislocation patterns has been the object of many extensive experimental studies. The essentials of pattern formation have been understood rather early (cf. the review [24]), but more recent reviews (cf. e.g. [25]) present a broader view, based on an increased variety of materials and deformation conditions.

The situation can be rationalized with the help of a few basic principles. By "dislocation pattern", one usually means "forest dislocation pattern". These arrangements necessarily involve local, athermal mechanisms which provide stable nuclei for the formation of a microstructure, such as the trapping of edge dislocations into dipoles and, above all, the formation of sessile locks through reactions between dislocations of different slip systems. Therefore, patterning is less likely to be observed in single glide than in double or multiple glide.The athermal component of the flow stress which stems from the long range elastic dislocation-dislocation interactions also contributes to pattern formation by controlling the interactions between the growing nuclei and by increasing the rate of storage of mobile dislocations around them. As soon as such nuclei begin to develop, a cooperative process sets in because the long range internal stresses can be viewed as being both the cause and a consequence of patterning.

This tendency is counteracted by the presence of strong and dense localized obstacles (surmountable by thermal activation) such as the Peierls relief, whose periodicity is that of the crystallographic lattice, or small particles or defect clusters of large densities. In such cases, the internal stress almost appears as a modulation of the flow stress which is essentially thermal in character and the spatial distribution of dislocations is rather uniform. During straining, however, the dislocation densities increase so that, sooner or later, the mutual interactions between dislocations become predominant leading to pattern formation. This explains the well-known differences in dislocation microstructures between the BCC and FCC metals strained at low temperatures (e.g. below 0.1-0.2 T_m, T_m being the melting temperature). In FCC metals like copper or aluminum, the main thermally activated mechanism, forest cutting, is rather weak, so that cell structures are formed as soon as more than one slip system is active. In BCC metals there is a strong lattice friction due to the Peierls relief and cell structures only emerge after a substantial amount of strain (e.g. 0.1-0.2). More generally, the ratio of the average athermal and thermal contributions to the flow stress (cf. eq.1) may, then, serve as an indicator for the tendency to patterning.

3.1.2. *Slip Patterning and Forest Patterning*. Slip patterning, i.e. the type of patterning associated with slip traces at the surface of crystals, is connected with the collective properties of the mobile dislocations. It should be distinguished from the forest patterning, described in the previous section, which is associated with a volumetric cellular arrangement of dislocations. It can be inferred from eqs. (2) and (3), however, that forest and slip patterning are not independent of each other. Each creation or loss mechanism influences both types of densities according to its possible stress, strain, strain rate, temperature dependence and through the material parameters it involves. By definition, slip patterning occurs when the mean-free path of mobile dislocations in their slip plane is of the order of the specimen transverse dimensions in single crystals or of the grain size in polycrystals. (This is usually the case at small strains, as shown in Section 2.) While multiplication by dislocation sources is essentially an athermal mechanism, any factor which contributes to a reduction of this mean-free path disrupts slip patterning and increases the tendency to volumetric (i.e. forest) pattern formation. A high stacking fault energy induces both cross-slip and mutual annihilation, and an increase in temperature or a decrease in strain rate enhances the latter tendency. A low stacking fault energy, on the contrary, promotes planar slip and the formation of dynamic pile ups which, owing to the stress concentration at their tips, can cross through obstacles impenetrable to isolated dislocations.The build-up of forest patterning, e.g. with increasing strain, effectively reduces the mean-free path of mobile dislocations to values of the order of one to a few micrometers.

The microstructure may also contain defects other than dislocations, not included in eqs. (2) and (3). For instance coherent and shearable precipitates become softer when repeatedly sheared by mobile dislocations, which induces a softening effect and leads to concentration of the strain within localized active lamellae, thus promoting slip patterning in a broad sense. The same holds for any other instability of type h, for instance the destruction of short range order by moving dislocations (see Neuhäuser, this volume) whose softening effect induces the formation of fine or coarse line or band patterns. On the other hand, unshearable precipitates may trap incoming dislocations and serve as anchoring points for the formation of a three-dimensional structure whith wavelength related to the average precipitate spacing. Grain boundaries are also impenetrable obstacles to dislocations and, obviously, patterning is no longer possible when the grain size is reduced to small values in the sub-micrometer range.

3.1.3. *The Influence of Straining Conditions*. The perfection of a forest pattern is controlled by the degrees of freedom of the stored dislocations, i.e. by the availability of paths allowing for rearrangement and/or mutual annihilation in cell walls and equivalent structures. The structures formed at low temperatures usually consist of fuzzy walls because recovery of redundant dislocations is not efficient enough. At moderate temperatures (0.2 to 0.4 T_m) or in materials with a high stacking fault energy, reasonably well organized subgrain structures with boundaries consisting of the "geometrically necessary" dislocations needed to accommodate the misorientations between the adjacent subgrains are observed. This rearrangement of walls into subgrain boundaries with increasing temperature is a consequence of local cross-slip events of screw dislocations and at higher temperatures (0.5 T_m and above) of climb processes which provide another degree of freedom to edge dislocations. Thus, as a rule, screw segments can annihilate more easily than edges at low and moderate temperatures and the forest contains a majority of non-screw segments trapped in form of tangles or multipolar configurations. At high temperatures, all species have a sufficient number of degrees of freedom so that recovery in subgrains predominates over the disturbing influence of incoming dislocations. Once cells or subgrains are formed, the geometry of the respective boundaries is essentially determined by their tendency to reduce the stored elastic energy and dynamics play a minor role. With increasing temperature, therefore, both the forest density and the long range internal stresses are strongly reduced by recovery processes.

It is possible to obtain well-organized patterns at moderate temperatures under straining conditions which favour the formation of stable dislocation arrangements. For instance, in creep tests, the strain rate is usually several orders of magnitude smaller than in constant strain rate tests. The role of time effects, such as diffusion and the local motion of edge dislocations by climb is enhanced with respect to that of glide processes. In cyclic deformation, the shuttle motion of dislocations and the possibility of achieving large cumulated strains favour the formation of configurations stable with respect to the applied stress of both signs. As a consequence, both the annihilation of screw dislocations and the mutual trapping of edges into dipolar and multipolar configurations are enhanced. This leads to various types of wall and channel structures, in particular to Persistent Slip Bands in pure FCC single crystals. The latter are well suited for modelling when cyclically deformed in quasi-steady state and with only one slip system activated.

3.1.5. *Scaling Laws*. Despite this large variety of mechanisms and straining conditions it seems that the arrangements of the microstructure obey, more or less closely, two very simple scaling laws in most materials. One relates the stress to the total dislocation density or to the forest density

$$\sigma/G = \alpha \, (b \, \rho^{1/2}), \tag{14}$$

This expression has already been used in Section 2.2. For individual dislocations, eq. (14) simply yields the stress necessary to reach the critical bowed out configuration between forest obstacles of spacing $\rho_f^{-1/2}$. As far as collective effects are concerned, the origin of this scaling law is not always understood. It can be derived from dimensionality arguments and virtually every model for strain hardening leads to such a form with numerical values for α ranging between 1/5 and 1/2, sometimes up to 1 [26]. It is however clear that eq. (14) does not apply to situations where the mobile density plays a prominent role. For instance, as noted in [24], dislocation pile-ups obey rather different scaling laws. The dislocation density involved in eq. (14) is, therefore, probably that of the forest and since the latter contributes to a large fraction of the total density, except at small strains, one may as well make use of the total density. Most probably also, friction stresses should not be included in the left-hand side of eq. (14) when they are significant. When they are not, i.e. in FCC metals, it can be shown that eq. (14) applies to the total stress, to the internal stress, and hence to the local effective stress with a different value of α [16] and with ρ denoting the density of the forest.

The second scaling law applies to many but not all materials [27]. It relates the characteristic length, d, associated with the forest pattern, i.e. a cell or subgrain size or the distance between parallel walls, to a stress, usually the flow stress minus the frictional stress.

$$(\sigma/G) = K \, (b/d), \tag{15}$$

The dimensionless constant K is frequently reported to be of the order of 20-30 for metals and alloys [27], with some notable exceptions, however. This expression is the cornerstone of the so-called mesh length theory of work hardening which assumes that dislocation patterns are in equilibrium and that their preferred arrangement is the one which minimizes the wall or subgrain energy. Since the elastic stress fields of dislocations are inversely proportional to the distance, eq. (15) directly follows from the assumption that dislocations are in equilibrium. This assumption, however, neglects all dynamic effects and, except perhaps at high temperatures, its applicability to such a highly dissipative process as plastic deformation is quite doubtful.

Nevertheless, the dynamic approaches discussed in the following sections have not, up to now, been able to yield the physical content of the two constants α and K. It is suggested that eqs. (14) and (15) actually apply to dynamic steady states, which could explain both their universality and the deviations which are sometimes observed. Figure 9, which is reproduced from the work of Raj and Pharr [27], shows a compilation of data on aluminum and iron and indicates that there exists at least a good correlation over several decades between the applied stress and the inverse of dislocation cell (or subgrain) size.

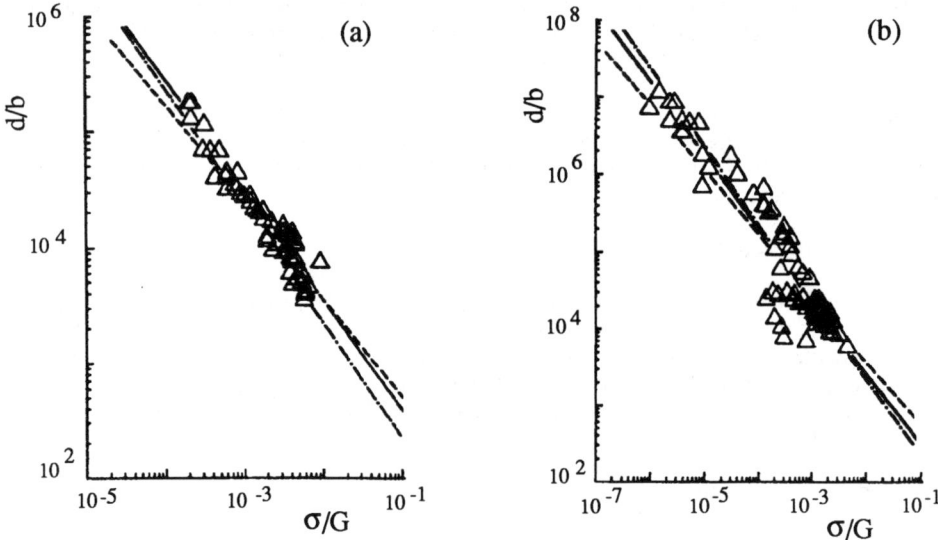

Figure 9. The relation between normalized cell size, d/b, and normalized stress, σ/G, in iron (a) and aluminum (b). (After Raj and Pharr [27]).

3.2. ANALYTICAL APPROACHES.

To take into account the distribution of dislocations in time and space, one has to express dynamic balance equations for the various coupled dislocation populations involved during plastic flow. Denoting by ρ_i one of these populations, we have:

$$\partial \rho_i / \partial t + \text{div} \underline{J}_i = g_i (\rho_i, \rho_j, \underline{v}_i, \underline{v}_j) \tag{16}$$

$\underline{J}_i = \rho_i \underline{v}_i$ is the dislocation flux associated with the species i moving with a velocity $\underline{v}_i$ and g_i is a creation-destruction term analogous to the terms at the right-hand side of eqs. (2) and (3) which involves creation and annihilation events and the couplings with other populations denoted by the index j. When the divergence term is omitted, eq. (16) reduces to the balance equations for uniform densities which are commonly used in models for strain hardening (cf. Section 2). When the conservation of the density ρ is assumed, i.e. when $g_i = 0$, eq. (16) reduces to a form wich has been used in many continuum models of dislocation theory. In its present form, the reaction-transport system, eq. (16), accounts for both long range and local interactions and appears suitable for the study of dislocation patterning. The development of such studies encounters, however, a few difficulties.

Strictly speaking, a dislocation population is defined as an ensemble of isolated dislocations or of elementary configurations which have in common the same creation-destruction mechanisms. In principle, thus, one has to distinguish between all dislocations of different Burgers vectors, and - given the Burgers vector - between individual dislocations of both signs (cf. D. Walgraef, this book) and to consider at least two different characters, screw and edge, which have quite different properties while being strongly coupled. To avoid dealing with an excessive number of coupled equations it is instructive, in a particular study, to reduce to a minimum the number of interacting populations and to carefully select the relevant creation, destruction and coupling terms.

For the moment, this has been done only for the case of fatigue patterning, whose study has been initiated by Walgraef, Aifantis and coworkers [28-32], and further by Kratochvil [33, 34]. Indeed, in the fatigue of FCC or BCC single crystals oriented for single slip, it is possible to reduce the number of populations to only two: mobile dislocations of essentially screw character and less mobile multipolar ensembles of edge dislocations.

A second difficulty resides in the fact that it does not always seem possible to bring the reaction-transport equation into the reaction-diffusion form

$$\partial \rho_i / \partial t + D_i \Delta \rho_i = g_i, \qquad (17)$$

where D_i is an effective diffusion coefficient associated with the species i. The scheme leading from eq.(16) to eq.(17) is as follows. The dislocation velocity involved in the definition of the flux is a function of the local effective stress σ^*, i.e. of both the applied stress and the internal stress (cf. eq. 1). This last quantity is the sum of all the long range interaction stresses between the local density at the position $\underline{r}$ and all the other dislocations (position $\underline{r}'$) in the crystal considered:

$$\sigma_i(\underline{r}, t) = \propto \int \underline{I} (|\underline{r}' - \underline{r}|) \, F(\underline{r}') \, \rho(\underline{r}', t) \, d\underline{r}' \qquad (18)$$

I is the pair-interaction function between two dislocations and F is a distribution function taking into account the possibility for the dislocations to have positive or negative signs. The integration is carried out over the volume of the crystal considered. For isolated dislocations, the pair interaction function is long ranged, $I \propto 1/(|\underline{r}' - \underline{r}|)$, so that the internal stress definitely has a nonlocal character. If, however, dislocations arrange themselves in such a way that their long range interaction stresses are screened off, eq. (18) can be expanded in Taylor series around $\underline{r}' = \underline{r}$ and the first term of this expansion leads to the reaction-diffusion form of eq. (17).

The first attempt to reduce the internal stress to a simpler form is due to Holt [4]. In his study Holt considers an ensemble of parallel, infinite screw dislocations of same Burgers vector and with both signs in equal density, in the absence of any dynamics (g = 0, no applied stress). The preferred wavelength of the first fluctuation which develops from an initially random distribution is estimated through an argument of energy minimization. By assuming that the long range internal stresses are negligible outside a certain cut-off radius, Holt obtains a preferred wavelength proportional to this screening distance by expanding a form analogous to eq. (18). The result obtained, a sinusoidal distribution of the total density is, however, not consistent with the assumption that self-screening occurs. In addition, it is well-known from experiment that no periodic arrangement of the microstructure is expected in such a situation.

In the case of fatigue patterning, Walgraef and Aifantis [28] take advantage of the fact that the less mobile dislocations are arranged in form of veins of multipolar arrays. The internal stresses are then effectively screened, at least after a preliminary cyclic hardening stage during which the vein structure is formed. The integral in Eq. (18) can then be expanded in the vicinity of $\underline{r}$ and, after a few approximations, two pseudo-diffusion coefficients are defined, one for the mobile dislocations and a second one for the less mobile multipoles. This model contains a few inaccuracies in the way it deals with local dislocation interactions and with dislocation mobility. Nevertheless, it yields a picture of patterning and of pattern evolution with stress, taken as a control parameter, which closely reproduces experimental observations in pure FCC crystals. When the cyclic stress amplitude is increased, a periodic vein structure emerges from the inital uniform dislocation density and, further, a ladder-like pattern of Persistent Slip Bands appears. Kratochvil [33] has elaborated in some detail the value of the "diffusion coefficient" associated with dipoles, noticing that a dipole can only drift under the effect of a stress gradient. His model is coherent with the types of vein structures observed in cyclically deformed BCC Fe-Si alloys. The reaction-diffusion models developed so far seem, therefore, to be able to describe patterning during cyclic deformation. It is not quite clear, however, why Persistent Slip Bands, characteristic of FCC crystals, are obtained in one case [28] and not in the other [33], since there is no or little crystallographic input in both of these models.

Future development of the reaction-transport or reaction-diffusion scheme could attempt at dealing with two particular situations in monotonic straining. (i) Patterning during secondary creep, i.e. another type of dynamic steady state but with several active slip systems and a long range pair-interaction. A possible objective should be then to investigate the formation of cell structures and their scaling laws, possibly within some self-screening approximation. (ii) Mobile dislocations and the transition between static and propagating types of behaviour. In this case, an extended version of Holts' problem could be dynamically investigated, with either screw or edge dislocations, and with two coupled populations corresponding to the two possible signs (see [35] and Walgraef, this volume).

In each case, the main difficulty arises from the need to self-consistently justify the occurrence of a critical screening radius for the long range interactions, or from dealing with the full nonlocality of the system. This is the main reason why numerical simulations have been developed, since they should in principle be able to reproduce the transition from a random to an ordered dislocation distribution, and to yield information on such quantitites as the cut-off radii for the internal stresses as well as on their evolution.

Besides the described approaches introducing diffusion-like terms at "microscopic" (dislocation) level, there were some attempts to account for nonlocal interactions at macroscopic scale. In particular, it has been suggested (see Aifantis, this volume) to include in a constitutive equation for the local stress a term proportional to the second derivative with respect to coordinate. The present authors proposed, instead, the idea [13, 20] that exchange of plastic activity (mobile dislocations) between adjacent slip planes via the cross-slip mechanism may produce a diffusion-like term in a constitutive equation, which term should be responsible for propagating solutions. This path will be pursued in studies on propagative instabilities of plastic flow.

3.3. COMPUTER SIMULATIONS OF DISLOCATION PATTERNING.

3.3.1. *2 D Simulations*.

The 2 D simulations presently available make use of two different techniques. The method used by Ghoniem (see [36] and Ghoniem, this volume) derives from molecular dynamics, while the one described in Refs. 35 and 37 derives from cellular automata. In this last case, parallel infinite straight dislocations of the same Burgers vector and character are considered and long range as well as local interactions are taken into account. Space and time are discretized and the dislocations, seen end-on, are viewed as point-like objects which move within a cellular space according to a stress criterion. The natural length of the elementary cells is twice the critical annihilation distance, which is denoted by y_e for edge dislocations and y_s for screw dislocations (cf. Section 2.1.). Approximate values for these quantities have been found in copper at room temperature by Essmann and Mughrabi [38] to be $y_s \approx 50$ nm and $y_e \approx 1.6$ nm.

In practice, this type of simulation is only reminiscent of cellular automata and it has to be adapted to deal with some specific static or dynamic dislocation properties. The reader is referred to Ref. 37 for more detail about the technique itself.

The 2 D simulations on screw or edge dislocations reveal that no periodic structure is formed both in the presence and in the absence of applied stress. This is in agreement with experiment but contradicts the conclusions drawn by Holt [4]. In the absence of applied stress (annealing experiment), a random initial configuration gets blocked after a few tens of generations into a metastable state consisting of clusters of dipoles or multipoles.

The study performed on screw dislocations in creep conditions casts some light onto the typical arrangements of mobile dislocations. Depending on the nature of the multiplication mechanism, either periodic arrays of active slip lamellae, with the spacing of a few y_s, or slip propagation with a constant velocity across the simulated crystal, are obtained. Since the critical annihilation distances vary as the inverse of the applied stress, so does the slip line spacing in the simulation. For edge dislocations in the same conditions, the simulated patterns consist of irregularly spaced walls perpendicular to the slip plane, which remain in dynamic steady state without exhibiting a well-defined periodicity.

Finally, a determination of the cut-off radius for the long range elastic interactions was performed on a metastable configuration after an annealing experiment. The mean value was found to be of the order of 5 μm, indicating that there is some self-screening effect but that, nevertheless, internal stresses remain long-range. These results are described in more detail in [35], along with other features not reported here.

Although this elaboration yields interesting results, in particular with regard to slip patterning, it leads to a conclusion that only 3 D simulations will be able to produce realistic types of forest patterning, with probably smaller values of the cut-off distance for the long range elastic interactions. The additional properties to be taken into account in an extended three-dimensional version of such simulations are the coupling between edge and screw segments, the interactions between different slip systems and, above all, the formation of reaction junctions and sessile locks when two segments of different Burgers vectors cross each other. In what follows, we report on preliminary results obtained with 3 D simulations.

3.3.2. *3 D Simulations: Simulation Technique*.

Although 3 D simulations are technically quite complex, they are conceptually simpler than 2 D simulations because they do not involve any arbitrariness in the way the dislocation-dislocation interactions are treated. Here, we discuss the basic principles underlying these simulations. Additional detail can be found in [39].

Space is, again, discretized but this time every dislocation loop or line is decomposed into a set of elementary dislocation segments (EDSs) of either edge or screw character which are distributed on a three-dimensional net. The character of the EDSs and the slip geometry are accounted for by attributing to this 3 D net the symmetries of the simulated crystal, as exemplified by Figure 10 in the case of an FCC lattice.

On Fig. 10, all the possible 1/2<110> Burgers vectors and screw directions are along the diagonals of the elementary cell, while the edge directions, <112>, are along the lines joining one corner to the middle of an opposite face. The usual slip planes are {111}.

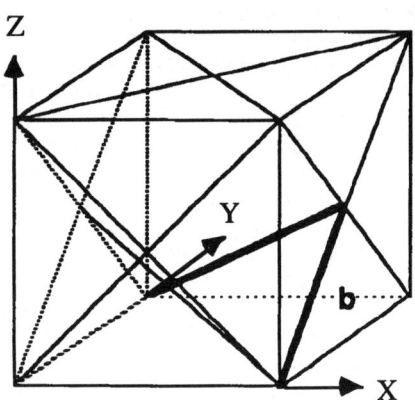

Figure 10. The FCC lattice of a simulated crystal and a portion of dislocation loop (bold lines). The screw segment is along [011], the edge one is along [211] and the slip plane is (11$\bar{1}$).

During its glide or cross-slip motion an EDS moves by a multiple of the elementary translations permitted by its slip geometry and by the local value of the effective stress. For instance, glide of a screw EDS involves translations parallel to the edge direction associated with the same Burgers vector and vice-versa. During such motion, new elementary segments are created to maintain the continuity of dislocation lines. This ensures, in addition, a coupling between edge and screw components and the operation of multiplication mechanisms.

The basis of the discretization is necessarily the smallest of the two critical annihilation distances defined in Section 3.1.3., i.e. y_e, the critical annihilation distance for edges. This defines the lengths of the elementary screw and edge segments, 2.44 nm and 4.23 nm respectively. The simulated crystal contains about 10^{12} elementary cells and has a side of 15 µm, with either periodic boundary conditions or image forces at its outer surface.

The long range interaction stresses between EDSs are computed using the stress fields of finite dislocation segments which may substantially differ from those of infinite segments used in the 2 D simulations. For instance, two parallel infinite dislocations with the same Burgers vector but one of edge and the other of screw character do not interact, while this is not the case for two finite segments. It is interesting to note that the segmented dislocation lines acquire a line tension whose value appears to be quite realistic. This is illustrated by Fig. 11 which compares theoretical and "experimental" (i.e. drawn from the simulation) estimates of the shear stress τ necessary to maintain at equilibrium a segmented loop of diameter D. The full line on Fig. 11 represents the prediction drawn from dislocation theory, $\tau = 2\Gamma/D$, where Γ is the line tension.

The long range pair interaction stresses are computed within a rather large cut-off distance. For each interacting pair, it depends on a criterion containing the lengths of the two segments (which may consist of one or several successive EDSs of the same nature) and the distance between their middles. The cut-off radius is a maximum for two segments whose distance is smaller than one of their lengths and, according to the results obtained for 2 D (cf. Section 3.3.1.), its value has been fixed to 5 µm.

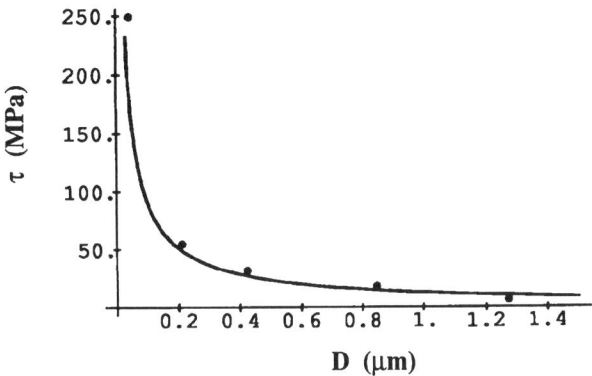

Figure 11. Shear stress τ needed to maintain at equilibrium a dislocation loop of diameter D as a function of D. The dots represent values obtained by the 3 D simulation and the full line represents the dependence predicted by dislocation theory.

The computation of the local effective stresses and the definition of the possible steps of motion of the elementary segments proceed as an extension of the methods previously used for 2 D. All stresses are expressed in tensorial form and a small frictional stress (3×10^{-4} G, a value typical for an FCC metal) is detracted from the resulting shear stresses in the glide or cross-slip plane. To simulate the effect of dislocation dissociation on cross-slip, this friction stress can be made larger in the cross-slip plane than in the glide plane. Together with the crystallography and the elastic constants, this represents, at the present stage, all the material parameters involved in the simulation.

The number and nature of active slip systems directly follow from the orientation of the crystal with respect to externally applied forces. Any type of straining conditions, uniaxial or multiaxial, monotonic or cyclic, can in principle be imposed on the simulated crystal.

Local processes such as direct annihilation, annihilation of dipoles, cross-slip and mutual trapping are naturally taken into account by the simulation and involve no input parameter. Regarding the interactions between two segments of different Burgers vectors, jog formation during cutting processes is neglected, as it is thought to have no influence on mechanical properties of FCC metals at room temperature. The formation of stable dislocation junctions or locks, as a result of attractive interactions and reactions during cutting processes, is accounted for by expressing that the meeting point of two segments has a zero velocity. Dislocation junctions, as well as cross-slipped segments are potential sites for dislocation sources, so that, unlike in the 2 D simulations, no input parameter is needed to account for multiplication mechanisms.

The most difficult step in all dynamic simulations of collective dislocation behaviour consists in accounting for the large dispersion of dislocation velocities, typically from a few Ås^{-1} to a few hundreds of ms^{-1}, particularly at small strains (cf. also Ghoniem, this volume). In its present stage, the 3 D simulation contains only athermal dislocation-dislocation interactions, the materials simulated being pure copper or aluminum at room temperature. The stress dependence of the velocity used is of viscous type, representing phonon and electron drag effects. Each time a segment has to move, a potential free-flight distance is defined, consisting of several substeps after which the possibility for further motion is reconsidered. The time scale of the simulation follows from the discretization in space and from the definition of the free-flight distances. Elementary time steps are typically of the order of 10^{-12} s.

3.3.3. 3 D Simulations: Preliminary Results. The output of the simulations consists of histograms and averages of various microscopic or macroscopic quantities, such as the dislocation densities and velocities, the internal stresses, strains and strain rates. The microstructure can also be imaged and we reproduce here two such simulated images which were obtained on one Apollo workstation at Metz University. Full scale simulations will be performed on the Cray 2 computer of the CCVR (Ecole Polytechnique Palaiseau).

Fig. 12 illustrates the mechanism of dislocation multiplication. The initial configuration consists of a segment pinned at its ends which develops into a Frank-Read source under the action of a constant imposed stress of 100 MPa. After 11 successive steps of simulation, the total strain is about 5×10^{-6} and the total dislocation density has increased by a factor of 20. The average dislocation velocity is of the order of 250 ms^{-1} and the total time elapsed is 1.2×10^{-11} s. These values are thought to represent the very fast initial dislocation multiplication stage in single crystals (see Neuhäuser, this volume). This clearly shows that to realistically deal with multiplication, one has to take into account very fast events and very small time steps, at least at the beginning of plastic flow.

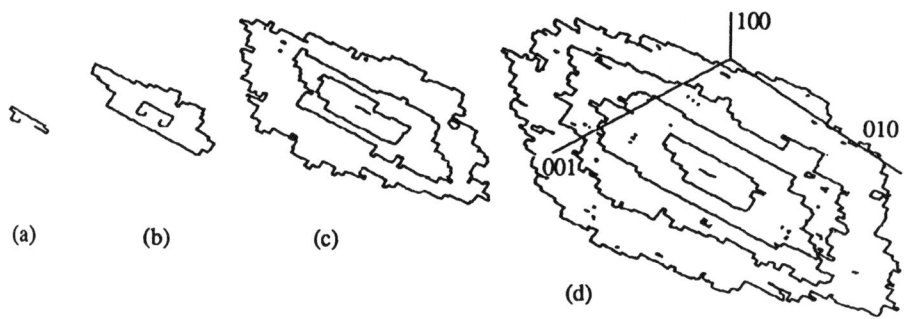

Figure 12. Several steps of operation of a Frank-Read source from an initial dislocation segment pinned at its ends. Number of time steps: (a): 1 (b): 3 (c): 7 (d): 11.

Figure 13 shows a simulation of easy glide under a constant tensile stress of 50 MPa parallel to [112]. The initial configuration is a representation of the Frank net of an annealed crystal; it consists of randomly distributed dislocation segments, of Burgers vector [$\bar{1}$01], pinned at their ends, with the density $\rho_i = 10^{11}$ m^{-2} and the average length $\rho_i^{-1/2}$. After a strain of 1.4×10^{-5} corresponding to the time of 1.75×10^{-8} s, the dislocation density has increased by one order of magnitude and a few dislocation sources have started operating. Closer inspection reveals that some edge dipoles have been formed and that some activity occurred in the cross-slip plane, although the applied stress has no component in this plane. A small internal stress starts building up, but, obviously, larger strains are necessary to observe the formation of a pattern.

This 3 D simulation is potentially able to give the answers to questions related to the modelling of dislocation patterning and of plastic instabilities. It can be adapted to deal with more complex microstructures, e.g. through the introduction of solute atoms or precipitates. Presently, the main objective is the reduction of the computation times to make possible simulations for strains in the range of percent.

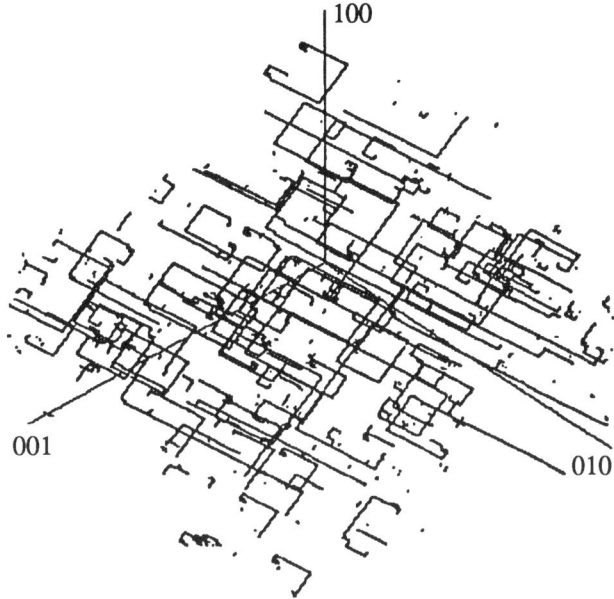

Figure 13. The development of easy glide on the (111) plane of a FCC crystal (copper) under a constant stress of 50 MPa, from an initial structure representing the Frank net in an annealed crystal.

References

[1] Lüders, W., *Dinglers Polytechnisches J.*, **156** (1860) 18.
[2] Le Châtelier, F., *Rev. de Métallurgie*, **6** (1909) 914.
[3] Portevin, A., Le Châtelier, F., *Trans. ASST*, **5** (1924) 457.
[4] Holt, D.L., *J. Appl. Phys.*, **41** (1970) 3197.
[5] *Mechanical Properties and Behaviour of Solids: Plastic Instabilities*, ed. by V. Balakrishnan and C.E. Bottani, World Scientific (Singapore) 1986.
[6] *Patterns, Defects and Microstructures in Nonequilibrium Systems*, ed. by D.Walgraef, Martinus Nijhoff (Dordrecht) 1987.
[7] *Non Linear Phenomena in Materials Science*, ed. by L. Kubin and G. Martin, Trans Tech Publications (Aedermannsdorf) 1988.
[8] Kocks, U.F., Argon, A.S., Ashby, M.F., *Thermodynamics and Kinetics of Slip*, Progress in Materials Science Vol 19, Pergamon Press (Oxford) 1975.
[9] Kocks, U.F., *J. Eng. Mat. Techn.*, **98** (1976) 76.
[10] Estrin, Y. and Kubin, L.P., *Acta Met.*, **34** (1986) 2455.
[11] Kubin, L.P. and Estrin, Y., *Acta Met.*, in press.
[12] Mughrabi, H., *S. Afr. J. Phys.*, **9** (1986) 62.
[13] Estrin, Y., in *Non Linear Phenomena in Materials Science*, ed. by L. Kubin and G. Martin, Trans Tech Publications (Aedermannsdorf), p. 417, 1988.
[14] Kubin, L.P. and Estrin, Y., *Rev. Phys. Appl.*, **23** (1988) 573.
[15] Estrin, Y. and Kubin, L.P., *Res Mechanica*, **23** (1988) 197.
[16] Friedel, J., *Dislocations*, Pergamon Press (Oxford) 1967.
[17] Kocks, U.F., *Progress in Materials Science*, Chalmers Anniversary Volume, Pergamon Press (Oxford) 1981.
[18] Kocks, U.F., in *Strength of Metals and Alloys* (Proc. ICSMA 5), ed. P. Haasen et al., Pergamon Press (Oxford), p. 1661, 1980.
[19] Kubin, L.P. and Estrin, Y., *J. Physique*, **47** (1986) 497.
[20] Kubin, L.P.and Poirier J.P., in *Non Linear Phenomena in Materials Science*, ed. by L. Kubin and G. Martin, Trans Tech Publications (Aedermannsdorf), p. 473, 1988.
[21] Kubin, L.P., Chihab, K. and Estrin, Y., *Acta Met.*, **36** (1988) 2707.
[22] Guillot, J. and Grilhé, J., *Acta Met.*, **20** (1972) 291.
[23] Räuchle, W., Vöhringer, O. and Macherauch, E., *Mat. Sci. and Engg.* **12** (1973) 147.
[24] Embury, J.D., in *Strengthening Methods in Crystals*, ed. A. Kelly and R.B. Nicholson, Applied Science Publishers (London), p. 331, 1971.
[25] Louchet, F., and Bréchet, Y., in *Non Linear Phenomena in Materials Science*, ed. by L. Kubin and G. Martin, Trans Tech Publications (Aedermannsdorf), p. 335, 1988.
[26] Lavrentev, F.F., *Mater. Sci. and Engg.*, **46** (1980) 191.
[27] Raj, S.V. and Pharr, G.M., *Mat. Sci. Engg.*, **81** (1986) 217.
[28] Walgraef, D. and Aifantis, E.C., *J. Appl. Phys.*, **58** (1985) 688.
[29] Aifantis, E.C., *Mat. Sci. Engg.*, **81** (1986) 563.
[30] Walgraef, D., Schiller, C. and Aifantis, E.C., in *Patterns, Defects and Microstructures in Nonequilibrium Systems*, ed. by D.Walgref, Martinus Nijhoff (Dordrecht) p.257, 1987.
[31] Walgraef, D., in *Non Linear Phenomena in Materials Science*, ed. by L. Kubin and G. Martin, Trans Tech Publications (Aedermannsdorf), p. 77, 1988.
[32] Aifantis, E.C., in *Non Linear Phenomena in Materials Science*, ed. by L. Kubin and G. Martin, Trans Tech Publications (Aedermannsdorf), p. 397, 1988.
[33] Kratochvil, J., in *Mechanisms and Mechanics of Plasticity*, ed. by J. Castaing, J.L.

Strudel and A. Zaoui, *Rev. Phys. Appliquée*, **23** (1988) 419.
[34] Kratochvil, J. and Libovicky, S., *Scripta Met.*, **20** (1987) 1625.
[35] Lépinoux, J., and Kubin, L.P., in *Strength of Metals and Alloys (ICSMA 8)*, ed. by P.O. Kettunen et al., Pergamon Press (Oxford) Vol 1, p.35, 1988.
[36] Ghoniem, N.M; and Amodeo, R., in *Non Linear Phenomena in Materials Science*, ed. by L. Kubin and G. Martin, Trans Tech Publications (Aedermannsdorf), p. 377, 1988.
[37] Lépinoux, J. and Kubin, L.P., *Scripta Met.*, **21** (1987) 833.
[38] Essmann, U. and Mughrabi, H., *Phil. Mag. A*, **40** (1979) 731.
[39] Canova, G., and Kubin, L.P., in *Continuum Models of Discrete Systems 6*, ed. by G.A. Maugin, Longman Scientific and Technical, in press.

NUMERICAL SIMULATION OF DISLOCATION PATTERNS DURING PLASTIC DEFORMATION

N. M. GHONIEM * and R. J. AMODEO **

* School of Engineering and Applied Science
University of California, Los Angeles
Los Angeles, CA 90024, U.S.A.

** Xerad Inc., 1526 14th Street, Suite 102
Santa Monica, CA 90404, U.S.A.

ABSTRACT. A new method for the numerical simulation of dislocation patterns in solids undergoing plastic deformation is introduced. The term dislocation dynamics (DD) is used to describe the technique and to distinguish it from traditional molecular dynamics (MD) simulations. The basis of the approach is the simultaneous solution of the equations of motion of aggregates of dislocations under the influence of self- and externally-applied stress fields. Dislocation climb and glide motions are given by phenomenological and empirical relationships. Forces on dislocations are computed by considering the long-range- and applied-stress fields of each dislocation in the simulation space. Short-range dislocation interactions are represented as events. These interactions are: generation, annihilation, pinning, junction, and dipole formation. Criteria for these processes are based on experimental observations. Two important computational aspects are emphasized for their uniqueness in the DD method. The first is that the stress field is tensorial and not scalar, thereby giving strong directional dependencies to dislocation motion. The second aspect is the need to appropriately select the simulation timestep in order to represent short-range reactions. Several examples are given to show the simulation of persistent slip bands and dislocation cells.

1. Introduction

The formation of regular patterns in nature has recently gained much attention in various disciplines of science and, over the last 20 years, theories for the phenomena of self-organization and bifurcation have been advanced. Many schools of thought have evolved as a result of this new perspective [1], and attempts are still being made to understand the driving force, or the fundamental basis behind the appearance of these organized structures.

Computer simulation has recently become an important tool in the field of scientific investigations and has gained equal significance to the traditional roles of theory and experiment. The advent of high speed supercomputers has allowed simulation of the time evolution of a large number of particles in molecular systems, both gaseous and liquid [2]. Only during the past decade have "computer experiments" in interacting dynamical systems emerged as a valuable means of understanding the complex behavior of physical systems without the need for the usual theoretical simplifications. The marriage of these two fields has offered an interesting approach to the solution of a problem which ocurs in materials subjected to engineering stresses, high temperatures, and irradiation.

During high-temperature deformation of metals (as in fatigue, creep, and hot forming), linear imperfections in the material (dislocations) tend to aggregate into spatially inhomogeneous patterns at critical values of external temperature and stress fields. This patterning behavior is a manifestation of the material's internal ability to resist further deformation and failure. A fundamental understanding of the mechanisms which control this phenomenon will enable us to design better structural alloys. The motivation for using computer experiments is to study the nature of dislocation pattern formation in metals. The purpose of these experiments is to identify critical stress/temperature/ material conditions for the onset and formation of spatially inhomogeneous dislocation patterns, and to identify fundamental dislocation processes that are responsible for this behavior. In general, computer simulations are not intended to substitute laboratory experiments, but rather to complement and interpret their results. Fundamental dislocation processes are describable in terms of basic material properties, applied stress, temperature, and irradiation conditions. Performing the simulation consists of implementation of these fundamental processes in the equations-of-motion of individual dislocations, and simultaneously solving these equations by developing dynamic particle simulation methods for dislocations. Particle methods applied to dislocations are unique, since they are based on the determination of vector rather than scalar fields. In general, application of these techniques eliminates the need to correlate data from specific experiments in order to solve for the resulting dislocation structure. In addition, results from a simulation can provide immediate visual insight into the fundamental nature of the formation of these patterns (something a theoretical description or realtime experiment may be incapable of achieving).

Two major theoretical approaches, appearing within the last century, have attempted to explain the physical basis for the formation of organized structures (Fig. 1). Around the turn of the century, Gibbs proposed a hypothesis that stable states are achievable through minimization of the free energy of the system in consideration [3]. Many years later, Cahn was the first to use the principles set forth by Gibbs, and he developed much work related to spinodal decomposition [4]. More recently, Heerman [5] has done extensive work in presenting an updated critique and analysis of spinodal decomposition and related topics. His work also includes a description of computer simulation of phase transitions.

The other school of thought on the formation of regular structures in nature is based on the phenomenon of self-organization. Prigogine [1] and Haken [6] independently developed a theory describing self-organization of dissipative structures. The principle involved is a

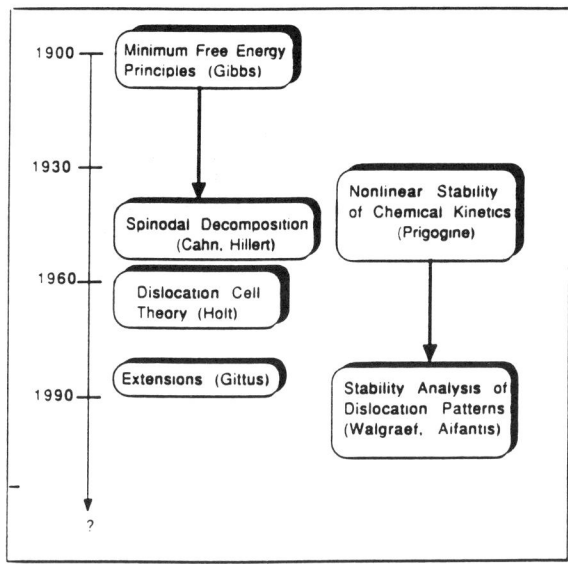

Fig. 1. Schematic of general methods to explain organized structures.

balance between the nonlinear reactive components of a process, and the diffusion of the particulate species through the medium. The result is, in general, a modulated structure which, at steady state, appears to be fixed in space. Both of these principles have been applied to study the formation of dislocation patterns in metals.

The earliest significant theoretical understanding of the critical phenomenon of dislocation pattern formation was developed by D. Holt [7] in 1970. Holt proposed that dislocations move to form a periodic structure of wavelength ℓ, characteristic of the diameter of the dislocation cell. Using analysis similar to spinodal decomposition, he showed that an array of dislocations is inherently unstable to small perturbations. In Holt's model, these small perturbations, driven by an unspecified source, consist of a local excess of concentration of dislocations of one character type, i.e., a dislocation density gradient. This gradient is responsible for the reduction in elastic energy which further drives the density perturbation.

Holt's analysis led to the prediction that the cell size (or perturbation wave length ℓ is related to the total dislocation density, ρ, as

$$\ell = K_c \rho^{-1/2} , \qquad (1)$$

where K_c is a constant. Since the applied stress is proportional to the square root of the dislocation density, the cell size was predicted to decrease with the applied stress (shown in Fig. 2). The experimental data falls within the band given by the values $K_c = 10 - 100$. The principle of minimizing the free energy to arrive at a stable configuration of the system was also used in an approach by Gittus [8-10] to correlate with experimental observations of dislocation cell size, and to determine the theoretical value of the constant K_c for metals.

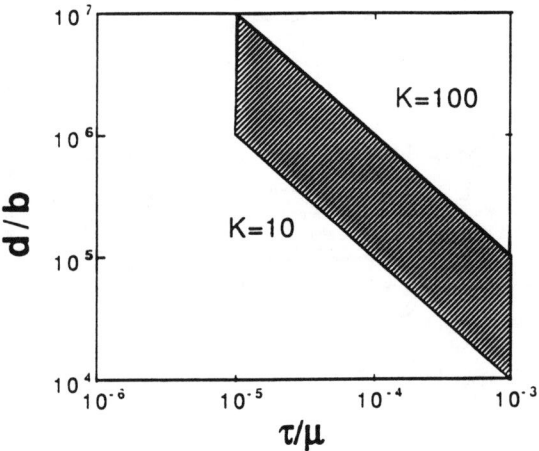

Fig. 2. Experimental range for the relationship between the dislocation cell size and the applied stress.

A recent conference on low-energy dislocation structure (LEDS) [11] has picked up where Holt and Gittus left off in the description of the driving force for the formation of these structures. Low-energy dislocation structures (LEDSs) are formally defined as any dislocation structure in which neighboring dislocations screen each other's stress field [12]. The term LEDS suggests that the driving force for the formation of these dislocation structures is the reduction of the elastic free energy of the system. This occurs when dislocations cluster into stress-screened arrays.

Sandstrom [13,14] and Li [15] developed a set of kinetic equations based on fundamental dislocation reaction processes occurring within the cell and at the cell boundaries. They basically described recovery as four possible processes: (1) annihilation of dislocations within the cell, (2) annihilation of dislocations in the cell boundary, (3) absorption of dislocations at the cell boundary, and (4) emission of dislocations from the cell boundary.

Rate theory descriptions of the physics of dislocation substructure formation and dislocation interactions include the kinetics of interstitial and vacancy interaction with dislocations. An example is the model developed by Ghoniem, Matthews, and Amodeo [16]. In this fomulation, the evolution of the average dislocation density in high strength steel is coupled with a phenomenological theory of dislocation creep.

The previous models were limited to a description of the rate processes of dislocation reactions, once the substructure is defined. Walgraef and Aifantis however [17-19], were the first to use Prigogine's concept of self-organization to arrive at a description of the formation and stability of dislocation patterns. In their approach, they solved reaction-diffusion equations to derive the conditions under which spatial modulations emerge in the dislocation system. Dislocation reaction mechanisms are quite complicated, and a simplified analytical model is prone to uncertainties if the underlying parameters are not well defined. Computer simulations can be used to define such parameters.

2. Numerical Simulation of Particle Systems

In physical systems comprising many identical particles which interact according to a force field, analytical descriptions are based on conservation principles, and are often simplified in order to obtain a solution. Although analytical solutions can be quite general and they can predict the response of the system to external variables, they must be based on certain assumptions. The validity of such analytical solutions is therefore dependent upon the accuracy of the underlying assumptions, which quite often ignore many physical details.

The advent of high speed computers has allowed introduction of a new way of thinking about the theoretical representation of physical systems. Methods which are based upon direct solutions to the equations of motion (EOMs) are increasingly becoming indispensible tools to understanding the physical origins of macroscopic phenomena. Numerical particle simulation methods can be very powerful when used to show the types of dominant mechanisms or interactions which lead to the particular set of observations. The basic idea behind all particle methods is simply to solve the coupled EOMs for all particles representing the system at small time intervals. However, the practical implementation of such a simple idea can represent a considerable challenge. The system size (number of interacting particles), and the effects of the outside "world" on its behavior usually represent sources of inaccuracy. This challenge is a physical one, and can be surmounted by experience and knowledge of the physical characteristics of the system. The other type of challenge is computational. There is no possible way presently available to actually mimic the dynamical interaction of a physical particle system without statistical averaging over time, space, and system size. The accuracy of the simulated behavior is determined by this sampling process. Coarse samples introduce more error than finer samples. Faster and larger computers increase the resolution of time, space, and system size, and hence the simulation accuracy. In this section, we will give a brief review of particle simulation methods, outlining their main characteristics. The purpose is to give the reader the minimum necessary background for our new method which is developed particularly for the numerical simulation of dislocation dynamics.

If the interaction force between the particles does not decay rapidly as the distance between them, then for each particle the influence of all other particles must be computed at every simulation step. This results in a computer algorithm with $O(n^2)$ complexity, where n is the number of particles in the system. For large n, the computational effort required rises rapidly and can exceed even the most powerful supercomputers available. An alternative to the $O(n^2)$ algorithms is the family of particle simulation methods: the particle-particle (PP) model, the particle-mesh (PM) model, and the particle-particle/particle mesh (P^3M) model [20]. In the first method, PP, the state of the physical system at a time t is described by a set of particle positions and velocities. The total force on a particle is computed, the EOMs are integrated, and the time counter is updated by a value dt.

In the PM method, the potential is assumed to satisfy certain elliptic equations. A computational mesh is overlaid on the domain, the source terms are approximated at the mesh points by local interpolation, and the potential is then computed by solving the elliptic problem with the given source terms by fast solvers. Finally, the forces on the individual particles are computed by interpolating the potential field to the particle positions.

The PP method is useful for small systems with long-range forces or for large systems where the forces of interaction are non-zero for only a few interparticle distances. The PM method is computationally faster but less accurate, because it can only handle smoothly varying forces. The P^3M method combines the two methodologies, and allows the simulation of large correlated systems with long-range forces. In this method, the interparticle forces are divided into a rapidly varying short-range force and a slowly varying long-range force. The complexity of the methods are often $O(n\log n)$, a big improvement over the $O(n^2)$ methods. However, the interpolations introduce errors and thus the accuracy is limited, although the accuracy can sometimes be improved via local corrections [21].

Another type of method is "clustering" algorithms [22]. Mayer and Mayer [23] introduced the cluster concept into the molecular theory of fluids. Their cluster has been identified as a "mathematical cluster" because it does not represent a physical cluster of particles, often referred to as a supermolecule. In clustering methods, subsets of particles are lumped together and a monopole (center of mass) expansion is computed which can be used to compute the potential at a far enough distance away from the selected subset. Using this basic idea, one can derive $O(n\log n)$ algorithms as well. The drawbacks are that sophisticated data structures must be used to keep track of which subsets of particles are sufficiently clustered to make the monopole expansion accurate enough, and the method is not so useful when relatively few clusters exist.

A major breakthrough was recently achieved when Greengard and Rohklin [24] announced an $O(n)$ algorithm which is optimal. They used multipole expansions [25] of clusters, together with techniques for shifting the center of expansion of a Taylor series, to determine the forces due to a collection of particles. Basically, they divided the Cartesian space into a regular (fine) grid consisting of cells, each containing a certain number of particles. The multipole expansion can then be computed for all cells on the finest grid in a systematic and efficient way.

Dynamical methods were formulated to predict the trajectories of a number of interacting physical entities as a function of time. The term molecular dynamics (MD) has been used to describe the early versions of such calculations, which were applied to the study of the motion of molecules in a fluid. Alder and Wainwright used hard-sphere and square-well [26] potentials to perform the first MD calculations. In 1964, Rahman [27] made the first MD simulations of fluids with continuous potentials. Verlet [28] made significant contributions to the applications of MD methods. In traditional MD, the total energy for a fixed number of atoms in a fixed volume is conserved as the dynamics of the system evolve in time. The dynamical equations are explicitly integrated yielding particle positions and momenta. Some form of velocity renormalization, or other constraints, are introduced to represent isothermal or isobaric thermodynamic processes.

In 1980, Andersen [29] proposed a mixed MC/MD algorithm for performing isothermal simulations where stochastic collisions from a heat bath are treated in accord with the MC process. A similar approach for a many-body system has been suggested by Abraham [2], where the EOMs are solved considering the stochastic component due to Brownian motion. All the methods described here have been applied primarily to atomic or molecular systems.

Additional constraints are usually applied to MD simulations of equilibrium systems. Constraints, such as a constant temperature, constant pressure, and conserved total momentum guarantee that the macroscopic thermodynamic properties of the system are well represented by the small statistical sample in an MD simulation. In the next section we present a new methodology for representing interactions between dislocations. Several characteristics of the approach are fundamentally different from particle methods described here.

3. The New Methodology of Dislocation Dynamics

3.1. BASIC DISLOCATION CHARACTERISTICS

Dislocations are line defects inside imperfect crystals which generally have 3-D characteristics. A single dislocation can therefore move on different slip systems, with components which may be screw, edge, or mixed. Primary dislocations move on primary slip systems, and they are the first to exhibit motion during plastic deformation. At critical values of applied stress and temperature, secondary dislocations are activated. They interact with primary dislocations leading to multiplication, junction formation, and other processes.

Mobile dislocations are responsible for the observed plastic deformation, and they can move in the crystal if the resolved shear stress on the slip plane is greater than a critical value (the friction stress). When a dislocation is immobilized, it provides further pinning points in the lattice for mobile dislocations. Clustering is attributed to the evolution of immobile dislocations, which ultimately result in various organized structures. Dipoles and multipole bundles are found to be important components of the organized dislocation substructure. In the following, we will summarize the basic characteristics of dislocation interactions which have a significant effect on the emergence of an organized substructure. Literature on dislocations is quite extensive, and we will make no attempt to be comprehensive.

3.1.1. Dislocation Motion. An edge dislocation can move on a glide plane in the direction of its Burgers vector (glide), and can also climb in a direction perpendicular to the glide plane (climb). Screw dislocations can additionally cross slip from one glide plane to another. We will focus here on edge dislocations, since it is relatively easy for screw dislocations of opposite Burgers vectors to mutually annihilate.

Dislocation glide is described by a phenomenological equation of the form [30]:

$$v_g = M_g F_g , \qquad (1)$$

where M_g is the glide mobility and F_g is the net force in the glide direction. The net force on the glide dislocation is expressed by

$$F_g = b\tau_{eff} , \qquad (2)$$

where τ_{eff} is the effective local stress on the dislocation computed from the applied and internal stress fields and b is the magnitude of the Burgers vector. Simple theoretical models for the glide mobility on the basis of self diffusion [13], kink propagation [31], and solute drag [32] give results which are independent of the stress. This is applicable only over a limited range of applied stress, however. It has been experimentally determined [33,34] that the relationship between velocity and stress is nonlinear over a wide range of applied stress. The nonlinear form of the glide velocity given by Eq. (1) is commonly expressed as [35]

$$v_g = v_0(\tau_{eff}/\tau_0)^m , \qquad (3)$$

where v_0 is the shear wave velocity, τ_0 is a material constant, and m is the stress exponent. Meyers and Chawla [36] have determined three regions for the relationship given by Eq. (3), where m can vary. For example, the stress exponent for iron in the first region is 35 and in the second and third regions it is, respectively, 1 and 0.67 [33]. Figure 3 shows our extrapolated calculations of the velocity-stress relationship for a single gliding dislocation in both copper and iron.

At high temperature, the process of climb constitutes the dominant mode of dislocation motion. Climb is thermally activated, and hence is dependent upon the diffusive properties of the material, particularly vacancies to the dislocation core. A phenomenological expression for climb is given by [37]:

$$v_c = v^*(\tau_{eff}/\mu) , \qquad (4)$$

where μ is the shear modulus and v^* is a characteristic thermal climb velocity.

During glide motion, kinks must form first on the dislocation and then propagate under the applied stress. On the other hand, climb motion proceeds by jog propagation along the climb direction under the local stress. It is interesting to note that while the stress dependence of the glide velocity can be highly nonlinear, dislocation climb is only linearly dependent on the stress. It has been found that the ratio of climb to glide velocity is negligibly small for copper for applied stresses in the range of 20 - 100 MPa and at operating temperatures < 100°C. These conditions will be further discussed in Section 4.2 for the conditions necessary to form persistent slip bands (PSBs) under fatigue loading. Figure 4 shows the ratio of the climb-to-glide velocities in iron in the temperature range of 400°-600°C and the stress range of 100-264 MPa. It is quite evident that in Fe, the climb velocity is a significant component of the overall motion. This situation is an important ingredient in the formation of the dislocation cell substructure, as will be discussed in Section 4.3.

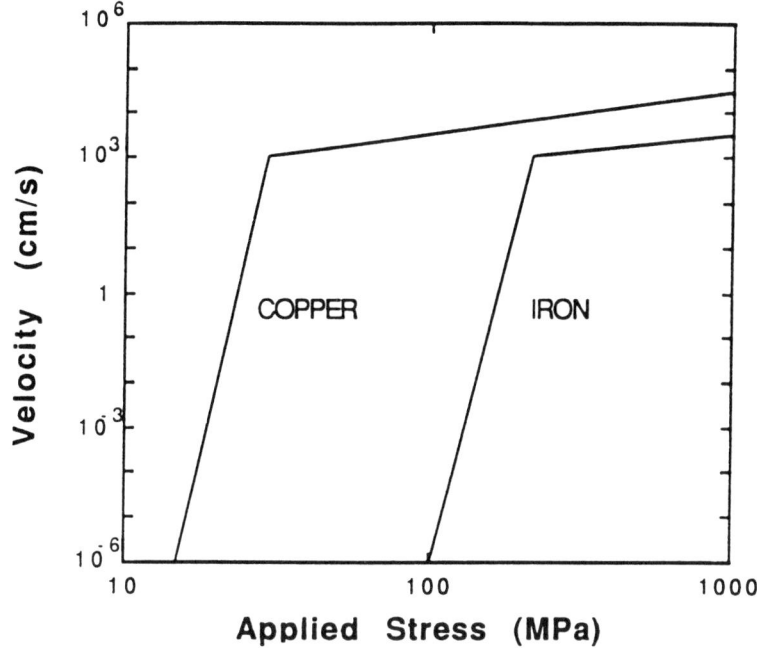

Fig. 3. Calculated velocity-stress relationship for gliding dislocations in Cu and Fe.

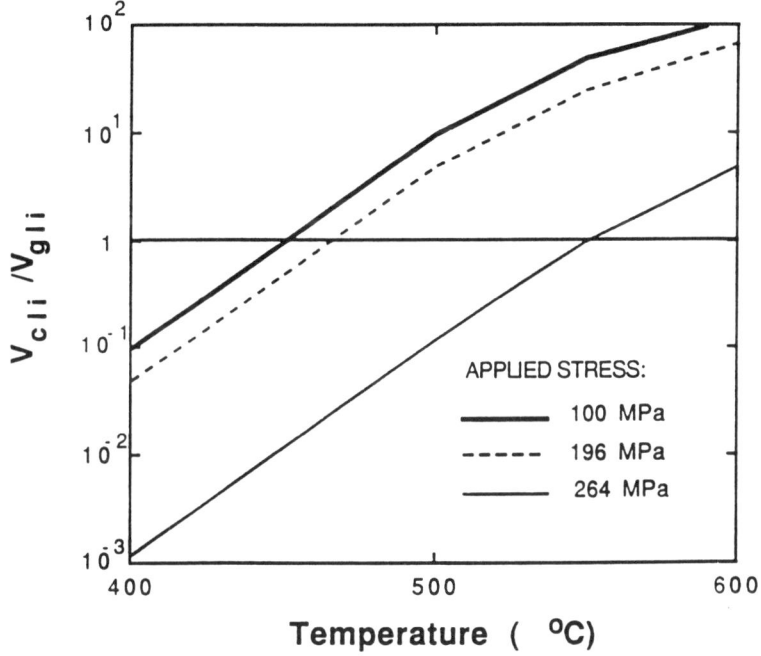

Fig. 4. Ratio of climb-to-glide velocities in Fe.

3.1.2. Long-Range Interactions. Let the x-direction represent the dislocation Burgers vector $\vec{b}$, the y-direction represent the dislocation climb vector $\vec{b}_c$, and the z-direction represent the sense vector of the dislocation $\vec{\xi}$. If, for example, the system consists of three coplanar dislocations, all traveling on different [111] glide planes, one of the dislocations represents the primary dislocation. The 3D characteristics of this dislocation can then be expressed in general terms as a tensor which consists of components

$$\bar{\bar{T}}_D = \begin{bmatrix} \vec{b} \\ \vec{b}_c \\ \vec{\xi} \end{bmatrix} = \begin{bmatrix} b_x & b_y & b_z \\ b_{cx} & b_{cy} & b_{cz} \\ \xi_x & \xi_y & \xi_z \end{bmatrix} . \tag{5}$$

This can be considered the dislocation tensor (DT). By convention, the DT of a primary dislocation is given by the identity matrix.

The stress tensor $\bar{\bar{\Sigma}}$ is determined by transforming the stress tensor of a test dislocation $\bar{\bar{\Sigma}}_t$ into the reference coordinate system. The stress tensor $\bar{\bar{\Sigma}}_t$ of an arbitrary test dislocation is given by

$$\bar{\bar{\Sigma}}_t = \begin{bmatrix} \sigma_{xx} & \sigma_{xy} & \sigma_{xz} \\ \sigma_{yx} & \sigma_{yy} & \sigma_{yz} \\ \sigma_{zx} & \sigma_{zy} & \sigma_{zz} \end{bmatrix} , \tag{6}$$

where, in Cartesian coordinates

$$\sigma_{xx} = -\frac{\mu b}{2\pi(1-\nu)} \left[\frac{y(3x^2 + y^2)}{(x^2 + y^2)^2} \right] ,$$

$$\sigma_{xy} = \sigma_{yx} = \frac{\mu b}{2\pi(1-\nu)} \left[\frac{x(x^2 - y^2)}{(x^2 + y^2)^2} \right] ,$$

$$\sigma_{yy} = \frac{\mu b}{2\pi(1-\nu)} \left[\frac{y(x^2 - y^2)}{(x^2 + y^2)^2} \right] ,$$

$$\sigma_{xz} = \sigma_{yz} = \sigma_{zx} = \sigma_{zy} = 0 ,$$

$$\sigma_{zz} = -\frac{\mu b}{\pi(1-\nu)} \frac{y}{x^2 + y^2} ,$$

for edge dislocations with the sense direction vector $\vec{\xi}$ coincident with the z-axis, and

$$\sigma_{xz} = -\frac{\mu b}{2\pi} \frac{y}{x^2 + y^2} ,$$

$$\sigma_{yz} = -\frac{\mu b}{2\pi} \frac{x}{x^2 + y^2} ,$$

$$\sigma_{xx} = \sigma_{xy} = \sigma_{yx} = \sigma_{yy} = \sigma_{zz} = 0 \quad,$$

for screw dislocations with $\vec{\xi}$ lying along the z-axis. ν is Poisson's ratio.

The rotational transform between two coordinate systems i and j is given by:

$$\bar{\bar{T}}_{rot} = \bar{\bar{T}}_{ij} = [\bar{\bar{T}}_{Di}] \cdot [\bar{\bar{T}}_{Dj}] \quad. \tag{7}$$

If system j is the primary coordinate system, the transformation becomes

$$\bar{\bar{T}}_{rot} = \bar{\bar{T}}_{ii} = [\bar{\bar{T}}_{Di}] \quad. \tag{8}$$

Transforming the components of $\bar{\bar{\Sigma}}_t$ (i.e., σ_{ij}; $i,j \in x,y,z$) into the primary coordinate system is accomplished by the following operation:

$$\bar{\bar{\Sigma}} = \bar{\bar{T}}_{rot} \cdot \bar{\bar{T}}_{rot} \cdot \bar{\bar{\Sigma}}_t = \sum_{\ell=x}^{z} \sum_{m=x}^{z} T_{i\ell} T_{jm} \sigma_{\ell m} \quad. \tag{9}$$

The force per unit length on the test dislocation can be computed by applying the Peach-Koehler equation [38]

$$\vec{F} = (\vec{b} \cdot \bar{\bar{\Sigma}}) \times \vec{\xi} \quad. \tag{10}$$

This force will have a glide component and another climb component. A 2-D velocity vector can then be formed with glide and climb velocities.

A simpler long-range interaction model can be readily constructed if the two dislocations are parallel. If the reference dislocation is located a polar point (R,θ) away from the test dislocation with Burgers vector along $\theta = 0$, we obtain:

$$\bar{\bar{\Sigma}}_t = \frac{\mu b}{2\pi(1-\nu)R} \begin{bmatrix} -\sin\theta & \cos\theta \\ \cos\theta & -\sin\theta \end{bmatrix} \quad. \tag{11}$$

The rotational transform, which rotates the stress tensor into the reference dislocation's coordinate system, is given by

$$\bar{\bar{T}}_{rot} = \begin{bmatrix} \cos\gamma & \sin\gamma \\ -\sin\gamma & \cos\gamma \end{bmatrix} \quad, \tag{12}$$

where γ is the angle between the reference dislocation Burgers vector and the (R,θ) vector. The stress in the reference dislocation's system is therefore given by

$$\bar{\bar{\Sigma}} = \bar{\bar{T}}_{rot} \cdot \bar{\bar{T}}_{rot} \cdot \bar{\bar{\Sigma}}_t = \begin{bmatrix} \sigma_{11} & \sigma_{12} \\ \sigma_{21} & \sigma_{22} \end{bmatrix} \quad, \tag{13}$$

where $\sigma_{11} = -K(\sin\theta - \sin 2\gamma \cos\theta)/R$,

$\sigma_{12} = \sigma_{21} = K \cos\theta \cos 2\gamma/R$,

$\sigma_{22} = -K(\sin\theta + \sin 2\gamma \cos\theta)/R$.

Substituting Eq. (13) into Eq. (10), we arrive at the expression for force components on a dislocation, which are given by the climb and glide components per unit length L, as

$$\frac{F_c}{L} = \frac{\mu |\vec{b}_t||\vec{b}_r|}{2\pi(1-\nu)} \frac{\sin\theta - \sin 2\gamma \cos\theta}{R} , \qquad (14)$$

$$\frac{F_g}{L} = \frac{\mu |\vec{b}_t||\vec{b}_r|}{2\pi(1-\nu)} \frac{\cos\theta \cos 2\gamma}{R} , \qquad (15)$$

where $\vec{b}_{t,r}$ are, respectively, Burgers vectors of test and reference dislocations.

3.1.3. Short-Range Interactions. When dislocations come in close proximity (i.e., $R \leq 1\text{-}2$ nm) the approximations of small-strain, linear elasticity break down. Dislocation-dislocation interactions at these distances involve atomic rearrangements which are driven by complex thermodynamic phenomena. It is impossible at the present time to accurately predict the probabilities of these rearrangements. Experimentally, however, conditions for the occurrance of some of these interactions have been determined. We will therefore describe short-range dislocation interactions as a set of "events" which are determined if certain conditions are satisfied, as discussed below.

1. Immobilization: If the stress on a mobile dislocation falls below the friction stress, the dislocation becomes immobilized. Once a dislocation is immobilized, the only consequence is that the total vector velocity is identically zero. An immobile dislocation can still contribute its elastic strain energy to the overall computation of internal stress on other dislocations. Immobile dislocations can be remobilized if the effective stress is raised above the friction stress.

2. Annihilation: Annihilation is the cancellation of two dislocations of opposite Burgers vectors which approach each other within a certain critical distance of separation. This annihilation has usually been treated as an average over the crystal of the recovery process [39,40]. Essmann and Mughrabi [41] have estimated a value for the critical distance for annihilation of two screw dislocations of opposite Burgers vectors as:

$$y_s \simeq \frac{\mu b}{2\pi \tau_g} , \qquad (16)$$

where y_s is the annihilation width for screw dislocations and τ_g is the shear stress required for dislocation glide. For copper, $y_s \simeq 1.8$ μm; for molybdenum, $y_s \simeq 0.19 - 2.25$ μm. For mixed or edge dislocations, annihilation will occur when the attractive elastic force between two dislocations exceeds the force required for dislocation climb [42]. The critical distance for annihilation of mixed dislocations is thus given by

$$y_m = \frac{\mu b^4}{2\pi \kappa \, U_f \sin\psi} \, , \qquad (17)$$

where y_m is the annihilation width for mixed dislocations, U_f is the energy of formation of atomic defects, ψ is the angle between the Burgers vector and the line vector, and κ is $1 - \nu < \kappa < \nu$. The critical distance for edge dislocation annihilation has been found to be on the order of 1.6 nm [43], which is much smaller than the critical distance for screw dislocations. This value is about an order of magnitude less than the average distance between dislocations within the cell boundary. Prinz and co-workers [44,45] determined this value to be greater than 1.6 nm, but their experiments were carried out at higher temperatures than those of Essmann and Mughrabi [41].

3. **Dipole Formation:** When two edge dislocations of opposite Burgers vectors approach each other, they can achieve a stable configuration if they remain at a distance greater than the critical distance for annihilation [41]. This configuration is known as a dipole and it can exist as a vacancy or interstitial-type configuration. Typical dipole lengths are on the order of tenths of microns [41,46]. Dipoles are composed only of edge dislocations since screw dislocations can easily annihilate by the cross-slip mechanism. Once formed, a dipole does not move as a whole if the external or internal stresses are changed, but it does change its configuration slightly. Application of an applied stress causes a slight change in the relative angle between the dislocations which constitute the dipole [47].

Two dislocations separated by a sufficient distance to form a dipole will not necessarily form a dipole configuration. It is found [46,48] that in order to form a stable configuration, the stress on a dislocation must be less than the passing shear stress for a dipole τ_p, i.e.,

$$\tau_p \simeq \frac{\mu b}{8\pi(1-\nu)y} \, , \qquad (18)$$

where y is the slip plane spacing.

A dipole does not necessarily represent the perfect alignment of two dislocations of opposite Burgers vectors. In fact, the relative angle between two dislocations in a dipole varies from 25 deg to 65 deg [46], with an average around 45 deg. The latter angle represents the position of minimum interaction energy of dislocations within a dipole.

4. **Junction Formation:** Attractive dislocations are those dislocations which experience a net attractive force with respect to each other. If the dislocations are parallel, then the difference in Burgers vectors between two attractive dislocations is between 90 deg and 270 deg. If two attractive dislocations approach each other, they will eventually intersect each other, the intersection being an annihilation event if the two dislocations are of opposite Burgers vectors. If they are not of opposite Burgers vectors, then one of three events can occur: (1) They lock together, forming a Lomer-Cottrell barrier; (2) They draw each

other out, forming a jog intersection; or (3) They intersect and then pass each other. The two dislocations will pass each other if the net stress on one of the dislocations is approximately greater than the pinning stress for dislocations which is given by

$$\sigma_p = \frac{\mu b}{2\pi \xi} \quad , \tag{19}$$

where ξ is the inter-dislocation separation distance.

5. Dislocation Multiplication: Dislocation multiplication is primarily attributed to the Frank-Read mechanism [49] for glide processes. During dislocation creep, the Frank-Read source is dominant and multiplication occurs by pinning of the dislocation, bowing out, and wrapping around the pinning points. Caillard and Martin [50] identified these points in aluminum as primarily small precipitates or impurity clusters and, to some degree, junction segments produced by two dislocations [51]. The latter mechanism is ruled out by the aforementioned authors as a major source, because the stress necessary to activate sources from these points is usually high enough to cause junction recombination [52]. On the other hand, Prinz and Argon [44] have identified anchoring points as dislocation dipoles and multipole bundles in the evolution of cell-wall-type structures [53,54]. The bowing out of free segments from these bundles is considered a major generation source [44].

Multiplication occurs if the total force on a dislocation exceeds the Orowan stress for dislocation reproduction. The Orowan stress criterion is given by:

$$\sigma_0 = \frac{2\mu b}{\lambda} \quad , \tag{20}$$

where λ is the inter-obstacle spacing. If this criterion is satisfied, dislocations are capable of multiplying at a rate

$$\dot{\rho} = \rho \frac{v_g}{\lambda} \quad , \tag{21}$$

where ρ is the total dislocation density. Dislocation multiplication is also possible by a process of climb, similar to the Frank-Read source, known as the Bardeen-Herring mechanism [55]. This process is dominant under conditions of vacancy supersaturation, such as irradiation or quenching conditions. The contribution to production due to the bowing of dislocation links by climb has been calculated by Nabarro [56]. It has been determined however, that this contribution is usually negligible compared to the recovery creep component [57] and is therefore not a major source of production of new dislocations.

3.2. COMPUTATIONAL ASPECTS OF DISLOCATION DYNAMICS

In order to produce a truly dynamical simulation of dislocation interactions, it is important to design the computations with a time resolution which is fine enough to capture the short-range interactions. Four criteria for timestep determination have been developed, as follow:

1. **Vector Timestep** (Δt_u): The maximum computational timestep for the system is limited by the minimum amount of time it would take two dislocations to experience a reaction (collision or annihilation). If we consider two dislocations of arbitrary Burgers vectors approaching each other, we define the universal vector timestep as:

$$\Delta t_u = \min\left[-\frac{|\Delta \vec{R}_{ij} \cdot \Delta \vec{v}_{ij}|}{|\Delta \vec{v}_{ij}|^2}\right] \quad , \quad (22)$$

where $\Delta \vec{R}_{ij} = (\Delta X, \Delta Y, \Delta Z)_{ij} = $ difference in position vector , (23)

$\Delta \vec{v}_{ij} = (\Delta v_x, \Delta v_y, \Delta v_z)_{ij} = $ difference in velocity vector . (24)

Positive values of the timestep are achieved if the two dislocations are relatively moving towards each other. For dislocations which move in two coordinate directions, this timestep represents the time until a collision between the two dislocations occurs. If the dislocations are of like sign, the collision will be a repulsive event. If the two dislocations are of opposite sign, the collision will be either an annihilation, a dipole formation, or a junction event.

For dislocations which only move in one coordinate direction, this timestep corresponds to the time to the distance of closest approach. The value of this distance is given by:

$$\Delta R_{min} = \frac{1}{|\vec{v}_{ij}|} \cdot (|\Delta \vec{R}_{ij}|^2 \cdot |\Delta \vec{v}_{ij}|^2 - |\Delta \vec{R}_{ij} \cdot \Delta \vec{v}_{ij}|^2)^{1/2} \quad . \quad (25)$$

If the two dislocations are on a collision course, then the discriminant in Eq. (25) for minimum separation is identically equal to zero. If the two dislocations are incapable of short-range interaction, then the value of the discriminant is positive.

2. **Advancement Timestep** (Δt_a): If two approaching dislocations are of opposite Burgers vectors and their interaction limits the system dynamics, then the calculated timestep should be the time it takes the dislocations to experience a short-range interaction, not the time it takes for the dislocations to occupy the same space. This advancement vector timestep is given by

$$\Delta t_a = \min(\Delta t_u)\left\{1 - \left[1 - \frac{|\Delta \vec{v}_{ij}|^2}{|\Delta \vec{R}_{ij} \cdot \Delta \vec{v}_{ij}|^2}(|\Delta \vec{R}_{ij}|^2 - \delta_0^2)\right]^{1/2}\right\} \quad , \quad (26)$$

where δ_0 is the characteristic short-range interaction distance. This is a more stringent timestep requirement than the previous one because the term in curley brackets is ≤ 1.

3. **Equilibrium Timestep** (Δt_e): If two dislocations of the same Burgers vector approach each other, they introduce an additional constraint on the timestep. If one of the two dislocations is immobile, it represents

an obstacle against which the mobile dislocation will come to equilibrium. Without viscous drag, a large timestep can introduce artificial numerical oscillations in the mobile dislocation trajectory. Our work on dislocation pile-up dynamics [58] indicated that if an equilibrium timestep is defined as

$$\Delta t_e = f \Delta t_u \quad . \tag{27}$$

then the average relative error in the position of the dislocation is found to be 36% for $f = 1/1.5$, 9% for $f = 1/5$, and 3% for $f = 1/15$.

4. **Multiplication Timestep (Δt_m)**: If the Orowan criterion is satisfied for a dislocation passing through obstacles on its glide plane, dislocation loops are produced. Each dislocation loop is then approximated as two edge dislocations separated by the inter-obstacle distance. The production rate equation is given by

$$\Delta \rho = \rho \frac{v_g}{\lambda} \Delta t \quad . \tag{28}$$

Within a multiplication strip, fractions of Orowan loops are not introduced. A cummulative multiplication timestep must pass before a number of loops ($N_{loop} \sim 1$) is introduced. This time is given by

$$\Delta t_m = \min(\frac{N_{loop}}{\nu' \rho}) \quad , \tag{29}$$

where ν' is the production frequency ($\nu' = v_g/\lambda$).

The system timestep is determined dynamically by surveying all dislocations and chosing the minimum timestep as

$$\Delta t = \min(\Delta t_u, \Delta t_a, \Delta t_e, \Delta t_m) \quad . \tag{30}$$

This procedure guarantees that the true dynamics of dislocation interactions are well simulated. For example, fatigue simulations for copper at 25°C give very short timestep ($\sim$ ns) until immobilization of the closest dislocations occurs. Thereafter, the timestep increases to the ms range. This wide dynamic range of timesteps is necessary for pattern simulations.

3.3. UNIQUE FEATURES OF DISLOCATION DYNAMICS

Dislocation dynamics methodology has unique features which are not necessarily embodied in conventional MD. Dislocations represent line defects which interact with each other through a long-range linear elastic field and a short-range nonlinear field. The system size in dislocation dynamics can represent an entire grain, for example, with 10,000 to 100,000 dislocations. In MD, on the other hand, systems containing up to 250,000 particles have been simulated and are only considered as a statistical sample of matter. Therefore, statistical thermodynamical principles can be used to re-normalize velocities if the studied phenomenon is near thermodynamic equilibrium. Periodic boundary conditions can be invoked in view of the statistical sampling argument. In DD, the application of thermodynamical concepts and statistical sampling is not yet established. The results of a dynamical simulation of a dislocation system represent a deterministic and not a sampled behavior of those dislocations contained within a grain. The presence of a long-range vector field for dislocations is not known in the majority of MD simulations. When short-range dislocation reactions such as annihilation and production are considered, particle conservation principles are not applicable. In MD simulations, particle conservation is a fundamental feature. In a similar way, energy conservation is more difficult to apply because of the various nonconservative dislocation reactions.

It is argued here that such fundamental differences between the DD and MD methodologies are strong enough to differentiate the two simulation approaches. The DD methodology is viewed as a possible new tool for the study of micromechanical problems where dislocations play a signficant role. A broader discussion of the specific differences between MD and DD simulations can be found in Ref. [59].

4. Numerical Simulation of Dislocation Pattens

4.1. PLANAR ARRAYS

Planar arrays arise from enhanced dislocation multiplication due to alloying elements [46]. Several nucleation sites develop, resulting in a set of parallel dislocations comprising groups of dislocation dipoles in band configurations. Once a single band forms, the nucleation process for an adjacent band is possible and appears to occur roughly 0-5 microns away.

We consider here a rectangular 2-D space ($4 \mu \times 2 \mu$) with an initial random distribution of 400 dislocations. A critical resolved shear stress is applied along the x axis, and the y axis is divided into strips where multiplication can occur. Periodic boundary conditions are invoked. Orowan-type dislocation multiplication is autocatalytic, and the mechanism is terminated at high dislocation densities within the strip. The parameters for this copper simulation at 25°C are given in Table I.

Table I. SIMULATION PARAMETERS FOR PLANAR-ARRAY FORMATION

Parameter	Value
Dipole width (Burgers vectors)	130
Annihilation width (Burgers vectors)	65
Applied stress (MPa)	30
Initial dislocation density (cm^{-2})	5×10^9
Friction stress (MPa)	5

Figure 5 shows the results of a simulation sequence at equal time intervals (~ 1/8 s). During early times (e.g., $t \leq 1/4$ s), the dislocation annihilation rate is substantial, as can be observed by comparing the first and second frames. At later times, random inhomogenieties develop leading to an oscillating pattern of planar arrays. High concentrations of dipoles are observed within each planar array. Many of the experimental features reported by Neuhauser [46] can be observed in these simulations, even though the overall model is quite simple.

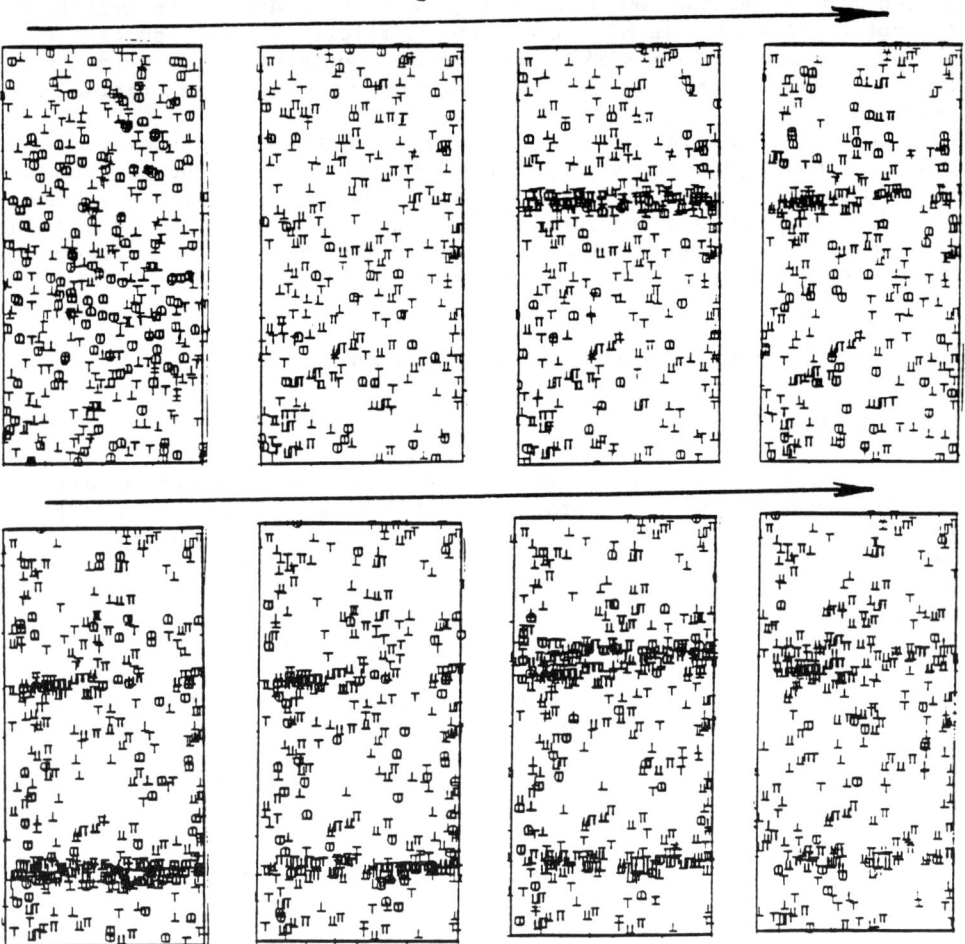

Fig. 5. Results of a simulation sequence for planar arrays in Cu at 25°C (parameters are in Table I).

4.2. PERSISTENT SLIP BANDS (PSBs)

Persistent slip bands (PSBs) are known to emerge from an original vein structure [60]. Climb is usually negligible, and the major component of dislocation motion is the glide process. It is experimentally found [61] that the major constituents of PSB walls are edge dislocation dipoles. Screw dislocations annihilate easily by the cross-slip process. We will therefore ignore the effects of screw dislocations.

The initial configuration of the vein structure in the simulation is given by two half-veins located at the edge of the simulation space (2.8 μ x 1 μ). Periodic boundary conditions are also used in this case. The parameters in Table I are again used for this simulation with a change in the dislocation density to 2.86 x 10^{10} cm^{-2}. The stress cycling frequency was taken as 1 Hz.

Figure 6 shows a simulation sequence over one second. The sequence represents frames (a total of 24) obtained at equal time intervals. Time progression is indicated by the arrows. Significant annihilation and reduction in the total dislocation density is observed inside the original vein structure. One can also see the development of a shear band at the bottom of the figure. This is interpreted as the mechanism by which slip lamellae occur within the material, which eventually drive the wavy band structure in PSBs.

4.3. DISLOCATION CELLS

Dislocation cell walls can comprise two or three sets of Burgers vectors [50]. The motion of dislocations of different Burgers vectors occurs in three separate directions. Since we are cutting a slice of the material and observing dislocations in two dimensions, it would appear difficult to simulate the true motion of dislocations. However, considering that the system of dislocation cells is isotropic in three Cartesian directions, we can represent the projection of the dislocation motion of two of the Burgers vectors in the plane of observation. Since the dislocation is a long line of discontinuity, and not actually a "link," a dislocation will not move out of the plane of observation. Hence the motion of the projection of the dislocation line will always be properly represented in the 2-D plane of observation. In the current biaxial simulation, we consider a system of two orthogonal glide directions for dislocation motion. This reduces the complexity of the simulation, and represents a more accurate interpretation of the projection of dislocation lines onto the observation plane.

The material is subjected to a cumulative strain within the primary range of dislocation creep, and it is assumed that a critical number of dislocations have already been produced. The following stage is the collapse of the dislocation structure into the cellular configuration. This happens within a short timespan (~ 1 s to 1 hr) depending upon initial conditions. Once the cells form, they exist as stable structures for a short period until coarsening occurs and they coalesce into subgrains. In the current study we consider only the initial stages of cell formation. The material considered in this study is alpha-iron. All of the cases studied in the following computer experiments consist of 1 square micron of the microstructure containing 400 dislocations (dislocation density: 4 x 10^{10}/cm^2). The dislocations are oriented with their Burgers vectors aligned with the corresponding diagonals of the square. Therefore there are a total of four different types of dislocations, or two sets of dislocation systems with orthogonal Burgers vectors. A monotonic stress is applied in the positive x direction.

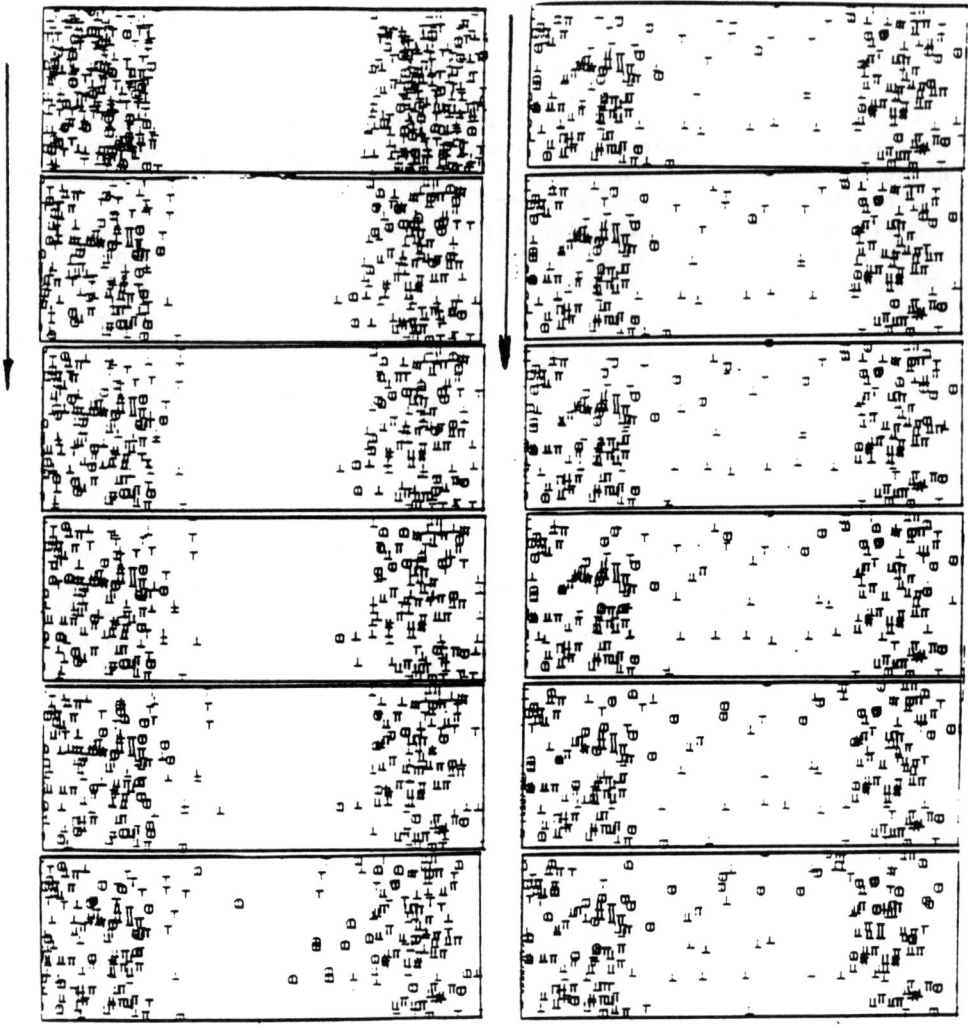

Fig. 6. Simulation sequence for PSB formation in copper at 25°C under fatigue conditions (sheet 1 of 2).

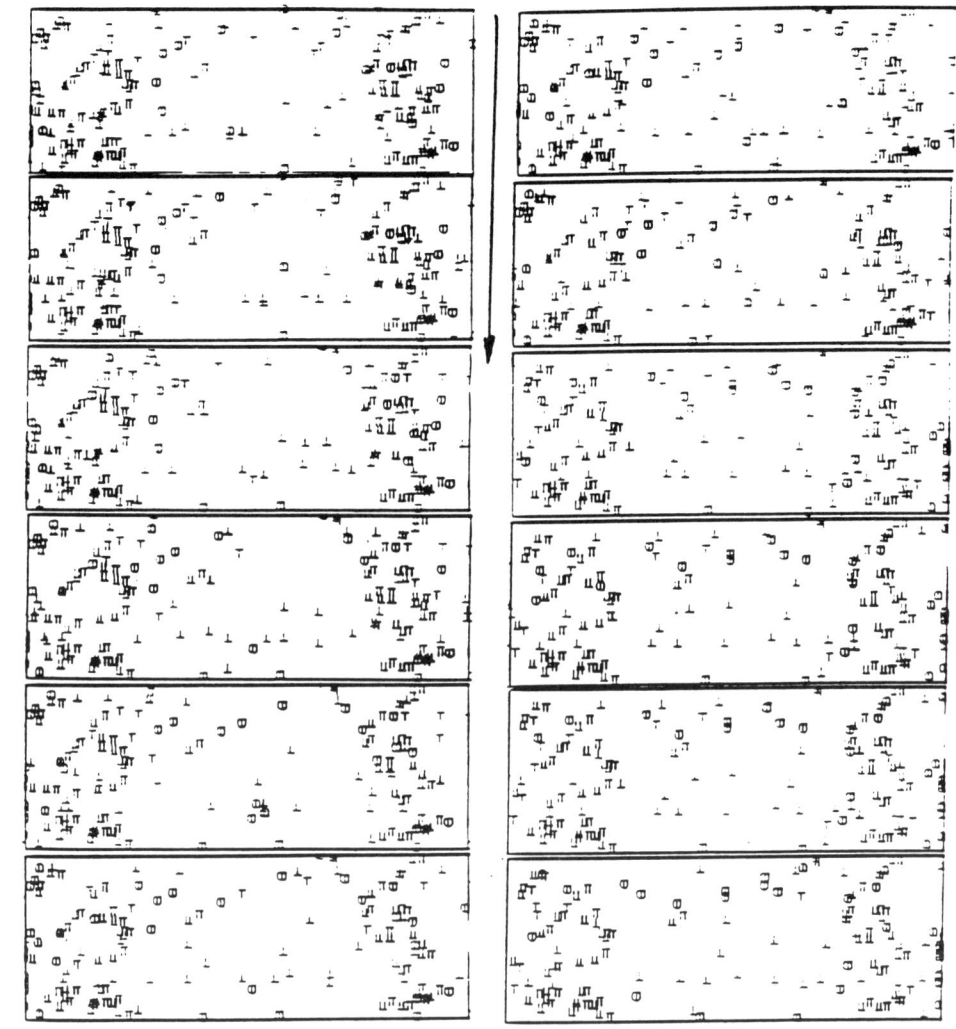

Fig. 6. Simulation sequence for PSB formation in copper at 25°C under fatigue conditions (sheet 2 of 2).

The temperature is varied between 400° and 600°C and the applied stress is varied between 100 and 400 MPa.

It is found that performing a simulation beyond the limits of these parameters does not produce organized structures. The computer simulation for the case of high temperature (600°C) and high stress (400 MPa) results in a cluster of dislocation multipoles which forms rapidly, while mobile dislocations which do not form dipoles experience annihilation by climb processes. On the other hand, a simulation at low temperature (400°C) and low stress (100 MPa) leads to the formation of a large number of immobile dislocations. This happens because the applied stress is much lower than the friction stress. Finally, at low temperature (400°C) and high stress (250 MPa), but with a much lower value of dislocation density ($\sim 1.4 \times 10^{10} cm^2$), the computer simulation produces a collection of mostly mobile dislocations in a loosely knit cellular configuration.

Figure 7 is a simulation of the formation of dislocation cellular structures under creep conditions [high temperature (600°C) and low stress (100 MPa)]. In this simulation, it is seen that a cellular structure rapidly forms. The cell sizes are on the order of 0.5 μ, which is somewhat lower than that expected from average conditions described by Holt's relationship between dislocation density and cell radius [7]. The experimental value of the average cell diameter corresponding to this equation is on the order of 1 micron for the given initial conditions. The resulting configuration contains a total of 210 dislocations, of which 206 are immobilized dislocations: 93 are immobile, 29 are dislocation dipoles, and 84 are junction dislocations. The boundaries of the cells are primarily composed of dipole or junction dislocations. It appears that the formation of the cellular structure is therefore critically dependent upon the total number of immobilized dislocations, in particular stable immobile dislocations.

Figure 8 shows the average internal and effective stresses of mobile dislocations as functions of time for a typical case. Initally, dislocations are randomly distributed throughout the medium. There is a general tendency for the reduction of the elastic stresses due to the rearrangement of dislocations into screened configurations. The effective stress follows the internal stress due to the averaging out of the applied and friction stresses for a random configuration. It should be recalled that the friction stress opposes the glide velocity in whichever direction the dislocation is gliding. This is why the friction stress averages to zero in the initial random configuration. The applied stress is actually dependent upon the angle of orientation of the dislocation with respect to the x-axis. For a single dislocation, the applied stress is given by $\sigma = \sigma_a \cos\alpha$. Thus integrating this over all angles results in an average applied stress of zero.

After about 90 timesteps of the simulation, it is interesting to observe that the internal stress bifurcates from the effective stress, and the two are separated by a nearly constant value until the end of the simulation. The steady-state value of internal stress approximately assumes the value of the effective critical resolved shear stress. The value of the effective stress approaches the difference between the internal stress and the friction stress. At this point, a cellular configuration has formed, and mobile dislocations are effectively screened from one another.

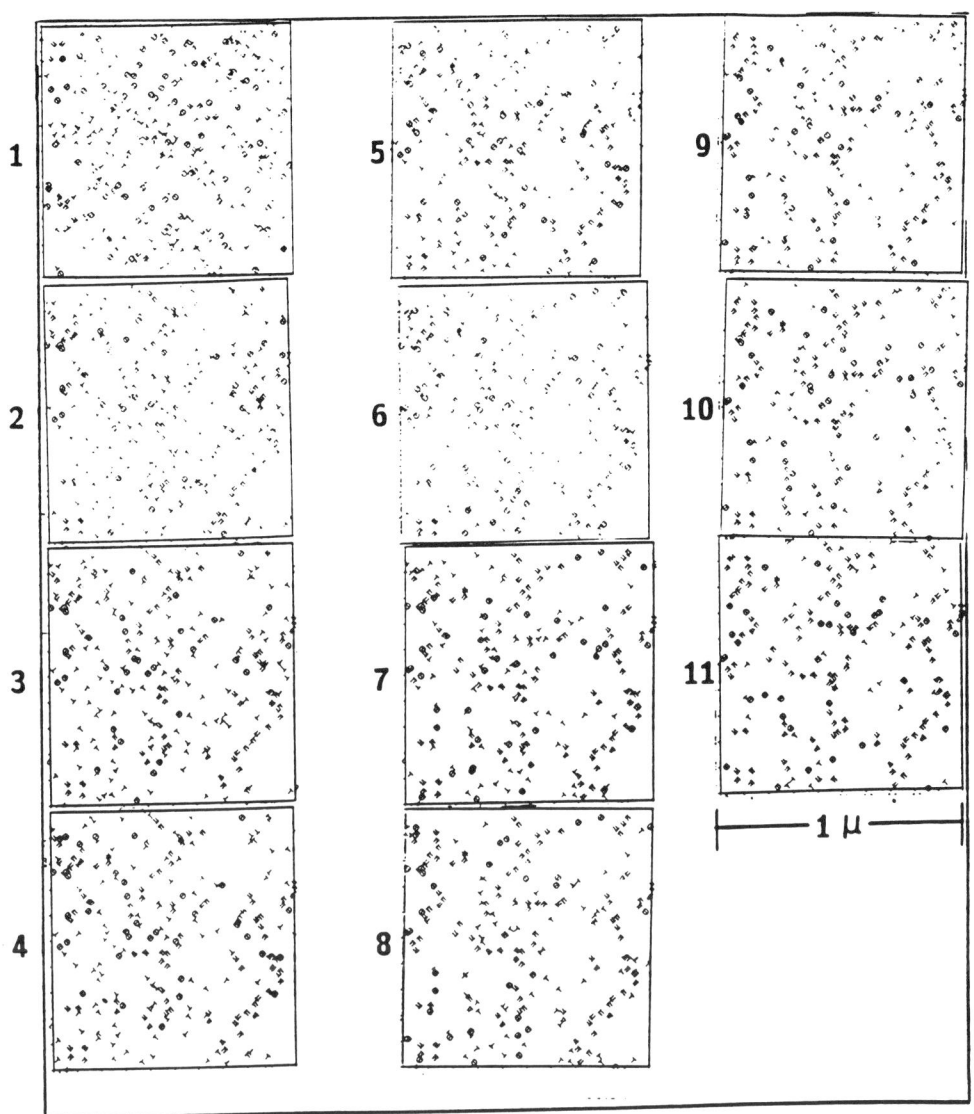

Fig. 7. Simulation of cell formation: Iron @ 600°C, 100 MPa, $4 \times 10^{10}/cm^2$ (~1 frame/s).

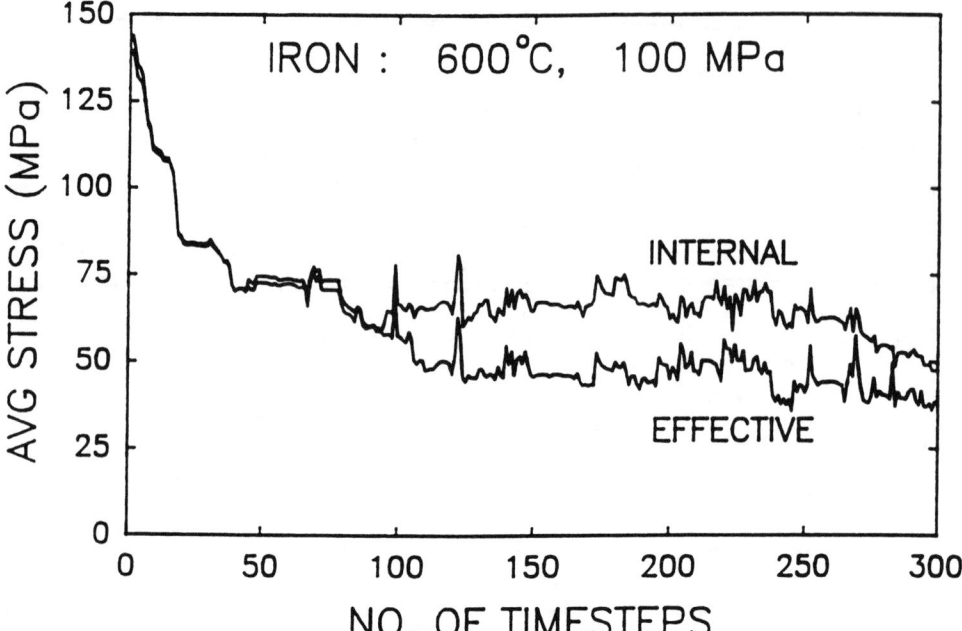

Fig. 8. Evolution of internal and effective stresses for the simulation of cell formation (total time = 10 s).

5. Conclusions

With a simple physical model of dislocation interactions, we have demonstrated that organized dislocation structures can clearly be numerically simulated. It is shown that relatively few physical mechanisms are necessary for the evolution of these organized structures.

The small value of climb velocity is the primary reason for the alignment of dipoles into banded structures. Since climb is negligible at low temperatures, dislocations in a stacked configuration are in a metastable condition. Because dislocations have formed dipolar configurations within the PSB, they have essentially stabilized after the alignment of the dislocations has occured. In general, however, the balance between annihilation and dipole formation is the key to the actual formation of these structures. The formation of planar arrays, on the other hand, is dependent upon heterogeneous multiplication processes and the stability of dipole structures along a slip plane.

It is found that the tendency to form one type of organized structure or another is dependent upon the material which is subjected to the conditions of applied stress and temperature. This is because the relative speed of dislocations in the climb and glide directions is an important contributing factor in the formation of organized spatial dislocation structures.

The formation of cellular structures is found to be critically dependent upon the total number of immobilized dislocations, in particular stable immobile dislocations. Initally, dislocations are uniformly distributed throughout the medium. The initial driving force for cell formation is the tendency of the system to reduce the total elastic strain energy through the reduction of the internal stresses by a rearrangement of dislocations into a more shielded configuration. The effective stress follows the internal stress due to zero averaging of both the applied and friction stresses for a stochastically random configuration. Nonlinear reactions, such as dipole and junction formation, do not permit this complete relaxation of the elastic strain energy. The average internal stress relaxes to a value on the order of the applied stress, which is consistent with a cell diameter that is inversely proportional to the applied stress, as observed experimentally. It is to be noted that the climb mobility of dislocations must still be sufficiently high for dislocation cells to form.

Although the emerging organized structure is undoubtedly of a lower total elastic energy as compared to the initial random configuration, it is constrained by nonlinearities which are inherent in dislocation reactions.

Acknowledgement

This work was supported by the U.S. Department of Energy, Office of Fusion Energy, Grant #DE-FG03-84ER52110, with UCLA.

References

[1] G. Nicolis and I. Prigogine, Self-Organization in Nonequilibrium Systems (Wiley Press, New York, 1977).
[2] F. F. Abraham, Adv. in Phys. 35 (1986) 1.
[3] J. W. Gibbs, Collected Works, v. 1 (Yale University Press, New Haven, CT, 1948) pp. 105-115, 252-258.
[4] J. W. Cahan, Acta Metall. 9 (1961) 795.
[5] D. W. Heerman, Phys. Rev. Lett. 52 (1984) 1126.
[6] H. Haken, Synergetics (Springer-Verlag, New York, 1977).
[7] D. Holt, J. Appl. Phys. 41 (1970) 3197.
[8] J. H. Gittus, Acta Metall. 22 (1974) 789.
[9] J. H. Gittus, Philos. Mag. 35 (1977) 293.
[10] J. H. Gittus, Philos. Mag. 39 (1979) 829.
[11] Mater. Sci. and Engrg. 81 (1986).
[12] D. Kuhlmann-Wilsdorf, Mater. Sci. and Engrg. 86 (1987) 53.
[13] R. Sandstrom, Acta Metall. 25 (1977) 905.
[14] R. Sandstrom, Acta Metall. 25 (1977) 897.
[15] J. C. M. Li, J. Appl. Phys. 33 (1962) 2958.
[16] N. M. Ghoniem, J. R. Matthews, and R. J. Amodeo, Res Mechanica, accepted.
[17] D. Walgraef and E. Aifantis, Intl. J. Engrg. Sci. 23 (1985) 1359.
[18] D. Walgraef and E. Aifantis, Intl. J. Engrg. Sci. 23 (1985) 1365.
[19] D. Walgraef and E. Aifantis, Intl. J. Engrg. Sci. 24 (1986) 1789.
[20] R. W. Hockney and J. W. Eastwood, Computer Simulation Using Particles (McGraw Hill, New York, 1981).

[21] C. R. Anderson, J. Comp. Phys. 62 (1986) 111.
[22] A. W. Appel, SIAM J. Sci. Stat. Comp. 6 (1985) 85.
[23] J. E. Mayer and M. G. Mayer, Statistical Mechanics (Wiley, New York, 1940) Chapters 14,15.
[24] L. Greengard and V. Rokhlin, "A Fast Algorithm for Particle Simulation," Yale University report YALEU/DCS/RR-459, 1986.
[25] A. Brandt, Math. Comp. 31 (1977) 333.
[26] B. J. Alder and T. E. Wainwright, J. Chem. Phys. 27 (1957) 1208.
[27] A. Rahman, Phys. Rev. A136 (1964) 405.
[28] L. Verlet, Phys. Rev. 15 (1967) 98.
[29] H. C. Andersen, J. Chem. Phys. 72 (1980) 2384.
[30] M. J. Turunen, Philos. Mag. 30 (1974) 1033.
[31] J. P. Hirth and J. Lothe, Theory of Dislocations (Wiley-Interscience, New York, 1982).
[32] M. W. Grabski and J. W. Wyrzykowski, Mater. Sci. and Engrg. 44 (1980) 229.
[33] D. F. Stein and J. R. Low, Jr., J. Appl. Phys. 32 (1960) 362.
[34] R. W. Rohde and C. H. Pitt, J. Appl. Phys. 38 (1967) 876.
[35] W. F. Greenman, T. Vreeland, Jr., and D. J. Wood, J. Appl. Phys. 38 (1967) 3595.
[36] M. A. Meyers and K. K. Chawla, Mechanical Metallurgy (Prentice-Hall, New Jersey, 1984).
[37] A. S. Argon, F. Prinz, and W. C. Moffatt, in: Dislocation Creep in Subgrain-Forming Pure Metals and Alloys, Eds. B. Wilshire and D. R. J. Owen (Pineridge Press, UK, 1981) p. 1.
[38] M. Peach and J. S. Kohler, Phys. Rev. 80 (1950) 436.
[39] R. Lagneborg, Met. Sci. J. 6 (1972) 127. or Metall.?
[40] R. Lagneborg, Met. Sci. J. 3 (1961) 161.
[41] U. Essmann and H. Mughrabi, Philos. Mag. 40 (1979) 731.
[42] J. Friedel, Dislocations (Pergamon Press, Oxford, 1954).
[43] U. Essmann and M. Rapp, Acta Metall. 21 (1973) 1305.
[44] F. Prinz and A. S. Argon., Phys. Status Solidi A57 (1980) 741.
[45] F. Prinz, H. P. Karnthaler, and H. O. K. Kirchner, Acta Metall. 29 (1981) 1029.
[46] H. Neuhauser, O. B. Arkan, and H. H. Potthoff, Mater. Sci. and Engrg. 81 (1986) 201.
[47] J. C. M. Li, Discuss. Faraday Soc. 38 (1964) 138.
[48] H. Mughrabi, in: Continuum Models of Discrete Systems (North-Holland Publ. Co., Amsterdam, 1981) p. 241.
[49] W. T. Read, Dislocations in Crystals (McGraw-Hill, New York, 1953).
[50] D. Caillard and J. L. Martin, Acta Metall. 31 (1983) 813.
[51] D. Caillard and J. L. Martin, in: Creep and Fracture of Engineering Structures, Eds. B. Wilshire and D. R. L. Owen (Pineridge Press, Swansea, 1981) p. 17.
[52] G. Saada, Acta Metall. 8 (1960) 841.
[53] H. Mughrabi, Philos. Mag., 23 (1971) 897.
[54] A. S. Argon, in: Physics and Strength of Plasticity, Ed. A. S. Argon (MIT Press, Cambridge, MA, 1970) p. 217.
[55] J. Bardeen and C. Herring, Imperfections in Nearly Perfect Crystals (Wiley Press, New York, 1952).
[56] F. R. N. Nabarro, Philos. Mag. A16 (1967) 231.
[57] B. Burton, Philos. Mag. A45 (1982) 657.

[58] R. J. Amodeo and N. M. Ghoniem, <u>Int. J. Engrg. Sci</u>. **26** (1988) 653.
[59] R. Amodeo and N. M. Ghoniem, "Dislocation Dynamics: Part I—A Proposed Methodology for Deformation Micromechanics," University of California Los Angeles report UCLA-ENG-8951 (June 1989) <u>Phys. Rev.</u>, submitted.
[60] C. Laird, P. Charsley, and H. Mughrabi, <u>Mater. Sci. and Engrg</u>. **81** (1986) 433.
[61] H. Mughrabi, <u>Mater. Sci. and Engrg</u>. **33** (1978) 207.

PATTERN FORMATION DURING CW LASER MELTING OF SILICON

K. DWORSCHAK, J.S. PRESTON *, J.E. SIPE and H.M. VAN DRIEL
Department of Physics
University of Toronto and Ontario Laser and Lightwave Research Centre
Toronto, Ontario, Canada, M5S-1A7

*Presently at: Department of Engineering Physics, McMaster University, Hamilton, Ont., Canada

ABSTRACT

We summarize our recent experimental and theoretical work on the molten-solid morphologies which form during irradiation of thin Si films by a cw, $\lambda = 10.6\mu m$ laser. We have observed a variety of ordered and disordered patterns which depend on laser intensity, spot size, polarization and angle of incidence.

Introduction

Nature is full of many examples of spontaneous symmetry breaking or self-organization in which new structures are formed from lower or higher symmetry states. The phenomena of Rayleigh-Benard instabilities, wind-driven pattern formation in clouds and the onset of laser action in an optically pumped material are common examples of how open systems can display new ordered states as a function of a controllable external variable.[1,2] Because of the wide-range of physical phenomena that can lead to pattern formation, the field is highly interdisciplinary in nature. In this article we describe pattern formation induced in silicon thin films during melting by cw lasers. A wide variety of ordered and disordered inhomogeneous states and "transitions" between these states have been observed as a function of angle of incidence, polarization, intensity and spot size.

The fact that inhomogeneous melting should occur at all is not unanticipated for two reasons. The first reason is related to the fact that the reflectivity of a bulk, molten semiconductor is typically three times greater than that of the solid semiconductor.

As one increases the intensity of a light beam on a semiconductor, therefore, at a certain intensity too much light will be absorbed for the material to remain a solid while too little light would be absorbed by the liquid to permit the melt to sustain itself. For a wide range of intensities, then, one might expect a semiconductor to form an inhomogeneous steady-state which possesses a macroscopic reflectivity intermediate between that of the uniform solid and uniform molten material. Indeed, several authors have noted that *incoherent* high intensity light can lead to such inhomogenous states. Celler et al.[3], Grigoropoulos et al.[4] and Jackson and Kurtze[5] have all considered or observed this spatially inhomogeneous state which naturally forms to satisfy the energy budget.

The second reason why pattern formation is not unanticipated when laser light is used to induce melting relates to an intrinsic *transverse instability* associated with the interaction of *polarized, coherent* radiation with all classes of absorptive matter.[6] Since even a microscopically rough surface can scatter radiation, the interference between incident and surface scattered light can lead to periodic energy deposition and periodic melting or vaporization. For linearly polarized light at normal incidence, "gratings" are therefore formed with their wavevector parallel to the light polarization. For picosecond and nanosecond pulsed-lasers, the induced structures depend solely on the surface electrodynamics and *transient* feedback processes which occur during a single pulse and from pulse to pulse.[7] For nanosecond and shorter pulses thermodynamic processes, such as heat flow, surface tension, etc., play little role, apart from a local restructuring of the surface profile. The interaction of cw laser beams with materials allows one to explore the *steady-state* development of periodic (and even disordered!) structures and consider the importance of feedback processes in a time-independent experiment.

Bosch and Lemons[8] and Biegelsen et al.[9] first considered the melting of silicon on sapphire samples by a linearly polarized, continuous laser beam. At the melting point the visible amount of blackbody radiation from the samples is sufficiently strong that one can observe the melt patterns through an optical microscope; the melt regions appear darker than the solid regions because of the lower emissivity of liquid silicon. Using a 10.6 μm laser Biegelsen and coworkers were able to observe the grating-type patterns seen earlier by pulsed lasers. However, in this case, true steady state patterns were observed. In addition, with shorter wavelength ($\lambda \simeq 0.5$ μm) lasers both groups had been able to observe solid and molten lamellae structures. Neither group considered detailed explanations of why the particular patterns formed or why they were stable.

We[10] have considered the melting of silicon on sapphire by a 10.6 μm cw beam in more detail. We have investigated the dependence of pattern formation on laser parameters such as wavelength, intensity, spot size, polarization and angle of incidence. In addition we have theoretically considered the electrodynamics associated with how a coherent, cw electromagnetic field interacts with an ordered or disordered structure in which metallic and dielectric regions are separated by distances of the order of a wavelength.

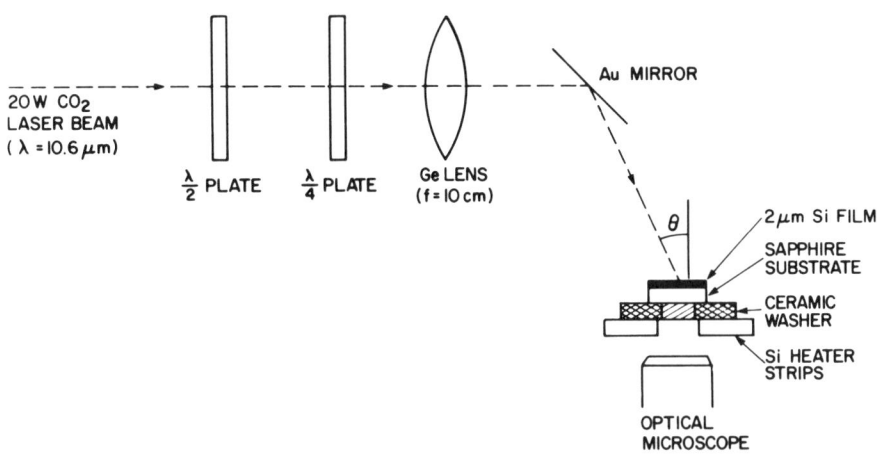

Figure 1. Diagram of the experimental setup used to observe pattern formation in silicon-on sapphire by a 10.6 μm, cw laser beam.

Experimental Details

Fig. 1 shows a schematic diagram of the experimental arrangement we have used to produce and observe the melt/solid patterns on silicon on sapphire. The laser is a cw, 14 W CO_2 linearly polarized laser operating at 10.6 μm. To control the polarization of the beam which reaches the silicon sample we use CdS quarter and half wave plates. The spot size on the sample was controlled with a 10 cm focal length lens. The samples are 2.5x2.5 mm² pieces of 2 μm thick polycrystalline Si on a 1 mm thick Al_2O_3 substrate. To observe the solid/melt patterns a microscope with a 10x objective was used to resolve the visible component of the blackbody radiation (T ≃ 1685K at the melting point) through the sapphire. The image was recorded by film or videotape. To reach the melting temperature we proceeded in two stages. Once the silicon film is brought to $\simeq$ 800 °C by resistive heating, thermally generated electrons and holes can absorb sufficient radiation at 10.6 µm to raise the temperature to the melting point. In various studies we have used film thicknesses in the range 0.5-3 µm, and at least in this range when the film begins to melt, it melts through the complete film thickness.

In experiments performed with light at non-normal incidence we had to consider the influence of s- and p-polarized light separately. Also, in studies of the solid-melt patterns formed under these circumstances we were forced to consider other paramters that change with the angle of incidence. In particular, for s-polarized light the reflectivity and spot size increase as the angle of incidence, θ, increases, requiring increased incident intensity to initiate melting. With our laser it was not possible to initiate melting for an angle larger than 60°. For p-polarized light, although the spot size increases with θ, the reflectivity decreases up to the Brewster angle. One could then simply vary both the spot size and intensity for large angles.

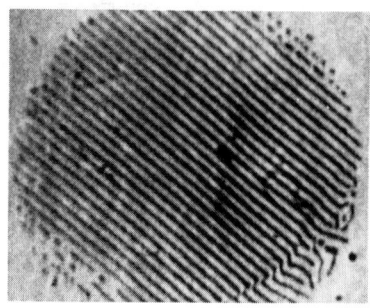

Figure 2. Solid-melt pattern formed at a power density of 2.7 kW/cm² with a spot size of 500 µm; grating spacing 10.6 ± 0.1 µm; dark regions are molten.

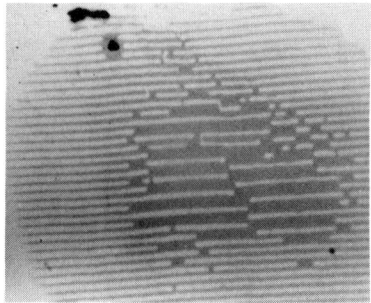

Figure 3. Solid-melt pattern formed by a normal incidence beam at a power density of 3.2 kW/cm² and a spot size of 500 µm; inner grating spacing 21 ± 0.4 µm.

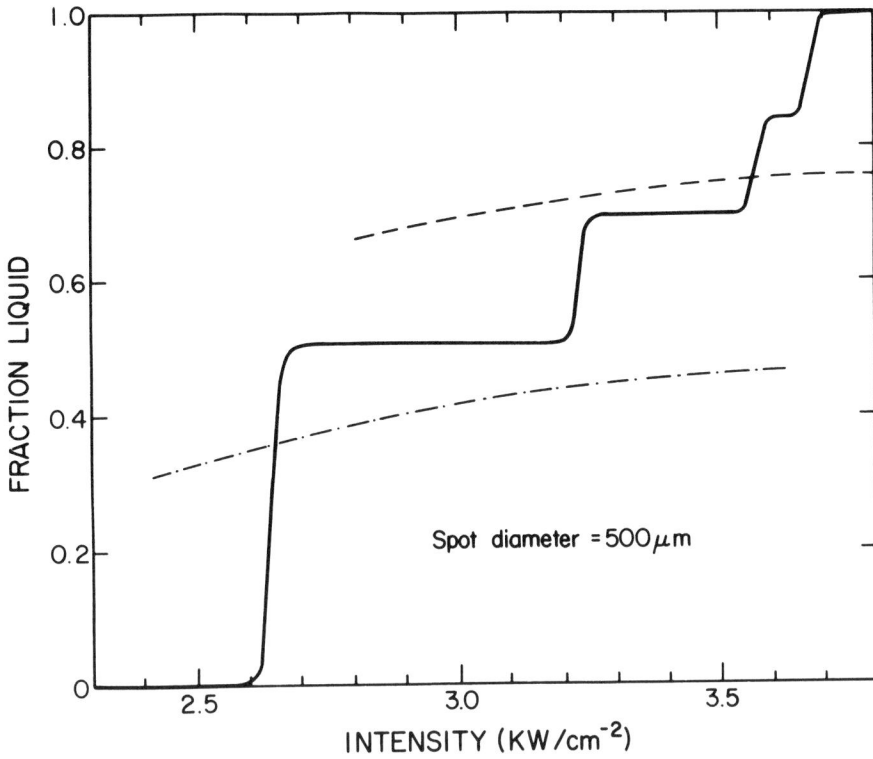

Figure 4. Fraction of liquid as a function of laser intensity illustrating the plateaus associated with "λ", "2λ" and "3λ" gratings; solid curve is a schematic of experimental results; dashed curve is based on a theoretical calculation.

Results

Let us first consider the types of patterns that form at normal incidence. Fig.2 shows a melt pattern generated with a laser spot size of 500 μm and an intensity of 2.7 kW/cm². Alternating strips of solid and liquid appear, with a spacing Λ = 10.6 μm ±0.1 μm, and oriented with the wavevector of the grating (κ_g) parallel to the polarization of the laser beam. Unlike the somewhat undulating gratings which form through pulsed laser-semiconductor interactions[6,7], in the cw case the strips making up the grating are virtually parallel illustrating the fact that feedback in the light scattering process has allowed the strongest component of the scattered light field distribution to dominate the pattern formation. On increasing the laser intensity to 3.2 kW/cm² the simple grating structure is replaced with a grating which has twice the period, as shown in Fig.3. Of course, as the intensity increases one would expect the amount of molten area to increase. What we have observed, however, is that the increase does not occur smoothly but rather occurs in discrete jumps, with the simple grating replaced by a doubled-period grating, which in turn is replaced with a triple-period grating. This is illustrated in Fig. 4 which shows the fraction of molten area as a function of laser intensity for a 500 μm spot size. Within a given structure, such as the simple grating, as the intensity increases there is simply a redistribution of the deposited power and local temperatures are changed without any additional melting. In this sense one has a very stable morphological "phase", which is somewhat insensitive to the spot size and laser intensity. The well-defined grating structure shown in Fig. 3, with the width of molten and solid strips constant across a spot illuminated with a Gaussian intensity profile, is evidence of this.

It is possible to generate other morphologies for other intensities and spot sizes. Fig. 5 shows that for an intensity of 1.7 kW/cm² and a spot size of 700 μm a disordered melt structure occurs with molten lamellae "dots" occuring in a solid background. For the larger spot size with Gaussian beam illumination, the lateral temperature gradient will be smaller as will be the lateral heat flow. In this case, one requires a lower intensity to initiate melting and the power flow through the system is lower. One therefore finds that the ratio of stored energy to power depositon in the system is lower. This ratio is a measure of the "thermodynamic time constant" over which the system can change. For the larger spot sizes the metastable "amorphous state" is preferred over the equilibrium ordered state that it may be trying to attain. Factors such as surface tension, viscosity, etc. therefore play a stronger role in defining the irregular patterns. These types of structures are observed upon melting for all spot sizes but for the smaller spot sizes, as noted above, as the intensity is increased, a grating structure very quickly develops. For the larger spot sizes, the structure remains irregular as the intensity increases, with solid dots in a liquid background replacing molten dots in a solid background. We have observed several different "morphological phases" which are stable within a range of laser spot sizes and intensities. These phases are summarized in Fig. 6. The dynamics of these patterns have also been studied albeit qualitatively. Near the "phase boundaries" it is possible, by rapidly scanning the beam or rotating the polarization, to initiate "defects" in the otherwise stable structures. It is also possible to change disordered structures into ordered ones. Well within the different phase regions the structures were much more stable.

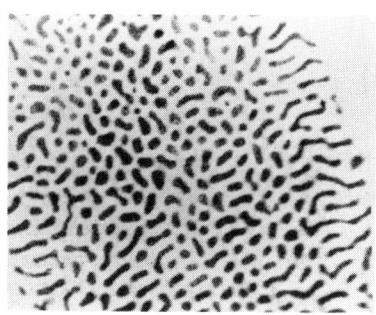

Figure 5. Solid-melt pattern formed by a normal incidence beam at a power density of 1.7 kW/cm² and a spot size of 700 µm; spacing between molten dots is ≃ 15 µm.

Let us now consider what patterns are observed for non-normal incident beams. For s-polarized incident light laser powers from 6W at normal incidence to 11W for $\theta = 60°$ were used for spot diameters ranging from 200 to 1200µm. As a function of θ two distinct types of disordered and ordered patterns were observed. For the lower powers and larger spot sizes disordered patterns could be observed for all θ but did not differ in any substantive manner from those observed at normal incidence. For smaller spot sizes grating patterns were observed with the grating wavevector oriented parallel to the incident polarization and with a period ,Λ, which varied as $\lambda/\cos\theta$. The fraction of melt did not vary substantially with θ, indicating that the solid and molten strips increased in width at the same rate with increasing θ. Period doubling was observed with increasing intensity for θ up to 50°, beyond which there was insufficient power to observe patterns other than the simple grating structure.

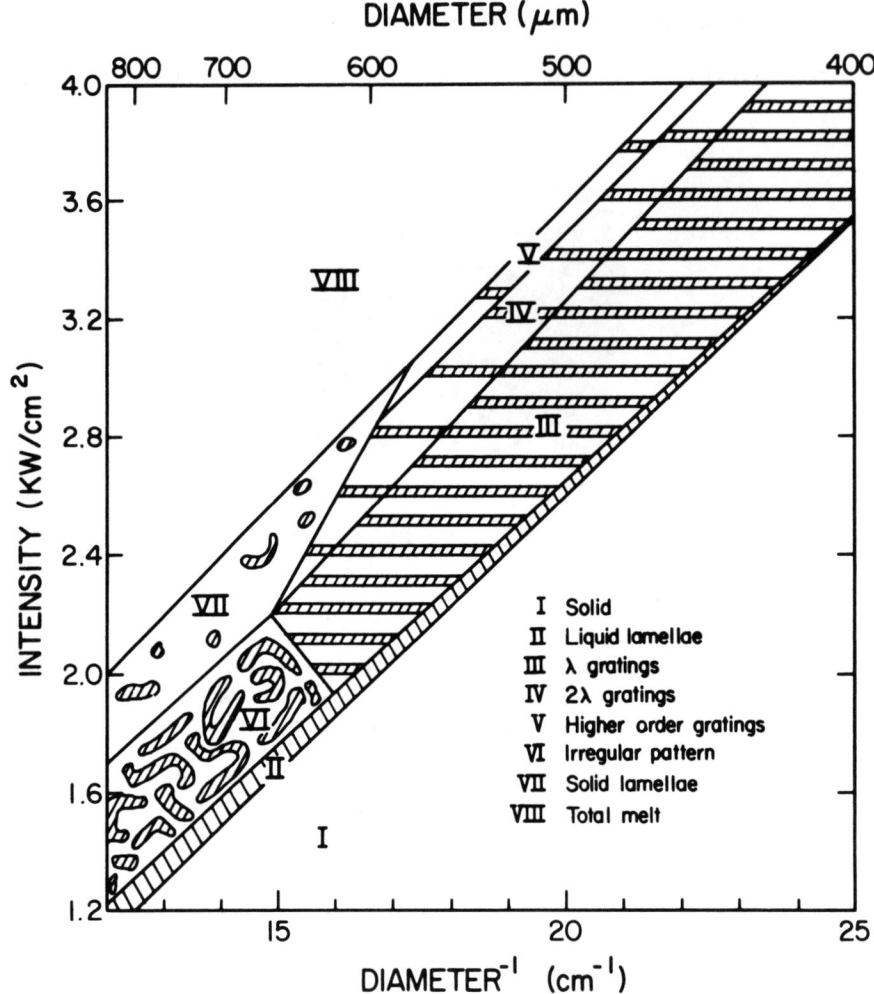

Figure 6. "Phase Diagram" illustrating the spectrum of solid-melt morphologies formed by a normally incident beam.

When p-polarized light is incident on the sample more complex behavior is observed. Three distinct regimes have been identified. The first regime is that for $\theta = 0°$ where s- and p-polarized light are equivalent. Between $\theta \simeq 2°$ and $30°$ only disordered patterns could be observed for all intensities and spot sizes. Above this angle and for smaller spot sizes coexisting but orthogonal gratings could be observed as shown in Fig. 7. The grating in the upper part of the Figure has a spacing which varies as $\Lambda = \lambda/(1-\sin\theta)$ and is oriented with its wavevector parallel to the κ_i, the component of the incident beam's wavevector parallel to the surface. The orthogonal grating has a spacing which varies as $\lambda/\cos\theta$. No period doubling was observed for either structure.

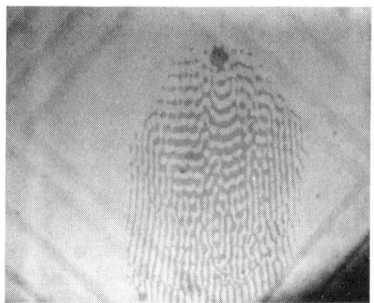

Figure 7. Example of an orthogonal grating pattern formed by a p-polarized beam incident at $\theta = 45°$.

Discussion

How can one explain the existence of these inhomogeneous phases and their stability? To be able to explain *ab initio* all the details of the ordered and disordered morphological phases we have observed is a difficult task. The problem of treating the interaction of light with a microscopically inhomogeous metal-dielectric is difficult enough.

The disordered inhomogeneous melt phases make the calculation of scattered light fields and inhomogeneous energy deposition intractable while surface tension and heat flow considerations further complicate the problem. At the present time we have set ourselves the modest goals of answering, on the basis of self-consistent arguments, the following questions. Can we account for the existence and stability of the simple and double grating structures when different fractions of the surface are molten, given that the reflectivity of bulk liquid silicon is higher than that of bulk solid silicon? Can we explain the different grating structures formed by s- and p-polarized light for off normal incidence? In the case of the disordered structures, can we explain the size of the molten and solid dots and their average separation simply by electrodynamic arguments? In essence then, initially we are assuming that thermodynamic and material properties allow certain types of structures to exist; we wish to determine the extent to which electrodynamics then dictates the existence of one type of grating versus another, and dot sizes and their separations. We will not discuss all of these questions here but merely give a flavor of the approach. For more detail the reader is refered to references 10 and 11.

The theoretical approach[10,11] we have used is based on the simple model of the inhomogeneous melt/solid structures shown in Fig. 8. We assume that induced polarizations in the liquid are essentially *uniform* and confined to the top and bottom layers for normal incidence. For non-normal incidence we assume the existence of uniform polarization fields confined to the walls of the melt-solid interfaces. A uniform polarization is assumed to exist in the bulk of the solid regions. These polarization densities are used with an electrodynamic Green's function for the thin film to calculate fields everywhere. The four (polarization density) parameters can be determined by the conditions that electric and magnetic fields vanish in the bulk of the liquid. Because of the simplicity of the model the problem is over specified, and it is only necessary to implement these conditions in a spatially averaged sense. Once the polarization parameters are calculated, the local power deposition can be calculated in the melt and in the solid regions. This of course depends on the structure of the inhomogeneous phase, and only becomes equivalent to that of bulk liquid or solid in the limit that the molten or solid areas have an extent which is large compared to the wavelength of light. To emphasize the importance of the microscopic structure in determining local power deposition we have found it convenient to express our results in terms of "structural absorption coefficents" for the liquid and solid regions. These give the ratio of the actual power deposition per unit area in a particular microscopic region to that of power deposition in an infinite bulk region of the same composition.

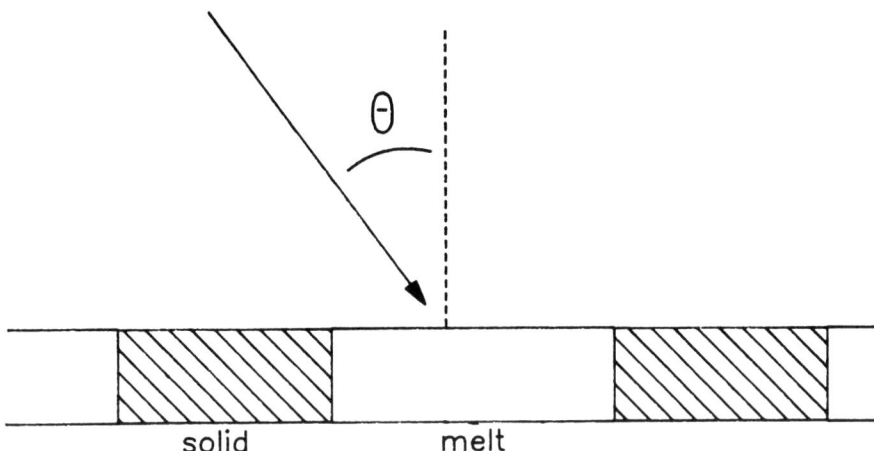

Figure 8. Schematic diagram of solid-melt cross-sections assumed to perform theoretical calculations.

In spite of the simplicity of the overall approach, we have found that many of the salient experimental features can be explained *in a self-consistent manner*. Taking the normal incidence problem first, we consider a melt/solid grating structure formed on a thin silicon film as a function of the grating spacing Λ as shown in Fig 9a,9b. The particular curves assume that 50% of the surface is molten, as is seen experimentally. When the grating wavevector is assumed to be parallel to the incident linear polarization, for Λ = λ one obtains a peak in the structural absorption coefficient for the liquid and a minimum for that of the solid region indicating that the relative energy deposition in the liquid versus solid is maximum at this spacing. It corresponds to the preferred grating spacing for the assumed melt fraction. Note as well that the structural absorption coefficient for the liquid (solid) at this spacing is well above (below) that for a grating with $\Lambda \rightarrow \infty$ corresponding to alternating infinite regions of solid and liquid. This illustrates that for the microscopic grating with $\Lambda = \lambda$, the absorption in the liquid region is enhanced over what a bulk liquid would give, while that of the solid is reduced. This is exactly the opposite to what one would expect from a geometric optics picture in which the liquid has the higher reflectivity. The actual structural absorption coefficients can be understood in terms of the large scattered fields which arise from a metal-dielectric grating. When these interfere with the incident field there is destructive interference in the solid but constructive interference in the liquid. In a sense the liquid shields the solid from the incoming radiation; we have referred to this effect as "interference shielding".[10] If we assume the grating wavevector is perpendicular to the incident polarization (Fig. 9b), the structural absorption coefficient is larger for the solid than the liquid, indicating that no grating should form for this configuration, consistent with observation.

If we now assume that 75% of the surface is molten one finds that the structural absorption coefficient of the solid is greater than that of the liquid for $\Lambda = \lambda$, indicating that the simple grating structure is no longer stable. However, for $\Lambda = 2\lambda$, the relative structural absorption coefficients are at a maximum. This is consistent with our observation of the double period grating when approximately 75% of the surface is molten. The dashed curves on Fig. 4 show the regions of predicted stable phases of simple and doubled gratings as a function of laser intensity. Good qualitative agreement is achieved. It is worth reminding the reader at this point that no attempt has been made to predict, as a function of the amount of surface that is molten, which particular structure will form. Without the benefit of a variational principle or some other principle to predict the most stable structure all we are attempting to do is show that our calculated results are consistent with what is observed.

Similar success has been achieved in explaining the formation of the various ordered structures for s- and p-polarized light for $\theta \neq 0°$. Fig. 9c and d show peaks in the molten structural absorption coefficient at $\theta = 60°$ for p-polarized light with the spacing and orientation of each of the orthogonal coexisting gratings that is observed experimentally. Appropriate results are also obtained for s-polarized light, confirming the existence of single grating structures.

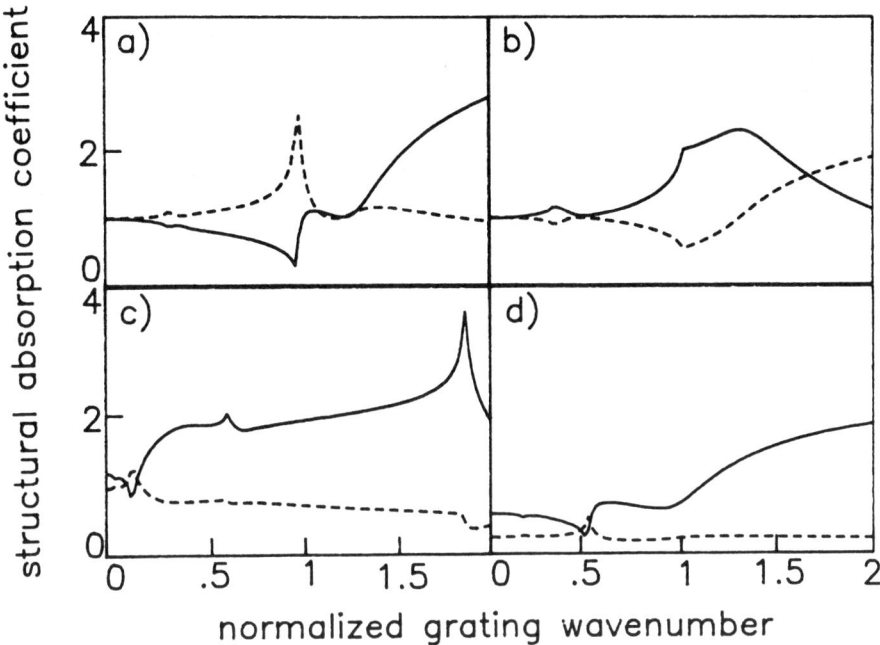

Figure 9. Structural absorption coefficient as a function of normalized grating wavenumber (λ/Λ); solid curve is associated with liquid, dashed curve with the melt. a) $\theta = 0$, E parallel to κ_g; b) $\theta = 0$, E perpendicular to κ_g c) $\theta = 60°$, p-pol., κ_i parallel to κ_g d) $\theta = 60°$, p-pol., κ_i perpendicular to κ_g.

The formalism has also allowed us to address question concerning the dot sizes and separation of solid or molten lamellae. As noted above, this is more difficult because of the random placement of a large number of such dots. One must therefore in principle treat a "many-body" problem. For simplicity, we initially considered an isolated circular solid or molten dot immersed in the complementary background. Calculations of the structural absorption coefficient as a function of disc size showed that the maximum (minimum) structural absorption coefficent for the molten (solid) disc occurred for a radius of $\simeq 3\mu m$, in agreement with experimental observation. As a next step we considered a pair of interacting molten lamellae. Fig. 10 shows the structural absorption coefficient for a pair of molten dots of optimum size as a function of their separation. One can see that this is maximum for an inter-disc separation of $\simeq 16$ μm indicating that at this separation the power deposited per unit area into the dots is highest and corresponds to the separation which is most favorable for their formation. This separation is in agreement with what is observed experimentally in Fig. 5 . These calculations show that, in spite of approximations related to the shape of the lamellae and their distribution, some of the key experimental results can be explained from a simple model which captures the salient features. They also indicate that the electromagnetic "interaction" between dots is probably important in determining the morphologies.

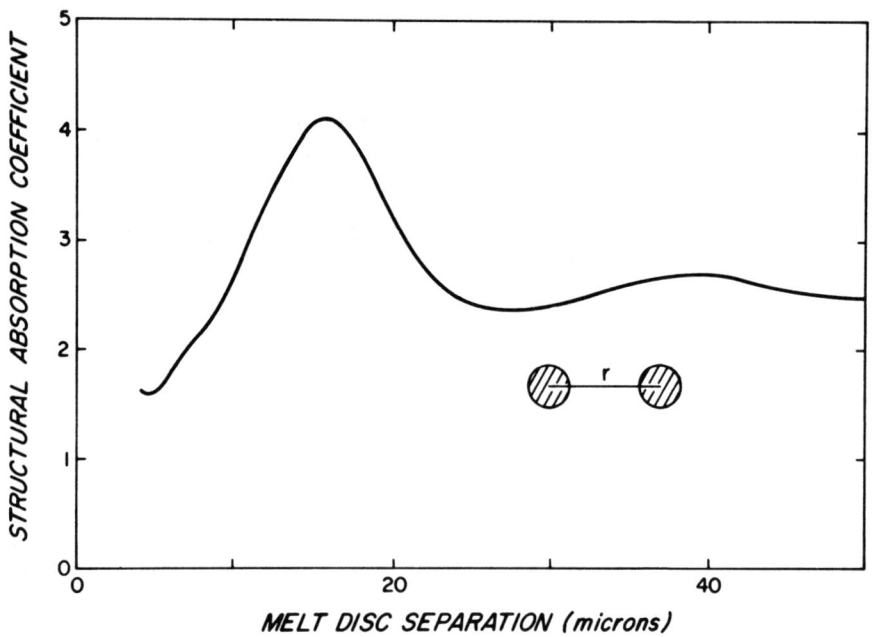

Figure 10. Structural absorption coefficient associated with two isolated molten discs in a solid background, as a function of disc separation; radius of discs is assumed to be 3 μm.

Conclusions

Much work remains to be done before we can fully understand the cw induced structures, in particular the stability of the disordered or "amorphous" structures, the nature of the transitions between amorphous and ordered strucutres, and the nature of those between different ordered structures. A key question, of course, is whether the spatial scales, especially in the disordered structures, are determined by electrodynamic rather than the thermodynamic or mechanical factors. In preliminary experiments with $\lambda = 1.06$ μm high power radiation we have observed only disordered structures with the sizes and separations of the lamellae not significantly different from what has been observed with $\lambda = 10.6$ μm laser light. This suggests that, for radiative induced melting in the absence of coherence, the spatial scales are determined by factors other than the incident light parameters and that 10.6 μm coherent light can more effectively couple and organize the molten and solid regions than light at significantly different wavelengths.

An additional important fundamental question is whether the symmetry breaking we see here is of the nature of the "true" Broken Symmetry[21] observed in many equilibrium systems, or merely characteristic of pattern formation often observed in dissipative systems.

Acknowledgements

We gratefully acknowledge support for this work from the Natural Sciences and Engineering Research Council of Canada and the Ontario Laser and Lightwave Research Centre.

References

1. G. Careri, *Order and Disorder in Matter*, Benjamin/Cummings(New York), (1984).

2. *Pattern Formation by Dynamic Systems and Pattern Recognition*, Ed. H. Haken, Springer Verlag (New York), 1979.

3. G.K. Celler, M.D. Robinson, L.E. Trimble and D.J. Lischner, *Appl. Phys. Lett.* **43**, p. 868 (1983).

4. C.P. Grigoropoulos, R.H. Buckholz and G.A. Donoto, *J. Appl. Phys.* **59**, 454 (1986).

5. K.A. Jackson and D.A. Kurtze, *Journal of Crystal Growth* **71**, 385 (1985).

6. H.M. van Driel, J.E. Sipe and J.F. Young, *Phys. Rev. Lett.* **49**, 1955 (1982).

7. J.F. Young, J.E. Sipe and H.M. van Driel, *Phys. Rev. B,* **30**, 2001 (1984).

8. M.A. Bosch and R.A. Lemons, *Phys. Rev. Lett*. **47**, 1151 (1981).

9. W.G. Hawkins and D.K. Biegelsen, *Appl. Phys. Lett* **42**, 358 (1983).

10. J.S. Preston, H.M. van Driel, and J.E. Sipe, accepted *Phys. Rev. B* (1989), J.S. Preston, J.E. Sipe, H.M. van Driel, and J. Luscombe, accepted *Phys. Rev. B* (1989).

11. K. Dworschak, J.E. Sipe and H.M. van Driel (unpublished).

12. P. Anderson in *Order and Fluctuations in Equilibrium and Non-equilibrium Statistical Mechanics*, Nicolis, Dewel, Turner Eds. (Wiley, New York, 1981) p. 289.

IRRADIATION-INDUCED CAVITY LATTICE FORMATION IN METALS

J.H. Evans
Materials Development Division
Harwell Laboratory
Oxon. OX11 0RA, UK.

ABSTRACT. The formation of cavity lattices in materials includes a large variety of experimental observations on void and bubble lattices following irradiation or ion implantation. The fundamental characteristic of the phenomenon is the close relationship between the cavity lattice crystallography and that of the host crystal. In this review, experimental results on metals will be highlighted, but analogous results on nonmetals, eg. calcium fluoride, alumina and silicon, will also be included. Recent theoretical models will be discussed with emphasis on those that explain the crystallography. Anisotropic self-interstitial diffusion, in two dimensions for metals and in one dimension for alumina and calcium fluoride, is a strong candidate to explain the majority of results.

1. INTRODUCTION

The phenomenon of void lattice formation in metals, together with analogous structures of inert gas bubbles, has received considerable experimental and theoretical attention since its discovery in the early 1970's [1,2]. For the three main metal structures (fcc, bcc and hcp), the common factor in all the observations has been the spatial alignment of cavities along crystallographic directions, leading to bcc and fcc cavity lattices in bcc and fcc metals respectively, and planar ordering of cavities parallel to the basal plane in hcp metals. It will be noticed that in the present context, void and bubbles are combined in the wider description of cavities as used in the article title. This is justified by the clear similarity between void and bubble lattices though the latter are always found on a finer scale.

The present article will be in three main parts. After a general introduction to voids and bubbles in metals, the experimental results available for irradiated or ion-implanted metals will be reviewed. In addition, similar results on defect alignment in other materials will be summarised, including the alignment of anion voids in calcium fluoride, cavities in alumina, and the interesting observation of a simple cubic lattice of silicon oxide precipitates in silicon. The second part of the article will introduce some of the theoretical models for cavity lattice formation, particularly those explaining the crystallography. The models based on the assumption that self-interstitial atoms (SIA) diffuse anisotropically seem applicable to all the results available to date. In the final part of the article, each system is discussed separately, bringing out results which allow models to be tested.

2. VOID AND INERT GAS BUBBLE GROWTH IN METALS

Since some readers may be unfamiliar with the topic of voids and bubbles in metals, this section will give a brief background to this area. Although both bubbles and voids are cavities, and thus from the host metal point of view they are 'constructed' from vacancies, their growth processes are very different.

2.1. Void Formation and Growth

This topic dates from 1966 when Cawthorne and Fulton [3] found voids in stainless steel following a high dose neutron irradiation in a fast reactor. Details of general interest in this topic can be found in conference proceedings in the early 1970's [4,5], but the mechanism had been anticipated earlier by Greenwood et al [6] who discussed how the vacancies and self-interstitials created in the irradiation damage process might be partitioned to the sinks in the system. The mechanism, known subsequently as bias-driven void growth, depends on the strain-induced interaction between the self-interstitials and the dislocations in the metal. The interaction biases the dislocations to trap interstitials preferentially, leaving the equivalent number of vacancies to be trapped at neutral sinks such as voids. The result is the growth of the void component in the system. This simple picture presupposes the existence of void nuclei; these are small vacancy clusters which are prevented from collapsing to vacancy loops by one or more associated gas atoms.

Generally speaking, under conditions where displacement damage creates vacancies and interstitials, eg. under neutron irradiation, ion implantation or high energy electron irradiation, voids are found to form in all metals over the temperature range between 0.2 - 0.3 and 0.5 of the metal melting temperature, T_m. The lower limits are related to vacancy mobility and build up of cascade damage (eg. vacancy loops) in bcc and fcc metals respectively. For $T \gtrsim 0.5\ T_m$, cavities will tend to maintain thermal equilibrium, $P = 2\gamma/r$, where P and r are the internal pressure and cavity radius and γ is the surface energy of the metal.

2.2. Growth of Inert Gas Bubbles

Inert gas atoms in metals are essentially insoluble and the resulting tendency for precipitation into bubbles is well known. Early studies involved low gas concentrations at high temperatures in the equilibrium bubble regime but for bubble lattice formation we are interested in high gas concentrations at relatively low temperatures. Although helium can be introduced into metals by (n,α) reactions under neutron irradiation or in some metals by tritium decay, all lattice observations have followed ion implantation. Under these conditions, ambient temperature implants lead to a very high concentration of small bubbles, usually of the order of $10^{19}/cm^3$. This appears to hold for all inert gases and all metals, while the temperature can vary up to about $0.35T_m$ without affecting the bubble parameters [7]. Lower limits have never been established.

Although there were early ideas that bubbles gained vacancies from the vacancy component of the displacement damage created during implantation, the athermal bubble behaviour eventually led to the suggestion [8] that bubble growth for the relatively low temperature implants was pressure-driven, with cavity vacancies being created by a process known as loop punching. Proposed by Greenwood et al [6], this process allows a bubble to relieve its overpressure by creating a plane of vacancies at its surface at the expense of creating an interstitial loop in the metal lattice. While atomistic details are uncertain in the high bubble density environment, the mechanism is accepted as a description of bubble growth under ion implantation in the temperature regime coincident with bubble lattice formation. The high predicted pressures in this mechanism are consistent with the discovery that the heavier inert gases precipitated into bubbles have been found in the solid phase [9,10].

3. EXPERIMENTAL OBSERVATIONS

3.1. Chronological Survey

In this section, we document the main steps in the evolution of void and bubble lattice results on metals and on other materials. Examples and references are also included in table I though this should not be taken as comprehensive. Most results up to 1982 are included in Krishan's review [11].

The first void lattice results on metals were made by the present author [1] and by Wiffen [2] on ion and neutron irradiated molybdenum, respectively. Rather remarkably, the voids were found to be aligned in a body-centred cubic (bcc) structure identical to and parallel with the parent metal lattice, but with a lattice parameter some two orders of magnitude greater. Acceptance of the results were undoubtedly helped by good lattice perfection, figure 1. Subsequent work has shown this phenomenon to occur on a far wider scale than initially might have been imagined. Within two years of the molybdenum observations, void lattices, or at least the tendency for voids to be spatially aligned, had been found in bcc niobium and tantalum [2], and the phenomenon extended to face-centred cubic (fcc) metals such as nickel [12] and aluminium [13], and to hexagonal-close-packed (hcp) magnesium [14]. Of special interest was the fcc nature of the void lattice in nickel where the structure again followed that of the parent crystal.

The exact matching of the void lattice with either the bcc or fcc

TABLE I

Selected Examples Of Cavity Lattice Observations

Material	Ordered defect (with examples of host materials and references)	Defect lattice structure
hcp metals	Voids ($Mg^{14,15}$) Helium bubbles (Ti^{20}) Krypton precipitates (Ti^{21}, Zr^{22})	ordering into 2-D layers parallel to basal plane
bcc metals	Voids ($Mo^{1,2}$, W^{33}, Nb^2, Ta^2, Fe^{35}) Helium bubbles (Mo^{16}, V^{19}, W^{19}, Ta^{19}, Fe^{19}) Neon bubbles (Mo^7)	bcc parallel to matrix
fcc metals and alloys	Voids (Ni^{12}, Al^{13}, st.steel^{34}) NiAl alloys^{83}, Ni-10%Cu^{84}) Helium bubbles (Cu^{17}, Ni^{17}, Au^{18}, Al^{19}, stainless steel^{17}) Krypton precipitates (Ni^{19})	fcc parallel to matrix but structural variants in bubble arrays^{23,24}
alkaline earth halides	Anion voids (CaF_2, $SrF_2{}^{26,27}$)	simple cubic lattice following matrix crystallography
alumina	Voids^{25}	linear alignment, perpendicular to basal plane
silicon	Silicon oxide precipitates^{28}	simple cubic lattice with major axes parallel to those of diamond-cubic Si

host lattice was broken in the results for magnesium. Here it was found that the alignment had a planar character, with voids aligned into rafts parallel to the basal plane [14,15]. Since there is no evidence in either the Mg results or the bubble results in hcp Ti and Zr (see below) for any ordering within bubble planes, the result clearly differs from the bcc and fcc results already outlined. Strictly speaking, cavity lattices are not formed in the hcp metals - cavity ordering would be a more accurate description. Nevertheless, the relationship between cavity and host lattice is still present and there seems little doubt that the hcp results should be included in the generic cavity lattice description.

The next significant result in metals was the finding by Sass and Eyre [16] that helium gas bubbles in molybdenum, formed by implantation of helium ions at ambient temperatures, were aligned in a bcc lattice, exactly analogous to the void lattice observation on the same metal. The main difference was one of scale, with lattice parameters of near 5 nm, consistent with the very high bubble densities observed. Mirroring the evolution of void lattice discoveries, helium bubble ordering has been subsequently found on a number of metals in all the three major metal structures, eg. copper, nickel, stainless steel [17], gold [18] and aluminium [19] in fcc metals, vanadium, tungsten, tantalum and iron [19] in bcc metals, and titanium [20,21] and zirconium [22] in hcp metals. Neon bubble lattices have also been found [7]. In all cases, the structure and alignment of bubbles has been identical to results for void lattices, though it is known that for bubble ordering in the fcc metals, variants in the structure are a common observation [23,24].

As mentioned in the previous section, the heavier inert gas atoms can precipitate within bubbles in the solid phase (due to pressures in the GPa range). However, the different phase of the inert gas atoms within the cavities has not in any way affected the tendency for alignment, at least for krypton in Ni [19], Ti [21] and Zr [22]. A good example of the planar cavity ordering found for krypton bubbles (or precipitates) in Zr is shown in figure 2.

In non-metals, the first observation of cavity alignment was reported in 1976. In neutron-irradiated alumina, Clinard [25] reported the formation of linear strings of cavities perpendicular to the basal plane of the parent hexagonal structure. In the same year Johnson et al [26] found that under 100 keV electron irradiation, simple cubic lattices of anion voids formed in calcium fluoride. In the other two alkaline earth halides examined, strontium fluoride gave a similar result with a less perfect lattice, but no lattice formation was seen in barium fluoride. Further details of this work, together with an early review of void lattice observations and theory are given in [27].

In this short survey of experimental landmarks, reference must also be made to the finding by van Ommen et al [28] of an oxide precipitate lattice in silicon after oxygen implantation at 575K. The lattice had a simple-cubic structure with major axes parallel to the parent diamond-cubic structure. The close crystallographic connection justifies including this observation here and in table I. In addition, from the

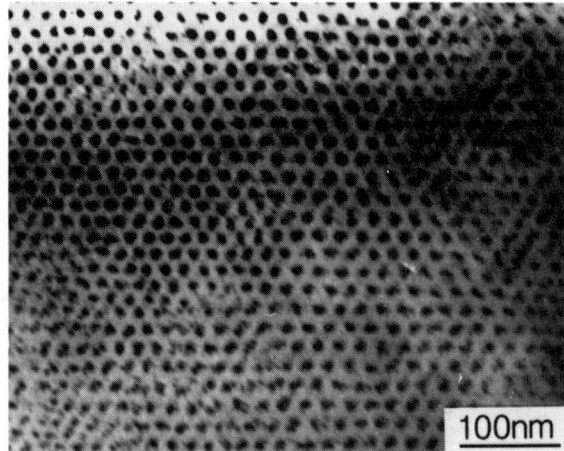

Figure 1. An almost perfect void lattice in molybdenum viewed along a <111> direction.

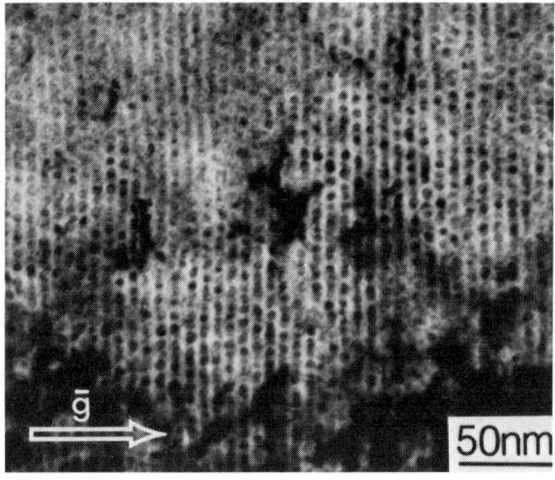

Figure 2. Electron micrograph showing the planar ordering of krypton bubbles parallel to the basal plane in zirconium. The g-vector is [0002].

silicon point of view the precipitate can be considered a cavity (cf. the krypton precipitates already mentioned).

3.2. General Observations

We now proceed to discuss void and bubble lattice formation in a more general way, starting with the evolution of the lattice. All observations agree that nucleation is random and that ordering occurs at a later stage. The mode of lattice formation, based on many examples where only partial or imperfect ordering has taken place, eg. [1,29,30], appears to be a long range process rather than one involving the nucleation and subsequent growth of small perfectly ordered regions. For inert-gas bubble arrays, diffraction information [7] shows the evolution of bubble lattice diffraction spots from an initial Debye-Scherrer ring, indicating that the sequence of lattice formation is similar to that for voids.

The tendency for void ordering appears to be improved at higher irradiation doses and with increases in void concentration. The latter effect can be clearly seen for voids in Nb and Ta where a variation in oxygen content can be used to influence void nucleation and markedly affect void lattice formation [31,32]. The lower perfection and less frequent observations of void lattices in fcc metals (especially under neutron irradiation) could well be attributed to the reduced rate of void nucleation in these metals relative to the bcc metals.

The influence of temperature on void lattice formation is difficult to assess since temperature always has a separate influence on nucleation parameters - void densities fall with increases in irradiation temperature. Assuming that the void lattice parameter will increase as the void density drops, then clearly the same parameter will rise with temperature. The data of Moteff et al [33] on both Mo and W are consistent with this picture; results for the two metals are identical for the same homologous temperature with void lattice parameters increasing by over a factor of three with irradiation temperature increase. In these metals, there is no step-wise change in void lattice perfection at temperatures where large changes in defect properties, eg. vacancy loop stability or vacancy evaporation from voids, are expected.

For inert-gas bubble formation in metals, where nucleation and growth appear to be independent of temperature over the range from ambient up to about 0.35 of the metal melting point (see section 2.2), no major effect of temperature on ordering parameters has been reported.

A discussion of the role of bombarding particle, ie neutrons, ions or electrons, on void lattice formation is subject to the same limitation as in the discussion of temperature effects above; void densities are again an important factor. As far as is known, where void densities are equal, the void lattice results in metals do not differ to a significant degree. At present there is little basis for following Krishan [11] in treating the electron irradiation void lattice results of Fisher and Williams [34] in stainless steel, and of Little [35] in iron, separately from other void lattice observations.

4. THEORIES OF LATTICE FORMATION

Before examining the models proposed for cavity lattice formation in detail, it seems worthwhile listing the common factors arising from the experimental survey in section 3 and table I. Three important properties stand out: firstly, in all cases there is a clear relationship between the cavity lattice crystallography and that of the host material; secondly, in all the metal results the bubble lattice symmetry follows exactly that of the void lattices; thirdly, in all cases displacement damage is present. Cavity lattice formation or its improvement during post-implant or post-irradiation annealing has never been reported.

The reason for listing these properties is to raise the question of whether more than one model is needed to explain the main cavity lattice results. While mechanisms could differ from voids to bubbles, and from one metal structure to another, the common factors above appear to be against this. It should be noted that one important result showing a difference in fcc metals for bubble lattices relative to void lattices, ie. the appearance of variants in the lattice structure [23,24], can be considered additional to any fundamental process.

Many proposals have been made to explain cavity lattice phenomena but models frequently are very selective, mostly being applied to void lattices and ignoring bubble lattices. Additionally, the hcp results are rarely considered. Details of models up to the early 1980's are included in the articles by Krishan [11] and Johnson and Chadderton [27] but the area has continued to be under active discussion. The major model areas are listed with references in table II and each will be discussed separately in the following sections.

TABLE II
Main Classes of Models for Cavity Lattices

Models	References
1. Elastic Interactions	36-39
2. Isomorphic Decomposition	42
3. Phase Instability Models	44-50
4. Anisotropic Self-Interstitial Diffusion	
a) One-Dimensional	51,53-55
b) Two-Dimensional	52,56-59
5. Cavity - Interstitial Loop interaction	61,62

4.1. Elastic Interactions

The first suggestion to explain void lattice formation came in the proposals of Malen and Bullough [36] where the cubic anisotropy of the elastic constants was used to provide an interaction between voids in metals. The discovery of void ordering in isotropic metals such as tungsten caused the model to be modified to include the faceting of the void as an important factor in the centering of the subsequent lattice [37]. The model gives both bcc and fcc lattices, with an energy minimum of about 0.5 eV per void, though Stoneham found no minimum for a simple cubic lattice [38]. The scale of the lattice has often been described by the ratio of void lattice parameter to void radius (A/r). At the energy minimum, Tewary [39] gives a value of 8 for this parameter, which lies in the wide experimental range of 6 to 15 deduced for voids from the compilation of Krishan [11]. However, it is larger than the range of 4 to 7 found for bubble lattices.

The main problem with the interaction models has been to assess whether the energy minimum is sufficient to lead to lattice formation rather than contributing to the separate question of its stability. Intuitively, an interaction model would be more likely to play a part in bubble lattice formation where bubble separation distances are very small. For large inter-void spacings where several cases greater than 100 nm have been reported, eg. [15,30], it has been suggested [30] that interaction of the required magnitude seems unlikely. Since the interaction models do not involve the radiation damage directly, apart from its effect on helping to induce surface migration of ad-atoms and hence cavity diffusivity, it might be expected that imperfect cavity lattices should be improved by post-irradiation annealing. As yet, no evidence has been presented to show this.

The various difficulties in applying the interaction model to void lattices has been compounded by the recent results of Finnis [40] who simulated the energy of a void lattice in molybdenum using a molecular dynamics program on a CRAY-2 computer. This was an exact calculation without the approximations present in the previous analytical approaches. Although an energy minimum was found, it was lower for an fcc lattice than the expected bcc lattice; furthermore, the void spacing for the minimum energy was unrealistically small. It is not yet known how these results could be altered by replacing voids by overpressurised bubbles.

4.2. Isomorphic Decomposition

An analogy has been made by Johnson and Chadderton [27] between void lattice formation and the ordering found during precipitation and ageing of certain alloy systems, eg. [41-43]. In this latter area it seems that the elastic interaction, through the anisotropy of the elastic constants, is responsible for the precipitate ordering. Theory shows that an energy minimum will occur at rather small (A/r) ratios, between 2 and 3, with the positive or negative value of the elastic anisotropy factor ($C_{11}-C_{12}-2C_{44}$) for the host material determining whether the lattice will be bcc or simple cubic respectively. The interest of Johnson and Chadderton in this mechanism arose from the fact that the anion void lattice in their CaF_2 results can equally be considered as a lattice of calcium particles. However, since the elastic anisotropy factor is positive for fluorite, the prediction of the mechanism, a bcc lattice, disagrees with the observed simple cubic lattice and can therefore be rejected. For bubble lattices generally, where the solid bubbles could be more correctly described as precipitates, the experimental A/r ratios exceed the values predicted on this theory. Fcc lattices are not predicted.

4.3. Phase Instability Models

Many authors have used general aspects of phase instability theories to explain the void lattice. (Bubble lattices are usually ignored). As summarised by Johnson and Chadderton [27], Martin and co-authors [44,45] first developed their theory using concepts from vacancy concentration fluctuation theory. The void lattice forms on this model from a certain mode of the collective vacancy concentration, with voids nucleating on the periodic peaks of defect density waves. Both fcc and bcc lattices are predicted. Krishan introduced a bifurcation analysis [46] in which the void lattice can develop from a vacancy concentration as a process of self-organisation in an open non-equilibrium dissipative system. These models are still under discussion [47-50].

One difficulty with some of these theories is that they use defect properties, such as vacancy loop density or thermal vacancy emission, which have a very large temperature dependence. As mentioned in 3.2, there is no evidence for any sudden change in void lattice formation that might coincide with expected changes in these defect properties. However, one major criticism of these theories is that they do not yet predict the various lattice crystallographies. This could be a considerable handicap to proposals that the models contain the key factor or factors in cavity lattice formation. The emergence of a calculated lattice wavelength might only point to the formation of a cavity structure with a liquid-like spatial distribution.

4.4. Anisotropic Self-Interstitial-Atom (SIA) Diffusion

4.4.1. One-Dimensional SIA Diffusion

Among one of the first models to explain void lattice formation was the approach of Foreman [51] which depended on the anisotropic movement of self-interstitial atoms, either through one-dimensional diffusion or via dynamic replacement sequences. It was pointed out that SIA moving one-dimensionally along a particular direction would tend to order voids into rows along that direction. The mechanism works because unaligned voids will always receive a larger flux of SIA than voids which are locally shadowed, while the greater arrival rate of SIA on unshadowed sides of voids relative to shadowed voids will always move voids towards linear alignment. In a metal, the most probable directions for one-dimensional SIA diffusion are those which are close-packed. If this is assumed, then the final stable lattice formed by the mechanism must be three-dimensional and have close-packed directions parallel to the host matrix. This property immediately gives a simple physical explanation for the coincidence of the structure of void and bubble lattices with the parent fcc and bcc metals. The model has not been applied to the hcp metals.

Generally, the model itself (leaving aside its initial assumptions) is regarded as plausible but one aspect has been criticised [52]. If, in three linearly aligned voids, the central void is off-centre, then the imbalance of its SIA capture volume between it and the two neighbours will cause it to drift (and accelerate) towards the nearest void. This drift could act against lattice formation unless balanced by alignment or partial alignment of voids in other directions of SIA diffusion. It seems that there could be a small potential barrier here against the nucleation of lattice formation. Intuitively, this barrier might be low for a high cavity density but only some form of modelling could assess its importance as the density parameter decreases.

With regard to the application of the model, the main difficulty is whether anisotropic SIA diffusion, or replacement collision sequences with sufficient length, can take place as required. The model has recently received attention in several papers by Woo and Frank [53-55] where a strong connection has been made with the postulated (and somewhat controversial) low temperature crowdion configuration of self-interstitials.

4.4.2. Two-Dimensional SIA Diffusion

A second model based on the anisotropic diffusion of interstitials was introduced by the present author in 1983 [56] and has been extended in subsequent papers [57-59,52]. The model was inspired partly by the Foreman model above, and by the suggestion made by Jacques and Robrock in 1982 [60] that their internal friction results on molybdenum could be explained by SIA diffusion in two dimensions. In exploring what effects such diffusion might have on cavity growth, it emerged that cavities could spatially order in a planar fashion, parallel to the plane of SIA diffusion [56]. Although somewhat analogous to the 1-D SIA model referred to earlier, the mechanism cannot rely on a direct shadowing process. The mode of operation can be followed by considering the simplest case of a metal having SIA diffusion on only one set of planes and being irradiated in the void formation regime. For voids of equal size positioned randomly in a metal, there will always be localities where voids are by chance aligned, fully or partially, on a plane common to that of the SIA diffusion. As a direct consequence of their positions relative to each other, they must receive a reduced SIA flux compared to planar regions where voids are isolated or where the planar void density is low. (In unit time, the same number of SIA are distributed among a greater or lesser number of voids respectively). Since the vacancy flux to equally sized voids must be equal using rate theory, the growth of aligned voids, based on the differences between vacancy and interstitial fluxes, must be faster than for isolated voids. The latter may well shrink.

In considering the 1-D SIA diffusion processes, Woo and Frank [54] have neatly described the alignment mechanism as one of Darwinian selection. Only the aligned voids, ie. the "fittest", survive. Exactly the same description applies in the 2-D SIA situation. The overall alignment is helped considerably by the non-uniform deposition of SIA on void surfaces; any single void will be cut by a number of planes depending on its size, and these planes all carry different SIA fluxes. Thus, any part of a partially aligned void protruding into a high flux region will receive an enhanced SIA flux. The net effect of this must be to move voids away from planes having high SIA concentrations and clearly aids their alignment into planes (planar ordering) parallel to the SIA diffusion planes. It can be seen that the normal rate theory approach in which the smeared out vacancy and interstitial concentrations can be represented by two equations is no longer valid. Although the vacancy concentration can still be represented by a single equation, each individual plane on which the SIA diffuse will see a different sink strength and thus need a separate equation. This clearly precludes any easy analytical approach to the spatial position of the voids.

To overcome this problem, the 2-D SIA diffusion model has been incorporated into a computer simulation as described in [57,58]. Briefly, the simulation first considered a cube of material containing voids in random positions and a smeared out dislocation density, and in which the self-interstitials were only allowed to move on one set of planes parallel to one of the cube faces. Each cycle of the computer program consisted of partitioning a small number of vacancies among the voids according to rate theory and then carrying out the same exercise for the SIA, but treating the SIA carrying planes separately, plane by plane. Since each of these planes received an equal number of interstitials which were then partitioned to the variable number of voids being cut by each plane, it was easy to see how partially aligned voids were able to grow faster than isolated voids. By examining the SIA fluxes reaching individual voids along the different planes bisecting the voids, it was possible to follow their growth (or shrinkage) and to quantitatively calculate void movements from high to low planar SIA concentrations. This movement was significant in perfecting the planar alignment of voids.

The results for this case of 2-D SIA diffusion on one set of planes are given in figure 3a, where the projected positions of random voids can be seen to become ordered on to planes parallel to the plane of SIA diffusion. This final structure is identical to that found in the hcp metals Mg, Ti and Zr where void and bubble alignment is found parallel to the basal plane. It is important to note that no wavelength emerges from the analysis. This is exactly the situation seen in void results on magnesium [14,15]; the example from [14] shown in figure 3b shows striking agreement with the computer simulation. Further evidence consistent with the model comes from the observation made by Jostsons and Farrell [14] that the voids in magnesium had aspect ratios very different from equilibrium bubbles; the latter were relatively equiaxed, but the former tended more towards a disc morphology with the disc faces parallel to the basal plane. This result follows naturally from the planar interstitials which must always erode parts of voids protruding into planes with higher SIA concentrations.

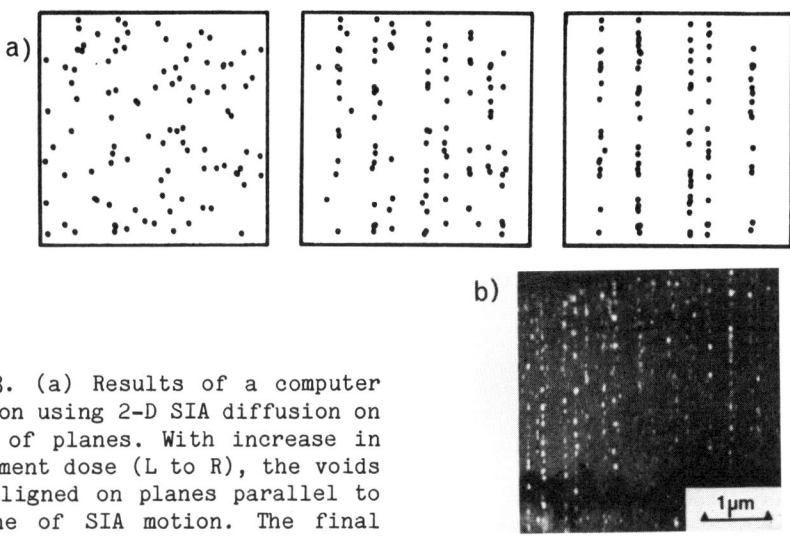

Figure 3. (a) Results of a computer simulation using 2-D SIA diffusion on one set of planes. With increase in displacement dose (L to R), the voids become aligned on planes parallel to the plane of SIA motion. The final result agrees remarkably with the experimental result (b) showing planar alignment of voids in Mg [14].

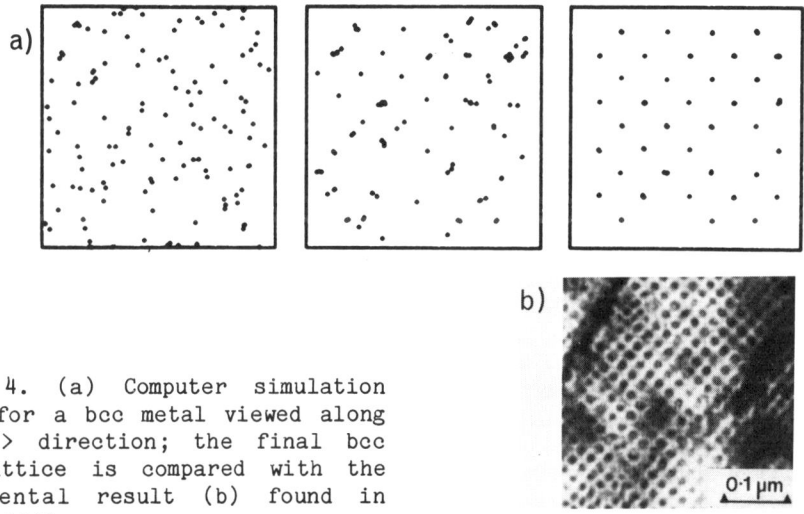

Figure 4. (a) Computer simulation result for a bcc metal viewed along an <001> direction; the final bcc void lattice is compared with the experimental result (b) found in niobium [31].

Following the application of the 2-D SIA diffusion model to explain (for the first time) the hcp results, the approach was extended to examine the cubic metals where the assumption of 2-D SIA diffusion on close-packed planes meant respectively, six sets of {011} planes in the bcc structure, and four sets of {111} planes in the fcc structure. Clearly, this is a complicated situation since voids will now receive local information about their neighbours via SIA fluxes from several different planes. However, on the basis that complete planar ordering would only be possible if a perfect lattice was formed, void lattices with close-packed planes defined by the planes of SIA diffusion were predicted, hence communicating the structure of the parent lattice to the voids, in agreement with experiment.

This important result was confirmed in further computer simulations [58] for both the bcc and fcc cases. As an example, figure 4 shows a <001> projection of the evolution of a random cavity distribution into an ordered bcc lattice. The final lattice is compared with an experimental result obtained for niobium [31]. Not surprisingly, the ordering in both the cubic simulations was significantly slower than the simpler hcp case. It was noted that the efficiency of the lattice formation was about an order of magnitude higher than that found in practice, a point easily explained by the moderating effect of any out-of-plane SIA jump-steps.

Both 4.5.1 and the present section have concentrated on void lattice formation. However, it is important to note that both anisotropic SIA diffusion mechanisms can apply equally well to bubble lattices. Only the properties of SIA diffusion are crucial, not the mode of cavity growth.

4.5. Cavity - Interstitial Loop Interaction

Recently, new models have been proposed separately for bubble and void lattices by Dubinko and co-authors [61,62]. In both cases the interaction between cavities and interstitial loops has been involved. Taking the bubble lattice model first, it will be recalled that the description of bubble growth given in section 2.2 pointed to a pressure-driven process, with additional reference to the well-known Greenwood, Foreman, Rimmer [6] loop punching model. Dubinko et al [61] have assumed that the loop punching operates within the fine scale of bubble formation and that loop glide occurs along the crystal close-packed directions. The presence of these small loops will cause an effective repulsion between two bubbles in the glide direction and balance a diffusive attraction between adjacent bubbles. Although it is easy to see that a model of this type must give the right bubble lattice crystallography (at least for the bcc and fcc lattices), the key processes which move the bubbles to their correct positions are not discussed. In fact, it would seem that a loop from one bubble punched out toward an imperfectly aligned bubble will be partially absorbed by that bubble and move it away from linear alignment.

Setting aside this serious difficulty in the model, its application for the cubic metals also rests on the atomistic details of loop punching. Dubinko et al use the observation of loop punching from isolated helium clusters (actually platelets, rather than bubbles [63]) to justify their approach but extrapolation to the high bubble density situation cannot be certain. Another difficulty must be the anisotropy induced into any loop punching process by the lateral stress usually associated with the relatively near-surface implants.

In their model on void lattice formation, Dubinko et al [62] use the gliding of self-interstitial loops formed during irradiation to form void lattices very much along the lines of the single interstitials in Foremans model. The result is a tendency for one-dimensional alignment along the loop gliding directions - usually the close-packed directions. Although the nucleation barrier mentioned in the description of the Foreman model could still apply, it is clear that the final lattice would fit the results for the cubic lattices. The main assumption must be that sufficient interstitial loops are continuously created. One interesting claim is that since the model requires the unfaulting of sessile loops to form perfect loops, metals with a high stacking fault energy will be favoured for void lattice formation. As suggested in [62], this could be tested using alloying additions to modify stacking fault energies.

5. DISCUSSION

Having summarised and critically appraised the mechanisms proposed for cavity lattice formation in the previous section, the question is which of these survive. Reasons have been given for eliminating the phase stability models (no crystallography), the isomorphic decomposition model (A/r far too small, no prediction of fcc lattices), the elastic interaction model for voids (Finnis simulation) and loop punching models for bubbles (loop punching will act against linear alignment). This leaves the models based on the anisotropic diffusion of interstitials, but we could also include the Dubinko model for voids based on the glide of interstitial loops. It should be emphasised for the models using the anisotropic diffusion of interstitials that we are not discussing quantitative diffusivities in different directions or on different planes, but the anisotropic nature of the interstitial jump-step.

That the two models based on anisotropic self-interstitial diffusion could work has been established beyond reasonable doubt; the question of their application is another matter. Clearly both models are not 'ab initio' but start with an assumption on the one- or two-dimensional migration of the self-interstitial. The questions then are whether the assumptions are reasonable, whether they can be validated in other ways, and which of the two variants of the overall model applies to the different experimental results.

For both the 1-D and 2-D SIA models, there are experimental situations where only one model can apply. These are the linear alignment of voids in alumina and the planar ordering of voids and bubbles in the hcp metals. These will be discussed in turn before going through the other systems listed in table I.

5.1. Alumina

In the results section we have noted that the ordering of voids in alumina is linear along the c-axis of the hexagonal structure. Clinard et al [25] in discussing their discovery suggested that the result was consistent with the 1-D SIA diffusion model of Foreman. Certainly the 2-D SIA mechanism cannot be applied to explain the result. Clinard et al supported their conclusion with the observation that the macroscopic swelling associated with the void formation was anisotropic, with most of the swelling being in the c-axis direction, entirely consistent with the presence of 1-D SIA jump-steps along the same axis.

5.2. Hcp Metals

The planar arrays of voids in Mg and bubbles in Ti and Zr are completely consistent with planar diffusion of interstitials in these metals along or within the basal plane. It is significant that no other model has been applied to these results though it might be less easy to eliminate 1-D SIA diffusion than appears initially. The linear ordering of cavities along directions within the basal plane could conceivably combine to give planar ordering. However, there would be a definite lattice structure within the planes of cavities. In fact the cavities within the planes are reported [21,22] to be randomly placed leaving on present evidence only the 2-D SIA model for these metals. The particular application of this model to Zr has been discussed recently [22] where attention was drawn to the fact that researchers trying to explain radiation creep and breakaway swelling in Zr and Zr alloys are increasingly invoking the anisotropic diffusion of interstitials in their modelling (see [64] for several examples). The fact that such diffusion will give the observed bubble lattices in Zr clearly gives the models considerable support.

Equally important are the theoretical studies on the structure and migration paths of SIAs in hcp metals. These have been reviewed recently by Bacon [65] and by Frank [66] with general agreement that for hcp metals with c/a ratio less than ideal (eg. Zr, Ti and Mg), the B_O interstitial has the most stable configuration. This interstitial sits on the basal plane midway between the two octahedral sites above and below the plane. Viewed on the basal plane the extra atom sits symmetrically between its three nearest neighbours and could be regarded as a planar crowdion. The significant property of this configuration is that its jump-steps are calculated to migrate on or within the basal plane, exactly fitting the requirement of the proposed 2-D SIA model.

When the agreement between the model and the experimental void results in magnesium (the variable lattice spacing and the shape of voids - see section 4.4.2) are added to the above, the case for applying the 2-D SIA model to Mg, Zr and Ti appears to be extremely strong. In passing, it is worth noting that for hcp metals with a greater than ideal c/a ratio, eg. zinc and cadmium, the stable SIA configuration is expected to change to the D_C configuration, a c-axis dumb-bell, which should have a strong tendency to migrate one-dimensionally in the c direction [66]. If correct, then the Foreman model should operate to give linear cavity alignment, exactly as for the alumina case discussed in 5.1. However, these metals have not yet been studied. Their low melting temperatures will require implantations at well below ambient to obtain homologous temperatures similar to those used in studying the high bubble densities and planar ordering in titanium and zirconium.

5.3. Bcc Metals

If self-interstitials migrate two-dimensionally on the close-packed {011} planes in the bcc metals, then the discussions in section 4.4.2 demonstrated that bcc bubble and void lattices would be a reasonable outcome. Such an outcome is also plausible with 1-D SIA. That one of these models is correct is almost certain since the idea of anisotropic SIA diffusion provides a straightforward explanation of a seemingly unconnected void phenomenon. This phenomenon, in which large voids shrink while small voids grow, is seen both in niobium [67] and molybdenum [68] and qualitatively is explained very simply by considering the behaviour of void sink strength with void radius for one or two dimensional SIA diffusion relative to three-dimensional vacancy diffusion. The relevant equations will not be repeated here but with the 2-D SIA assumption, the niobium results of Loomis and Gerber [67] can be quantitatively described with surprising accuracy [52]. Furthermore, the sink strength arguments used in explaining these results also explain the long standing association of void swelling saturation with lattice formation besides giving lower limits of the values of the (A/r) ratios for void lattices (about 7 and 9 for bcc and fcc lattices respectively [52]), in good agreement with experiment. The different growth mode of bubbles prevents a similar analysis so that a different range of A/r ratios is not too surprising. Other implications of the model have been discussed [69], including the several observations of enhanced void swelling at low irradiation doses.

The question remains as to which SIA diffusion model is more likely for bcc metals. We believe there is sufficient evidence to prefer the 2-D SIA model. In the first place there is the internal friction evidence of Jacques and Robrock, section 4.4.2, that led to the first application of 2-D SIAs to void lattice formation. Secondly, the interstitial configuration has been shown experimentally to be a <011> dumb-bell and it is easy to see a plausible jump-step, figure 5, that leads to the required planar migration. Theoretical modelling of SIA migration paths in bcc metals with interatomic potentials available at present do not yet support this mode of diffusion [70,71], but the differentials in jump energies are small. Thirdly, in deciding between 1-D and 2-D ordering, we believe it is possible to use the early stages of ordering. In the 1-D SIA case, the initial stages must involve nearest-neighbour cavities because of the direct shadowing effects; thus they must be short range. However, the information carried by the local concentrations of 2-D SIAs can come from a far larger volume and thus can be a far longer range effect.

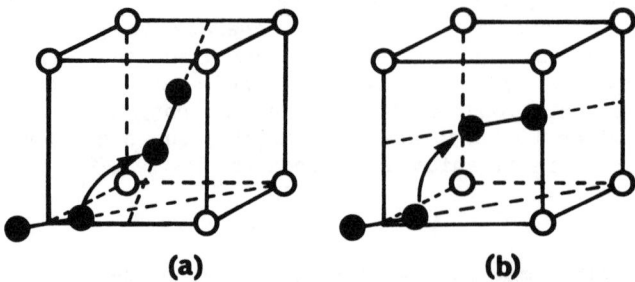

Figure 5. In (a) and (b) the two possible jump-steps for an <011> dumbbell self-interstitial configuration in a bcc structure are given. While the jump-step in (a) involves a rotation of the dumbbell to another direction, the jump-step in (b) maintains its orientation and must move two-dimensionally in one plane.

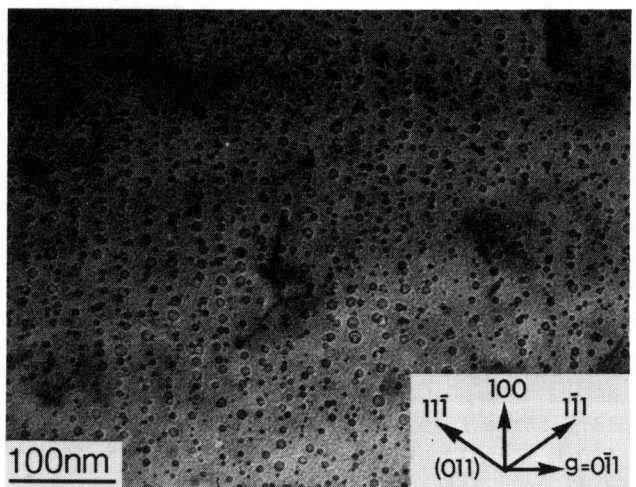

Figure 6. An electron micrograph showing the early stages of void alignment in molybdenum. The partial north-south alignment could be consistent both with 2-D SIA diffusion in the (0$\bar{1}$1) planes or 1-D diffusion along the [111] and [$\bar{1}$11] directions. However, the absence of any alignment in the marked [1$\bar{1}$1] and [11$\bar{1}$] directions rules out this possibility.

This latter situation fits the results in molybdenum where Liou et al [29] first pointed out the evolution of the void lattice via planar ordering. Here we show how the early stages of void lattice formation in molybdenum can be used directly as evidence against the short range nearest-neighbour alignment (along a <111> direction in a bcc structure) required both by the 1-D SIA model and in the Dubinko void model. In figure 6, an imperfect void lattice in Mo is seen imaged along the [011] direction. The north/south void alignments perpendicular to the [110] direction are clearly consistent with planar ordering on {011} planes. However, this void aligment is also expected on the 1-D SIA model from the two <111> directions lying perpendicular to [110]. The key factor in rejecting this possibility is the absence of any alignments corresponding to the other two <111> directions marked on the micrograph. These would have to be present for any model based on one-dimensional, or nearest-neighbour ordering. The visible alignments in figure 6 therefore must be planar and cannot be linear.

In addition to the above, 2-D SIA migration has been discussed with respect to resistivity recovery results in Mo, W [72] and Nb [73].

5.4. Fcc Metals

To create an fcc cavity lattice using anisotropic SIA diffusion requires the SIA either to move one-dimensionally along the close-packed <011> directions or two-dimensionally along the close-packed {111} planes. Separate support in applying the models to fcc metals is less than for the hcp and bcc metals, but fast void growth at low irradiation doses has been noted [69] while excellent evidence of void swelling saturation coincident with void lattice formation is available for Ni [11]. Certainly, when the experimental evidence for cavity lattices in metals is taken as a whole, it seems implausible that the fcc metals could have a different explanation to that for hcp and bcc metals.

The 1-D SIA diffusion model has received attention in several papers by Woo and Frank [53-55], where the model (both for fcc and bcc metals) is strongly linked to the controversial metastable low temperature crowdion configuration for self-interstitials and the associated recovery of stable interstitials in Stage III. Frank [55] argues that the crowdion model has experimental support and that the formation of void lattices effectively proves its correctness. However, there are many techniques detecting vacancy migration in stage III (eg. helium desorption spectrometry [74], positron annihilation [75] and perturbed angular correlation [76]), which challenge the viewpoint on the crowdion model. Certainly the formation of void lattices does not uniquely depend on the crowdion; as shown in section 4.4.2, the 2-D SIA model works equally well. On the other hand, it is possible that 1-D migration of stable interstitials might occur, unconnected with the low temperature crowdions. Woo and Frank have not applied their overall model to bubble lattices.

For fcc metals, discussion between 1-D and 2-D SIA migration has to resort to the arguments made in the previous section. The published micrographs of voids in fcc metals appear to fit the longer range ordering of the 2-D SIAs. eg [13,30], but additionally there is evidence for the absence of ordering along the traces of close-packed directions, analogous to the figure 6 situation for molybdenum. As the reader can verify, the (011) micrograph shown for voids in Al [30] would only appear to fit the planar case; one expected crystallographic trace appropriate to linear ordering is absent. We conclude that in fcc metals the evidence is marginally in favour of planar SIA diffusion.

One argument that could be used against 2-D SIA on the {111} planes in the fcc structure is that it is accepted that the stable SIA configuration is a <001> dumbbell. Experimental evidence on this configuration has been obtained at low temperatures but there is no a priori reason why this has to be maintained to high temperatures. Alternative configurations with the required migration property are available. Evans [56] pointed out that a <112> dumbbell belongs uniquely to a single (111) plane and could then diffuse two-dimensionally within that plane. More plausible is the planar crowdion configuration described by Beeler [77] which, in computer calculations, migrated within the (111) plane. It is interesting that the planar structure of this interstitial was identical to the B_0 interstitial referred to in the hcp discussions.

5.5. Calcium Fluoride

To explain the simple cubic lattice of anion voids in calcium fluoride using anisotropic SIA diffusion ideas, one would require either 2-D anion interstitials moving along {001} planes or 1-D migration along <001> directions. Johnson and Chadderton [27] have been strong proponents for the 1-D SIA diffusion mechanism. A feature of their proposal was that the defect with linear diffusion properties was the V_k centre, a precursor of the anion interstitial, rather than the interstitial itself. The V_k centre is known to diffuse linearly in CaF_2, and for 90° jumps to become more probable along the sequence $CaF_2 \to SrF_2 \to BaF_2$ [78].

It appears significant that the tendency for ordering was found to exactly follow this sequence, with no ordering being found in the BaF_2 case. Thus the 1-D SIA model appears to fit the available evidence. As discussed in 4.2, the possibility of any elastic interaction mechanism playing a part was excluded [27].

5.6. Precipitates in Oxygen-Implanted Silicon

The simple cubic precipitate structure found found in silicon [28] was explained on the basis of elastic anisotropy (similar to the isomorphic decomposition model, section 4.2). The value of the elastic anisotropy factor, $(C_{11}-C_{12}-2C_{44})$, was consistent with the simple cubic lattice. Although this may be true, we have to ask whether it is probable that the ordering in silicon has a different mechanism to the other systems covered in this review. Clearly a definitive answer is not possible but it is significant that attempts to form helium bubble lattices in silicon were unsuccessful [79], even though bubble precipitation occured on the same fine scale as in the bcc, fcc and hcp metals. If the elastic anisotropy model were correct for the precipitate lattice, it is difficult to see why it would not apply to the bubble results.

In the context of the present review, it is important to explore if the results for Si could fit one of the anisotropic SIA diffusion models. From the previous section on CaF_2, this must be possible in principle. However, although an examination of the diamond cubic structure shows no obvious SIA configuration for either 1-D or 2-D migration, the SIA configuration in Si is still a matter for some debate, eg. [80,81].

The conclusion from the above discussion is that neither the elastic anisotropy model nor the SIA models can apply, at least in a straightforward way. If they did, then the helium bubble lattice and the precipitate lattice should form with equal ease. The only clear difference between the two cases is the use of oxygen as an implanting ion for the latter. Since there are suggestions that SIA and oxygen atoms in Si form complexes [82], we propose here that one of these complexes must migrate anisotropically, either one-dimensionally along ⟨001⟩ directions or two-dimensionally within {001} planes. At present this appears a possible answer to explain the precipitate lattice in silicon.

6. SUMMARY

This paper has provided an experimental and theoretical review of the void and bubble observations at present available. Of the models which include the vital crystallography common to all the results, only those based on the anisotropic diffusion of interstitials appear viable. If we do not distinguish between the 1-D and 2-D SIA migration modes, then we can satisfy the simple and intuitive viewpoint that only one generic model should be involved in cavity lattice formation. Although the growth modes of voids and bubbles are different, the mobility of self-interstitials is a common factor in both cases. Of additional importance is the fact that there are separate experimental results which are also explained by anisotropic SIA migration. This is particularly so for the bcc metals Nb and Mo, but is also true for the hcp metal results.

In discussing the partition of the results among the two SIA diffusion models, it seems clear that 1-D SIA diffusion is applicable to alumina and probably to calcium fluoride while the evidence for 2-D SIA diffusion is very clear for the hcp results and almost certainly for the bcc metals. The fcc metal case is less clear but 2-D SIA migration appears to be favoured. For the precipitate lattice in silicon a new suggestion is made based on the anisotropic diffusion of oxygen-self-interstitial complexes.

Acknowledgement. This work has been supported as part of the UKAEA Underlying Research Programme. The author thanks Drs. Alan Foreman, David Mazey, Peter Johnson and Susan Murphy for useful comments on this paper.

REFERENCES

[1] Evans, J.H., Nature 1971 229, 403; Rad. Effects 1971 10, 55.
[2] Wiffen, F.W., ref.5, p.386.
[3] Cawthorne, C. and Fulton, E.J., Nature 1966 216, 575.
[4] Proc. Int. Conf. on Voids Formed by Irradiation of Reactor Materials, Reading, UK., 1970, Brit. Nucl. Energy Society.
[5] Proc. Int. Conf. on Radiation Produced Voids in Metals, Albany, 1971 (Eds. J.W. Corbett and L.C. Ianniello), USAEC, 1972.
[6] Greenwood, G.W., Foreman, A.J.E. and Rimmer, D.E., J. Nucl. Mater. 1959 4, 305.
[7] Mazey, D.J., Eyre, B.L. Evans, J.H., Erents, S.K. and McCracken, G.M., J. Nucl. Mater. 1977 64, 145.
[8] Evans, J.H., J. Nucl. Mater. 1978 76/77, 228.
[9] Templier, C., Jaouen, C., Riviere, J-P., Delafond. J. and Grilhe, J., Comptes Rendus 1984 299, 613.
[10] Evans, J.H. and Mazey, D.J., J. Physics F. 1985 15, L1; J. Nucl. Mater. 1986 138, 176.
[11] Krishan, K., Radiation Effects 1983 66, 121.
[12] Kulcinski, G.L., Brimhall, J.L. and Kissinger, H.E., J. Nucl. Mater. 1971 40, 166.
[13] Mazey, D.J., Francis, S. and Hudson, J.A., J. Nucl. Mater. 1973 47, 137.
[14] Jostsons, A. and Farrell, K., Radiation Effects 1972 15, 217.
[15] Risbet, A. and Levy, V., J. Nucl. Mater. 1974, 50, 116.
[16] Sass, S. and Eyre, B.L., Phil. Mag. 1973 27, 1447.
[17] Johnson, P.B., Mazey, D.J., Nature 1978 276, 595; Rad. Effects 1980 53, 195.
[18] Johnson, P.B., Mazey, D.J. and Evans, J.H., Rad. Effects 1983 78, 147.
[19] Mazey, D.J. and Johnson, P.B., to be published.
[20] Johnson, P.B., Mazey, D.J., J. Nucl. Mater. 1980 93/94, 721.
[21] Mazey, D.J. and Evans, J.H., J. Nucl. Mater. 1986 138, 16.
[22] Evans, J.H., Foreman, A.J.E. and McElroy, R.J., J. Nucl. Mater. in press.
[23] Johnson, P.B., Mazey, D.J., J. Nucl. Mater. 1985 127, 30.
[24] Johnson, P.B., Malcolm, A.L. and Mazey, D.J., Nature 1987 329, 316; J. Nucl. Mater. 1988 158, 108.
[25] Clinard, F.W., Bunch, J.M. and Ranken, W.A., Proc. Conf. on Radiation Effects and Tritium Technology for Fusion Reactors, Gatlinburg, 1975, ERDA CONF-750989, p.498.
[26] Johnson, E., Chadderton, L.T., and Wohlenberg, T., Rad. Effects 1976 28, 111.
[27] Johnson, E., and Chadderton, L.T., Rad. Effects 1983 79, 183.
[28] Van Ommen, A.H., Koek, B.H. and Viegers, M.P.A., Applied Physics Letters 1986 49, 1062.
[29] Liou, K., Smith, H.V., Wilkes, P. and Kulcinski, G.L., J. Nucl. Mater. 1979 83, 335.

[30] Horsewell, A. and Singh, B.N., Rad. Effects 1987 102, 1.
[31] Loomis, B.A., Gerber, S.B. and Taylor, A., J. Nucl. Mater. 1977 68, 19.
[32] Loomis, B.A. and Gerber, S.B., J. Nucl. Mater. 1978 71, 377.
[33] Moteff, J., Sikka, V.K. and Jang, H., Proc. Conf. on The Physics of Irradiation Produced Voids, Harwell 1974, AERE Report 7934, p.181.
[34] Fisher, S.B. and Williams, K.R., Rad. Effects 1977 32, 123.
[35] Little, E.A., unpublished results.
[36] Malen, K., and Bullough, R., Ref. 4, p.109.
[37] Tewary, V.K. and Bullough, R., J. Physics F. 1972 2, L69.
[38] Stoneham, A.M., J. Physics F. 1971 1, 778.
[39] Tewary, V.K., J. Physics F. 1973 3, 1275.
[40] Finnis, M.W., UKAEA Annual Underlying Report, 1987/88, p.101.
[41] Ardell, A.J., Nicholson, R.B. and Eshelby, J.D., Acta Met. 1966 14, 1295.
[42] Khachaturyan, A.G. and Airapetyan, V.M., Phys. Stat. Solidi A 1974 26, 61.
[43] Polat, S., March, C., Little, T., Ju, C.P., Epperson, J.E. and Chen, H., Scripta Met. 1986 20, 1739.
[44] Martin, G., Phil. Mag., 1975 32, 615.
[45] Benoist, P. and Martin, G., Fundamental Aspects of Radiation Damage in Metals, Gatlinburg, 1975 (Eds. M T Robinson and S Young), p.1236.
[46] Krishan, K., Phil. Mag.A 1982 45, 401.
[47] Krishan, K., Proc. Meeting on Non Linear Phenomena in Materials Science, Aussois 1987 (eds. L P Kubin and G Martin) Trans Tech Publications 1988, p.267.
[48] Koptelov, E.A. and Semenov, A.A., Phys. Stat. Solidi A 1985 89, 117
[49] Sugakov, V.I. and Selishchev, P.A., Sov. Phys. Solid State, 1986 28, 1641.
[50] Koptelov, E.A. and Semenov, A.A., J. Nucl. Mater. 1988 160, 253
[51] Foreman, A.J.E., AERE Report 7135 (1972).
[52] Evans, J.H., Materials Science Forum 1987 15-18, 869.
[53] Woo, C.H. and Frank, W., J. Nucl. Mater. 1985 137, 7; 1986 140, 214; 1987 148, 121.
[54] Woo, C.H. and Frank, W., Materials Science Forum 1987 15-18, 875.
[55] Frank, W., Proc. Meeting on Non Linear Phenomena in Materials Science, Aussois 1987 (eds. L P Kubin and G Martin) Trans Tech Publications 1988, p.315.
[56] Evans, J.H., J. Nucl. Mater. 1983 119, 180.
[57] Evans, J.H., Proc. 12th ASTM Symposium on the Effects of Radiation on Materials, Williamsburg, June 1984, ASTM STP 870, p.525.
[58] Evans, J.H., J. Nucl. Mater. 1985 132, 147.
[59] Evans, J.H., Proc. Meeting on Non Linear Phenomena in Materials Science, Aussois 1987 (eds. L P Kubin and G Martin) Trans Tech Publications 1988, p.303.
[60] Jacques, H. and Robrock, K-H., Proc. Yamada Conf. on Point Defects in Metals, Kyoto, Japan, 1981, Eds: J. Takamura, M. Doyama, and M. Kiritani, (Univ. of Tokyo Press, 1982) p.159.

[61] Dubinko, V.I., Slezov, V.V., Tur, A.V. and Yanovsky, V.V., Radiat. Effects 1986 100, 85
[62] Dubinko, V.I., Tur, A.V., Turkin, A.A. and Yanovsky, V.V., J. Nucl. Mater. 1989 161, 57.
[63] van Veen, A., Caspers, L.M. and Evans, J.H., J. Nucl. Mater. 1981 103/104, 1181.
[64] Proc. Int. Conf. on Fundamental Mechanisms of Radiation-Induced Creep Growth (Eds. C H Woo and R J McElroy), J. Nucl. Mater. 1988 159.
[65] Bacon, D.J., J. Nucl. Mater. 1988 159, 176.
[66] Frank, W., J. Nucl. Mater. 1988 159, 237.
[67] Loomis, B.A. and Gerber, S.B., J. Nucl. Mater. 1981 102, 154.
[68] Bentley, J., Eyre, B.L. and Loretto, M.H.: Proc. Conf. on Fundamental Aspects of Radiation Damage in Metals, Gatlinburg, 1975, US-ERDA Conf-7510006, p.925.
[69] Evans, J.H. and Foreman, A.J.E., J. Nucl. Mater. 1985 137, 1.
[70] Harder, J.M. and Bacon, D.J., Phil. Mag.A 1986 54, 651.
[71] Ackland, G.J. and Thetford, R., Phil. Mag.A 1987 56, 15.
[72] Schultz, H., Materials Science Forum 1987 15-18, 727.
[73] Petzold, J. and Schultz, H., Materials Science Forum 1987 15-18, 733.
[74] van Veen, A., Materials Science Forum 1987 15-18, 3.
[75] Hautojarvi, P., Materials Science Forum 1987 15-18, 81.
[76] Sielemann, R., Materials Science Forum 1987 15-18, 25.
[77] Beeler, J.R., Radiation Effects Computer Experiments (North-Holland, 1983) p.219.
[78] Norgett, M.J. and Stoneham, A. M., J. Phys. C. 1973 6, 238.
[79] Evans, J.H., van Veen, A. and Griffioen, C.C., Nucl. Instr. Methods B, 1987 28, 360.
[80] Mastri, P., Harker, A.H. and Stoneham, A.M., J. Phys. C. 1983 16, L616.
[81] Car, R., Kelly, P.J., Oshiyama, A. and Pantelides, S.T., Physica 1984 127B, 401.
[82] Deak, P., Snyder, L.C., Corbett, J.W., Wu, R.Z., and Solyom, A., Materials Sci. Forum 1989 38-41, 281.
[83] Chen, L.J. and Ardell, A.J., J. Nucl. Mater. 1978 75, 177.
[84] Mazey, D.J., Electron Microscopy Studies of Void Formation in Metals, Ph.D Thesis, Univ. of Salford, 1975.

THE FORMATION OF CLUSTERS OF CAVITIES DURING IRRADIATION

S.M.MURPHY
Materials Development Division,
Harwell Laboratory,
Didcot, Oxfordshire, OX11 0RA
U.K.

ABSTRACT. This paper discusses the formation of cavity clusters in irradiated metals. A simple rate theory model is used to describe the growth of cavities during irradiation. A linear stability analysis of this model shows that cavities in some regions can grow at the expense of cavities in other regions, thus leading to the formation of cavity clusters. Numerical results from this model are presented and compared with experimental results on a Fe-15%Ni-12%Cr alloy.

1. Introduction

The irradiation of metals and alloys can have a profound effect on their microstructure. For example, cavities may form and dislocation loops may nucleate and grow, causing swelling and hardening of the material. In many cases, the presence of the irradiation induces an instability which produces inhomogeneities in the microstructure [1]. As a result of this, the precipitation behaviour in alloys may be modified significantly by the irradiation. Even in pure materials which have been carefully annealed before irradiation, the microstructure after irradiation contains regions with a high dislocation density separated by regions with very few dislocations [2-4]. In other circumstances, periodic structures are formed; void and bubble lattices have been observed in a wide range of materials [5] and arrays of dislocation loops have also been observed in a number of pure materials [6-8]. In alloys, in addition to variations in the cavity or dislocation density, variations in composition can develop from one region to another [9-10].

The subject of irradiation-induced instabilities in metals and alloys has been studied theoretically by a number of authors. Martin [11] discussed the formation of oscillations in composition in an irradiated alloy and Martin et al. [12] showed that an irradiation-induced instability was responsible for the formation of zinc-rich precipitates in irradiated Al-Zn alloys. Krishan and Abromeit [13] and Abromeit and Martin [14] discussed irradiation-induced instabilities in concentrated alloys. Krishan [15] and Koptelov and Semenov [16] used a similar approach to study void lattices and Murphy [17] discussed fluctuations in the density of vacancy loops.

This paper discusses a model for cavity growth which displays an irradiation-induced instability. Thus, the cavities in some regions grow at the expense of cavities in the surrounding areas. This study was motivated by a recent observation of cavity clusters in an irradiated Fe-15%Ni-12%Cr-1%Si alloy [18]. It was also found that these cavity clusters were associated with regions rich in nickel and silicon. Thus the formation of cavity clusters may explain the observations of fluctuations in composition in irradiated Fe-Ni-Cr alloys. Nickel segregates towards point defect sinks, so once the cavity clusters have formed, nickel will migrate towards the cavities and iron and chromium will migrate away, giving rise to fluctuations in composition. Evidence in support of this proposal comes from ion-irradiations in materials with and without pre-injected helium (as discussed in the next section, in many circumstances the growth of small cavities is dependent on the presence of helium). Fluctuations in composition were observed in the material containing helium but not in the material containing no helium [19].

The mechanism discussed here is based on the preferential growth of cavities in some regions and the shrinkage of cavities in neighbouring regions. This model is similar to that described by Krishan [15] for the formation of void lattices. However, in the model discussed here the role of helium is included and the formation of vacancy loops is neglected. Thus this model is appropriate for high temperature irradiation conditions where the lifetime of vacancy loops is short and hence their concentration is very low.

2. Description of the model

During the irradiation of a metal, atoms are displaced from their normal lattice sites to form self-interstitial atoms and vacant lattice sites in the material. These interstitials and vacancies migrate through the lattice until they recombine, aggregate to form interstitial or vacancy clusters or disappear by absorption at sinks such as dislocations and grain boundaries. During neutron irradiation, helium atoms are also produced in the material by transmutation reactions. Helium is almost insoluble in metals and tends to aggregate to form helium bubbles. These bubbles grow during irradiation by absorbing both gas atoms and vacancies, and if there is a strong flux of vacancies to the bubbles, they eventually grow into voids. Here, the term void indicates that the pressure of gas within the void is too low to balance the surface tension forces. In this paper, the term cavity will be used to include both bubbles and voids, although most of the discussion will be concerned with small cavities which contain a high gas pressure.

The model presented in this paper includes vacancies, interstitials and gas atoms, and their absorption and emission by cavities. The model also includes dislocation lines which can absorb interstitials and vacancies. For simplicity, it is assumed here that gas atoms do not interact with these dislocation lines, although there are many observations of dislocation lines decorated with small gas bubbles. It is also assumed here that these dislocation lines are uniformly distributed through the material, and that their density is constant during the irradiation. Thus the concentrations of vacancies and interstitials, c_v and c_i, are given by the equations

$$\frac{\partial c_v}{\partial t} = K + D_v \nabla^2 c_v - D_v(c_v - \bar{c}_{vC})4\pi r_C C_C - D_v(c_v - \bar{c}_{vN})Z_{vN}\rho_N , \qquad (1)$$

$$\frac{\partial c_i}{\partial t} = K + D_i \nabla^2 c_i - D_i c_i 4\pi r_C C_C - D_i c_i Z_{iN}\rho_N , \qquad (2)$$

where K is the production rate of the interstitials and vacancies, and D_v and D_i are the diffusion coefficients of the vacancies and interstitials. For simplicity, bulk recombination of vacancies and interstitials is neglected in these equations. ρ_N denotes the dislocation density, and Z_{vN} and Z_{iN} denote the bias factors of the dislocations for the absorption of vacancies and interstitials, respectively. In general, dislocation lines absorb interstitials more readily than vacancies, so $Z_{iN} > Z_{vN}$. r_C and C_C denote the radius and concentration of the cavities, and $\bar{c}_{vC}$ and $\bar{c}_{vN}$ the equilibrium concentrations of vacancies adjacent to the surface of a cavity and a dislocation line, respectively. These concentrations are given by the relations

$$\bar{c}_{vC} = c_v^e \exp\left(\frac{\Omega}{kT}\left(\frac{2\gamma_S}{r_C} - p_g\right)\right), \tag{3}$$

$$\bar{c}_{vN} = c_v^e, \tag{4}$$

where c_v^e is the concentration of vacancies in the material under thermal equilibrium conditions. Here Ω is the atomic volume of the material, k is Boltzmann's constant and T is the absolute temperature. γ_S is the surface tension and p_g is the gas pressure within the cavity. It is assumed here for simplicity that the gas pressure is given by the Ideal Gas Law, i.e.

$$p_g = \frac{3n_{gC}kT}{4\pi r_C^3}, \tag{5}$$

where n_{gC} is the number of gas atoms in each cavity. The concentration of vacancies in cavities as a fraction of the number of lattice sites, which is denoted by q_{vC}, is given by

$$q_{vC} = 4\pi r_C^3 C_C / 3. \tag{6}$$

The concentration of vacancies trapped in cavities increases as the cavities absorb vacancies, so q_{vC} satisfies the equation

$$\frac{\partial q_{vC}}{\partial t} = 4\pi r_C^2 C_C \frac{\partial r_C}{\partial t} = 4\pi r_C C_C (D_v(c_v - \bar{c}_{vC}) - D_i c_i). \tag{7}$$

Here the concentration of cavities, C_C is assumed to be constant. The number of gas atoms in each cavity, n_{gC}, is given by the equation

$$\frac{\partial n_{gC}}{\partial t} = 4\pi r_C D_g c_g / \Omega - K n_{gC}, \tag{8}$$

where c_g is the concentration of gas atoms migrating through the material, and D_g is the diffusion coefficient of these gas atoms. The term Kn_{gC} describes the loss of gas atoms from cavities as a result of direct 'knock-on' interactions between the gas atoms and the bombarding high energy particles. The concentration of free gas atoms, c_g, is given by

$$\frac{\partial c_g}{\partial t} = K_g + D_g \nabla^2 c_g - D_g c_g 4\pi r_C C_C + K n_{gC} \Omega C_C, \tag{9}$$

where K_g is the production rate of gas atoms.

If the cavities are small and surface tension is important, i.e. $\bar{c}_{vC}$ is large, then the cavities grow by absorbing gas atoms. In this case, the growth of the cavities is limited by the supply of gas atoms, and not by the absorption of vacancies.

3. Instability in cavity size

The simple model presented in the previous section displays an instability which leads to spatial variations in the size of the cavities. In other words, cavities in one region grow at the expense of cavities in a neighbouring region. This process can be explained as follows. Suppose one region contains slightly larger cavities than other neighbouring regions. This situation could arise because the concentration of gas in this region was particularly high, or because at some stage in their history the cavities in this region were able to grow more rapidly (e.g. a number of gas bubbles could nucleate at a precipitate or at an interstitial dislocation loop where the diffusion of gas atoms into the cavities is rapid). Then this group of larger cavities absorb gas atoms more strongly (see equation (8)). This in turn causes the cavities to grow even larger by reducing the thermal emission of vacancies from the cavities (see equations (3), (5) and (7)). Because the absorption of gas atoms by the cavities reduces the local concentration of free gas atoms, further gas migrates into the region from neighbouring regions. So the cavities in this region continue to grow at the expense of cavities in a neighbouring region. This process is limited to some extent by the time required for the cavity radius to respond to an increase in the number of gas atoms in the cavity, and for the gas to migrate into the region containing larger cavities. The loss of gas atoms from cavities also tends to reduce the inhomogeneities in the cavity distribution.

This argument can be tested by using linear stability theory to determine the conditions for an irradiation-induced instability to occur. Thus, it is assumed that the concentrations of vacancies, interstitials and gas atoms are given by the relations

$$c_v(x,t) = c_v^0(t) + c_v^1(x,t) ,$$

$$c_i(x,t) = c_i^0(t) + c_i^1(x,t) ,$$

$$c_g(x,t) = c_g^0(t) + c_g^1(x,t) , \qquad (10)$$

and the cavity radius and the number of gas atoms per cavity are given by

$$r_C(x,t) = r_C^0(t) + r_C^1(x,t) ,$$

$$n_{gC}(x,t) = n_{gC}^0(t) + n_{gC}^1(x,t) , \qquad (11)$$

where c_i^0, c_v^0, c_g^0, r_C^0 and n_{gC}^0 are the solutions of the equations (1), (2), (7), (8) and (9) above assuming that there is no spatial variation in these quantities. For simplicity, it is also assumed here that the microstructure varies only in the x direction and any variations in other directions are ignored. The next step is to investigate whether the magnitude of the spatially varying components of the concentrations, $c_v^1(x,t)$ etc. increase or decrease with time. It is more useful to study the evolution of the fluctuations as a function of wavelength rather than position, and therefore the Fourier Transforms of these fluctuations (which are denoted by a tilde) are used in the remainder of this section. The Fourier Transform, $\tilde{c}_v(\lambda,t)$, of the vacancy concentration $c_v(x,t)$ is given by

$$\tilde{c}_v(\lambda,t) = \int c_v^1(x,t) e^{i\lambda x} dx , \qquad (12)$$

with similar equations for $\bar{c}_i$, $\bar{c}_g$, $\bar{r}_C$ and $\bar{n}_{gC}$. By taking the Fourier Transform of equations (1), (2) and (9), equations for the Fourier Transforms of the vacancy, interstitial and gas concentrations are obtained, and these are given by

$$\frac{\partial \bar{c}_v}{\partial t} = -D_v\lambda^2\bar{c}_v - 4\pi r_C^0 C_C D_v(\bar{c}_v - \bar{c}_{vC}) - 4\pi \bar{r}_C C_C D_v(c_v^0 - \bar{c}_{vC}) -$$
$$- Z_{vN}\rho_N D_v \bar{c}_v , \qquad (13)$$

$$\frac{\partial \bar{c}_i}{\partial t} = -D_i\lambda^2\bar{c}_i - 4\pi r_C^0 C_C D_i \bar{c}_i - 4\pi \bar{r}_C C_C D_i c_i^0 - Z_{iN}\rho_N D_i \bar{c}_i , \qquad (14)$$

$$\frac{\partial \bar{c}_g}{\partial t} = -D_g\lambda^2\bar{c}_g - 4\pi r_C^0 C_C D_g \bar{c}_g - 4\pi \bar{r}_C C_C D_g c_g^0 + K\bar{n}_{gC} C_C \Omega , \qquad (15)$$

where $\bar{c}_{vC}(\lambda,t)$ is the Fourier Transform of $\bar{c}_{vC}(x,t)$. Similarly, $\bar{r}_C$ and $\bar{n}_{gC}$ satisfy the equations

$$\frac{\partial \bar{r}_C}{\partial t} = \frac{1}{4\pi(r_C^0)^2 C_C}\left(\frac{\partial \bar{q}_{vC}}{\partial t}\right) - \frac{2\bar{r}_C}{r_C^0}\frac{dr_C}{dt}$$
$$= -(D_v(c_v^0 - \bar{c}_{vC}) - D_i c_i^0)\bar{r}_C/(r_C^0)^2 + (D_v(\bar{c}_v - \bar{c}_{vC}) - D_i\bar{c}_i)/r_C^0 , \qquad (16)$$

$$\frac{\partial \bar{n}_{gC}}{\partial t} = \frac{4\pi r_C^0}{\Omega} D_g \bar{c}_g + \frac{4\pi \bar{r}_C}{\Omega} D_g c_g^0 - K\bar{n}_{gC} . \qquad (17)$$

The vacancies, interstitials and gas atoms migrate rapidly through the lattice, and therefore their concentrations are always at steady-state. Thus, it can be assumed that

$$\frac{\partial \bar{c}_v}{\partial t} = \frac{\partial \bar{c}_i}{\partial t} = \frac{\partial \bar{c}_g}{\partial t} = 0 , \qquad (18)$$

and therefore the Fourier Transforms of the concentrations are given by the equations

$$\bar{c}_v = -\frac{(c_v^0 - \bar{c}_{vC})4\pi \bar{r}_C C_C}{\lambda^2 + 4\pi r_C^0 C_C + Z_{vN}\rho_N} + \frac{\bar{c}_{vC} 4\pi r_C^0 C_C}{\lambda^2 + 4\pi r_C^0 C_C + Z_{vN}\rho_N} , \qquad (19)$$

$$\bar{c}_i = -\frac{c_i^0 4\pi \bar{r}_C C_C}{\lambda^2 + 4\pi r_C^0 C_C + Z_{iN}\rho_N} , \qquad (20)$$

$$\bar{c}_g = \frac{K\bar{n}_{gC} C_C \Omega - 4\pi \bar{r}_C C_C D_g c_g^0}{D_g(\lambda^2 + 4\pi r_C^0 C_C)} . \qquad (21)$$

From equations (16) and (17), the Fourier Transform of the cavity radius, $\bar{r}_C$, satisfies the equation

$$\frac{\partial \tilde{r}_C}{\partial t} = X_1 \tilde{r}_C + X_2 \tilde{n}_{gC} \quad , \tag{22}$$

where

$$X_1 = -\frac{K(Z_{iN}-Z_{vN})\rho_N}{(r_C^0)^2 k_v^2 k_i^2}\left(1+\frac{4\pi r_C^0 C_C(\lambda^2+k_v^2+k_i^2)}{(\lambda^2+k_v^2)(\lambda^2+k_i^2)}\right) -$$

$$-\frac{D_v(\bar{c}_{vN}-\bar{c}_{vC})Z_{vN}\rho_N}{(r_C^0)^2 k_v^2}\left(1+\frac{4\pi r_C^0 C_C}{(\lambda^2+k_v^2)}\right) -$$

$$-\frac{D_v \bar{c}_{vC}(\lambda^2+Z_{vN}\rho_N)\Omega}{(r_C^0)^2(\lambda^2+k_v^2)}\left(\frac{9 n_{gC}^0}{4\pi(r_C^0)^3}-\frac{2\gamma_S}{r_C^0 kT}\right) \quad , \tag{23}$$

and

$$X_2 = \frac{3\Omega\, D_v \bar{c}_{vC}}{4\pi(r_C^0)^4}\frac{\lambda^2+Z_{vN}\rho_N}{\lambda^2+k_v^2} \quad , \tag{24}$$

where the total sink strengths in the material for vacancies and interstitials, k_v^2 and k_i^2 respectively, are given by

$$k_v^2 = 4\pi r_C^0 C_C + Z_{vN}\rho_N \quad , \tag{25}$$

$$k_i^2 = 4\pi r_C^0 C_C + Z_{iN}\rho_N \quad . \tag{26}$$

Similarly, $\tilde{n}_{gC}$ satisfies the equation

$$\frac{\partial \tilde{n}_{gC}}{\partial t} = Y_1 \tilde{r}_C + Y_2 \tilde{n}_{gC} \quad , \tag{27}$$

where

$$Y_1 = \frac{4\pi D_g c_g^0}{\Omega}\frac{\lambda^2}{\lambda^2+4\pi r_C^0 C_C} \quad , \tag{28}$$

and

$$Y_2 = -\frac{K\,\lambda^2}{\lambda^2+4\pi r_C^0 C_C} \quad . \tag{29}$$

These equations can be written in the form of a matrix equation

$$\frac{\partial}{\partial t}(\tilde{r}_C, \tilde{n}_{gC}) = \begin{pmatrix} X_1 & X_2 \\ Y_1 & Y_2 \end{pmatrix} \begin{pmatrix} \tilde{r}_C \\ \tilde{n}_{gC} \end{pmatrix} = \mathbf{M} \begin{pmatrix} \tilde{r}_C \\ \tilde{n}_{gC} \end{pmatrix}. \tag{30}$$

The next step is to determine the eigenvalues of this matrix. If one of these eigenvalues has a positive real part, then the magnitude of the fluctuation represented by $(\tilde{r}_C, \tilde{n}_{gC})$ will tend to increase, i.e. a fluctuation in the radius of the cavities or in the number of gas atoms they contain will tend to grow instead of decay. Thus, some regions will contain a number of relatively large cavities, and the cavities in a neighbouring region will shrink and disappear. In most cases, both of the eigenvalues of the matrix $\mathbf{M}$ are real, with one large and negative, and the other small and either positive or negative. The value of this second eigenvalue gives the rate of growth of a fluctuation in the cavity population and is called the amplification factor [14].

4. Results

This section discusses the results of numerical calculations of the eigenvalues of the matrix $\mathbf{M}$ discussed in the previous section. In these calculations values for the initial cavity radius, r_C^0, and the cavity concentration are chosen. It is assumed that the concentrations of vacancies, interstitials and gas atoms, c_v^0, c_i^0 and c_g^0, and the concentration of vacancies in cavities, q_{vC}^0, are constant in time. This condition determines the values of c_v^0, c_i^0, c_g^0 and n_{gC}^0. Because gas is continually produced in the material, $\partial n_{gC}^0/\partial t > 0$, i.e. the concentration of gas trapped in cavities is increasing.

Figure 1 shows the results for the amplification factor as a function of the wavelength, $w = 2\pi/\lambda$, for the case of neutron irradiated steel. The radius and concentration of the cavities are taken from experimental measurements on a Fe-15%Ni-12%Cr-1%Si alloy irradiated in a fast reactor at 644°C [20]. The parameters used this calculation are listed in Table I. This figure shows that fluctuations with wavelengths $\leq 1\mu m$ are likely to develop in the material. There is no peak in the amplification factor, and this indicates that there is no preferred wavelength for these fluctuations. In other words, cavity clusters of differing sizes $\leq 1\mu m$ will develop in the material. (It is assumed here that the cavities which begin to shrink eventually disappear altogether.) This agrees with the observations of compositional changes in this material which show an irregular pattern of nickel-rich and nickel-poor regions, with clusters of cavities in the nickel-rich regions. A study of the cavity distribution indicated that the separation of the cavities is typically about 350nm [18].

Figure 1 also shows that the amplification factor falls with increasing dislocation density. Similarly, the value of the amplification factor tends to increase as the concentration of cavities falls because of the increased concentrations of vacancies and interstitials. However, in both cases the range of wavelengths for the cavity clustering is only slightly altered.

The initial size of the cavities is also important, see figure 2. It is clear from this figure that there is an optimum radius where the fluctuations develop most rapidly. However, the initial radius has little effect on the size of the cavity clusters. As the radius increases, the surface tension becomes less important and this tends to increase the amplification factor.

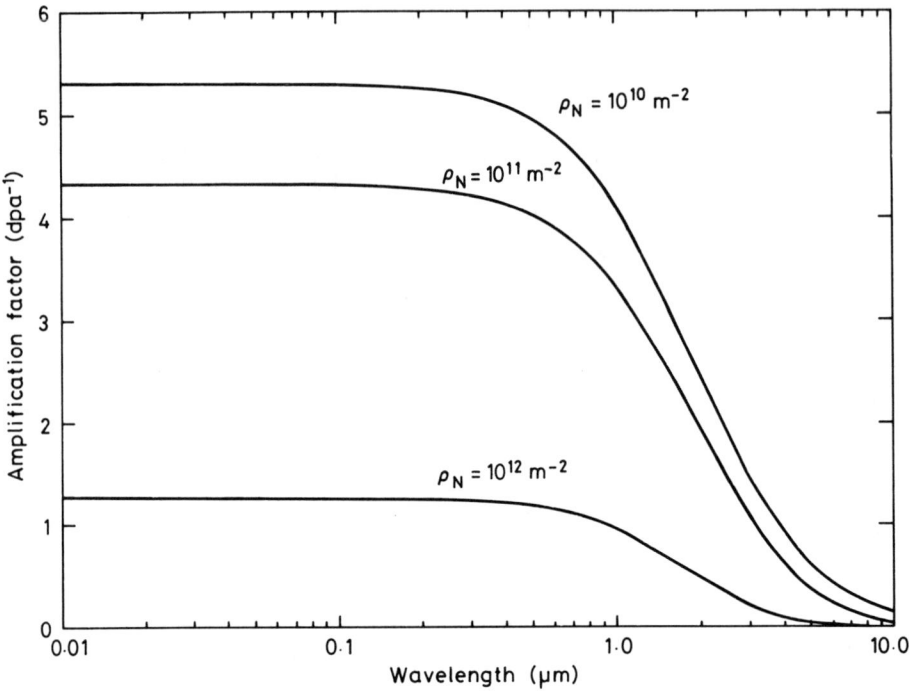

Figure 1. Amplification factor for fluctuations in cavity size in a steel irradiated at 644°C. Initial cavity radius, $r_C^0 = 3$ nm, and cavity concentration, $C_C = 3 \times 10^{20}$ m^{-3}. Other parameters are given in Table I.

TABLE I. Parameters used for numerical calculations

Displacement damage rate, $K = 10^{-6}$ dpa/s.
Production rate of gas, $K_g = 10^{-13}$ atoms / lattice site / second.
Irradiation temperature, $T = 644°C$.
Diffusion coefficient of vacancies, $D_v = 6 \times 10^{-5} \exp(-1.3\text{eV}/kT)$ m^{2}s^{-1}.
Diffusion coefficient of interstitials, $D_i = 10^{-7} \exp(-0.15\text{eV}/kT)$ m^{2}s^{-1}.
Diffusion coefficient of gas atoms, $D_g = 10^{-7} \exp(-0.15\text{eV}/kT)$ m^{2}s^{-1}.
Thermal equilibrium concentration of vacancies, $c_v^e = \exp(-1.6\text{eV}/kT)$.
Atomic volume, $\Omega = 1.206 \times 10^{-29}$ m^{-3}.
Surface energy, $\gamma_S = 2.0$ Jm^{-2}.
Dislocation bias factor for interstitials, $Z_{iN} = 1.10$.
Dislocation bias factor for vacancies, $Z_{vN} = 1.00$.
Cavity concentration, $C_C = 3 \times 10^{20}$ m^{-3}.

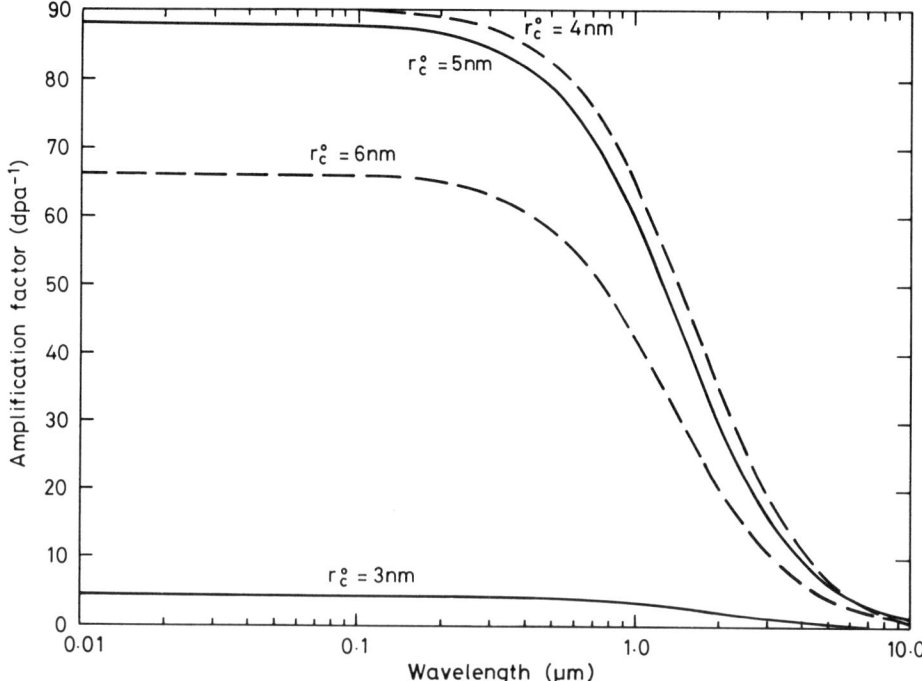

Figure 2. Amplification factor for fluctuations in cavity size in a steel irradiated at 644°C for a range of initial cavity radii. Cavity concentration, $C_C = 3 \times 10^{20} \, \text{m}^{-3}$ and dislocation density, $\rho_N = 10^{11} \, \text{m}^{-2}$. Other parameters are given in Table I.

At larger sizes there is a relatively small increase in the cavity radius for a given increase in n_{gC} and this tends to reduce the amplification factor.

Further work is needed to obtain results for lower irradiation temperatures. It is known from experiment that the concentration of cavities tends to increase and the size of the cavities tends to decrease as the irradiation temperature falls, although detailed experimental data are not available for the Fe-12%Cr-15%Ni alloy discussed above. These changes must be taken into account when predicting the fluctuations in cavity size under discussion here. Also, there are a number of approximations in the model which may influence the temperature dependence of the results. In particular, the recombination of vacancies and interstitials is neglected, and this will become more important at lower temperatures. It is hoped to extend the model to include the recombination of vacancies and interstitials, and to investigate the predictions of the theory at lower temperatures.

5. Summary and Conclusions

This paper has described a simple model for cavity clustering in irradiated metals. The growth of these cavities is largely determined by the concentration of helium in the material and the clustering is caused by the interactions of the helium atoms in the irradiated metal with the cavities. The results of numerical calculations show that this mechanism produces clusters of cavities for typical irradiation conditions, and that the predicted size of the clusters at 644°C agrees approximately with experimental observations [18].

Acknowledgements

The author wishes to thank Dr.T.M.Williams and Dr.C.M.Shepherd for many helpful discussions. This work is part of the longer term research programme funded by the Underlying Programme of the United Kingdom Atomic Energy Authority.

References

1. *Nonlinear Phenomena in Materials Science*, edited by G.Martin and L.P.Kubin, also published as *Solid State Phenomena*, Vols. 3&4, (Trans Tech Publications, Aerdermannsdorf, Switzerland, 1988).
2. Muncie,J.W., Eyre,B.L. and English,C.A., *Phil.Mag.A*, 1985, **52**, 309.
3. Jaeger,W., Ehrhart,P., Schilling,W., Dworschak,F., Gadalla,A.A. and Tsukuda,N., *Materials Science Forum*, 1987, **15-18**, 881.
4. Singh,B.N., Leffers,T. and Horsewell,A., *Phil.Mag.A*, 1986, **53**, 233.
5. Evans,J.H., this conference.
6. Hulett,L.D., Jr., Baldwin,T.O., Crump,J.C., III and Young,F.W., Jr., *J.Appl.Phys.*, 1968, **39**, 3945.
7. Kulcinski,G.L. and Brimhall,J.L., in: *Effects of Radiation on Substructure and Mechanical Properties of Metals and Alloys*, ASTM-STP 529, American Society for Testing and Materials, 1973, p.258.
8. Steigler,J.O. and Farrell,K., *Scripta Metall.*, 1974, **8**, 651.
9. Brager,H.R. and Garner,F.A., in: *Optimizing Materials for Nuclear Applications*, edited by F.A.Garner, D.S.Gelles and F.W.Wiffen, 1984, The Metallurgical Society of AIME, p. 141.
10. Williams,T.M., Boothby,R.M. and Titchmarsh,J.M., in *Proc.Int.Conf. on Materials for Nuclear Reactor Core Applications*, BNES, London, 1987, p.293.
11. Martin,G., *Phys.Rev.B.*, 1980, **21**, 2122.
12. Martin,G., Cauvin,R., Bocquet,J.-L. and Barbu,A., in: *Phase Stability During Irradiation*, Proceedings of a Symposium, Pittsburgh, Pennsylvania, October 5-9, 1980, edited by J.R.Holland, L.K.Mansur and D.I.Potter, 1981, The Metallurgical Society of AIME, p. 43.
13. Krishan,K. and Abromeit,C., *J.Phys.F: Metal Physics*, 1984, **14**, 1103.
14. Abromeit,C. and Martin,G., ref. 1, p. 321.
15. Krishan,K., *Phil.Mag.A*, 1982, **45**, 401.
16. Koptelov,E.A. and Semenov,A.A., *J.Nucl.Mater.*, 1988, **160**, 253.
17. Murphy,S.M., ref.1, p. 295.
18. Shepherd,C.M. and Murphy,S.M., Harwell Report AERE-R.13636 (1989).
19. Shepherd,C.M., unpublished work.
20. Shepherd,C.M., Murphy,S.M. and Williams,T.M., Harwell Report AERE-R.13538 (1989).

A MESOSCOPIC THEORY OF IRRADIATION-INDUCED VOID-LATTICE FORMATION

P. Hähner and W. Frank
Max-Planck-Institut für Metallforschung
Institut für Physik, and Universität Stuttgart
Institut für Theoretische und Angewandte Physik
P.O. Box 800665
D-7000 Stuttgard 80 (FRG)

Void-lattice (VL) formation in crystals under high-temperature irradiation is an example of self-organization resulting from highly dissipative processes. There is strong experimental evidence that anisotropic migration of self-interstitial atoms (SI) is a prerequisite [1,2]. The present note summarizes a recent mesoscopic treatment of VL formation based on the same physical picture as the Woo-Frank theory [1,3,4], which operates on the level of macroscopic averages. The radiation-induced defects featuring in this theory are (1) metastable 1-dimensionally migrating SIs (crowdions), bringing in the anisotropic migration mentioned above, (2) stable 3-dimensionally migrating SIs (dumbbells), and (3) 3-dimensionally migrating vacancies.

Expressions for the growth rates and drift velocities of voids in terms of the fluxes of the above-mentioned defects are derived. By means of Green's functions of stationary reaction-diffusion equations we calculate the fluxes of defects entering any particular void, which depend on the surrounding void configuration. In this way the dynamics of point defects is eliminated adiabatically and the dynamical void-void interaction may be modelled. These results enable us to set up two non-linear coupled integro-differential equations for suitably defined fields modelling the sink strength of voids on a mesoscopic scale.

From a linear stability analysis the following conclusions may be drawn :

1. The control parameters governing the transition to ordered void arrays are the void-number density ρ and the average void radius
2. The lattice parameter of the VL is not an intrinsic property of the material but depends on ρ in the pre-ordering regime.
3. For bcc (fcc) metals the mode marking the onset of VL formation is a planar ordering on {110} ({111}) planes. The perfect 3-

dimensional VL emerges as a superposition of all 6 (4) equivalent sets of planes.

4. For bcc, fcc, and simple cubic structures the VLs are isomorphic to the host lattices and possess the same crystallographic orientation. Further it may be shown that.

5. VLs are dynamically stable.

Conclusions (1)-(5) are in full agreement with experiments. Conclusion (3), which goes beyond the predictions of the Woo-Frank version of the theory, shows that, contrary to what has been done in [2], it is inadmissible to quote the initial formation of planar VLs as evidence against this theory.

Acknowledgements

The authors are grateful to Professor Dr. A. Seeger for his collaboration and to Dr. J. Evans for making available a preprint of his contribution to these proceedings. One of us (P.H.) would like to acknowledge the financial support by NATO for the participation in this Advanced Study Institute.

References

1. W. Frank, "Solid State Phenomena", **3/4**, 315 (1988)
2. J.H. Evans, these proceedings
3. C.H. Woo and W. Frank, J. Nucl. Mater. **137**, 7 (1985); 140, 214 (1986); **148, 121** (1987)
4. C.H. Woo and W. Frank, Materials Science Forum **15-18**, 875 (1987).

INDEX

Absolutely unstable 83
Acoustic emission 254,266
Activation volume 257,269
Active slip volume 251,268
Adiabatic shear band 264,265
Alloy solidification 124,180
Amorphous alloys 266
Amplitude equations 4,7,8,124,128,230
Balance equations 227
Benard instability 3,7,27,116
Bifurcation 4,35,66,124,127,142,303,356
Binary fluid convection 85
Binary mixtures 123,135
Bubbles 347,348,351,371
Cavities 347,351,360,371,372
Cavity clusters 371,376
    lattice 347,354,361
Cellular pattern 1,4,127,139,145
Climb 75,309
Collective dislocation effects 259,277
Condensed matter 25,187,185
Confined state 84
Conserved order parameter 111
Constitutional supercooling 136
Constitutive equation 255
Convection-diffusion equation 166
Convective instability 83
Convectively unstable 83
Cottrell solute atmosphere 245,246,258,271
Creep 324
Cross-slip 75,78,244,260,265,266,292,309
Crystal growth 147,159
    plasticity 221,242
Defects 6,7,9,12,16,44,73,74,128,195,337,381
Deformation bands 231,264
    of metals 73,241,304
Dendrite 126,130,203
    growth 180
Directional solidification 123,127,128,129,135,147
Dislocation 242,277,303,348,371
    annihilation 74,223,280,303,314,

breakaway	245,258,269
cell structure	251,252,310,321
dipole	289,303,315,326
dynamics	77,316
forest	79,277,278
front	206,213
generation rate	223
mobility	245
motion	243,290
multiplication	74,247,256,298,316
patterns	289,294,303,304,306
pinning	74,80,303
reaction	248,289,327
source	247
velocity	243,267
Dissipative structure	64,304,345
Domain walls	36,55
Dynamic recrystallization	258
External noise	83
Evaporation	109,112
Fatigue	304
Fe-Cr-Ni alloys	371
Fluctuations	2,186
Fracture	221,238
Freedericksz	35,49
Front propagation	69,203
Generation rate approach	268
Ginzburg-Landau equation	6,7,8,36,38,64,74,83,84,85
Glide	75,78,309
Growth of needle crystals	159
H-type instabilities	256
Hart's criterion	256
Hydrodynamics	26,113
Hydrodynamic equation	25
Immobile dislocations	80,81,225,227,309
Inert gases	348
Inhomogeneous melting	331
Instability	2,4,109,112,118,135,203,204,221,225,228,356,373
Interface	138,142,148
instability	138
Internal stress	257
Interstitials	81,348,358,360,361,372
Irradiation	347,348
effects	74,265
induced instability	75,371

Ivantsov solution	159,160
Kuramoto-Sivashinsky equation	30,86
Laser melting	331,332
Lattices	73
Lyapunov functional	64,70
Load serrations	253,262
Local shear transformations	266
Localized structure	63,66,72,266
Localization of shear	264,266
Localized solution	28,30,63
Lüder-band	247,250,254,265
M-type instabilities	255
Marangoni effect	109,111,113
Martensitic transformations	262
Mechanical twinning	261
Metallic glass	267
Metals	73,74,241,243,304,339,347,351,371
Microwaves	109,112
Mobile dislocations	79,80,225,227,277,278,309
Mullins-Sekarka instability	125,126,136
Necking	141,256
Nonlinear gradient term	85
Non-variational effects	71
Noise-sustained structure	83
Open-flow systems	83
Ordered-disordered transition	201
Parabolic tip	126,163
Patterns	1,4,7,9,20,26,63,73,109,123,129,195,335,337
Pattern formation	35,49,67,73,96,105,147,203,221,305,345
Pattern selection	41,83
Peeling	233
Periodic melting	332
Persistent slip bands	221,225,226,291
Phase diffusion	27
dynamics	2,25,128
Pipe flow	84
Plastic deformation	77
flow	277,286,292
instabilities	241,252,255,277,278,280
Porous medium	203,204
Portevin-LeChatelier effect	258,270,283,284
PLC-band	222,231,265
Propagating mode	28,264,265
Pseudoelasticity	262

Rate theory	75,306
Reaction-Diffusion	6,63,66,67,74,102,105,222,277,292, 293,381
front	203,206,207,217
Transport	203,206,217,226
Recovery	258
Saffman-Taylor fingering	129,149,204
Self-organization	1,203,221,222,356,304
S-type instability	258,283
Shape memory effect	262
Shear bands	222,264,266
Sidebranches	
Silicon thin films	331
Simulation	2,10,11,13,14,15,66,207,213,272, 294,295,303,307,316,319
Slip	
bands	221,249,264,291
band formation	253,257,260
(growth, development)	
band bundles	221,250,265
lines	221,249,282,289
localization	249
Slugs	84
Solid solutions	245
Solitons	85
Spatial structure	1,91
Spatio-temporal intermittency	83
Spinodal decomposition	36
Stick-slip	233
Structural invariant	186,188
Succinonitrile	135,137
Suzuki segregation	245
Strain hardening rate	255
Strain rate changes	269
Strain rate sensitivity	255,269
Stress relaxation	269
Structural softening	256
Systems far from equilibrium	1,26,123,129,203
Taylor-Dean instability	89
T-type instabilities	262
Thermal convection	27,109,110,124
Tip splitting	208,211,214,215
Transition	156,195,331
Traveling rolls	89
waves	12,68,128

Turbulence	6,16,17,28
Turing instability	63,64
Undercooling	163
Uniaxial deformation	281
Vacancies	75,81,348,358
Vacancy clusters	372
Viscous glide approach	267
Vapor differential recoil instability	112,121
Voids	73,347,348,353,381
Void lattice	73,347,349,381
Wavelength selection	138,145,147
Wavy instability	90
Work hardening	246
Work hardening rate	248

PARTICIPANTS

ACOMB, S.   Mathematical Institute, 24-29 St Gilles, Oxford OXI 3LB, England.

AHLERS, G.   University of California, Dept. of Physics, CA-93106 Santa Barbara, USA.

AIFANTIS, E.C.   Michigan Technology University, Dept. ME-EM, Houghton MI 49931, USA.

ALTINTAS, S.   Bogazici University, Dept. of Mechanical Engineering, 80815 Bebek, Istanbul, Turkey.

BARYAKHTAR, V.G.   Institue of Metal Physics, Academy of Sciences Ukr SSR, 36 Acad. Vernadsky Bd, 252142 Kiev 42, USSR.

BEN AMAR, M.   Ecole Normale Supérieure, Rue Lhomond 24, F-75231 Paris Cedex 05, France

BERTRAND, G.   Université de Bourgogne, Laboratoire de Réactivité des Solides, F-21004 Dijon Cedex, France.

BORCKMANS, P.   Université Libre de Bruxelles, Service de Chimie Physique, C.P.231, Bd du Triomphe, B-1050 Bruxelles, Belgium.

BRAND, H.   University of Essen, Dept of Physics, D-43 Essen 1, FRG.

BRATTKUS, K.   Caltech, Dept of Applied Math., Pasadena CA 91125, USA.

CASADEMUNT, J.   Universidad de Barcelona, Facultat de Fisica, Av Diagonal 647, E-08028 Barcelona, Spain.

CELIK, M.   Middle East Technical Univ. METU, 06531 Ankara, Turkey.

COLLET, J.F.   Indiana University, Dept of Chemistry, Bloomington, IN 47405, USA.

COULLET, P.   Université de Nice, Laboratoire de Physique Théorique, Parc Valrose, F-06034 Nice Cedex, France.

DE BRUYN, J.	Memorial University, Dept of Physics, St John's, NF, AIB3X7 Canada.
DEISSLER, R.J.	National Center for Atmospheric Research, PO Box 3000, Boulder, Colorado 80307-3000, USA.
DEWEL, G.	Université Libre de Bruxelles, Service de Chimie Physique, C.P.231, Bd du Triomphe, B-1050 Bruxelles, Belgium.
EVANS, J.	Harwell Laboratory, Materials Devel. Div. B393, UK Atomic Ener. Auth., Oxfordhire OX011 ORA, GB.
FAIVRE, G.	Université de Paris 7, Groupe de Physique des Solides, Place Jussieu 2, Tour 23, F-75005 Paris, France.
GECIM, H.S.	Hacettepe University, Electrical & Elect. Eng. Dept, 06532 Eytepe, Ankara, Turkey.
GESHEF, D.	Vrije Universiteit Brussel/FMAT, Pleinlaan 2, B-1050 Brussel, Belgium.
GHONIEM, N.	University of California, School of Engineering, CA-90024 Los Angeles, USA.
GIL, L.	Univesité de Nice, Laboratoire de Physique Théorique, Parc Valrose, F-06034 Nice Cedex, France.
GLEESON, J.	ATT Bell Laboratories, 600 Mountain Ave, MH 1E440, NJ 07974, USA.
GUTHMANN, C.	Université de Paris 7, Groupe de Physique des Solides, Place Jussieu 2, Tour 23, F-75005 Paris, France.
HÄHNER, P.	Max-Planck Institüt für Metallforschung, Heisenbergstr.1, D-7000 Stuttgart 80, FRG.
HANSELER, M.F.	Université Pierre et Marie Curie, I.E.S.C., Tour 15, Place Jussieu 4, F-75230 Paris Cedex, France.
HERNANDEZ-GARCIA, E.	Universitat de les Illes Balears, Dept de Fisica, E-07071 Palma de Mallorca, Spain.
HOUCHMANDZADEH, B.	Université de Grenoble 1, Laboratoire de Spectroscopie Physique, B.P.53,

	F-38041 Grenoble Cedex, France.
JOETS, A.	Université de Paris Sud, Laboratoire de Physique des Solides, F-91405 Orsay, France.
KAISER, M.	Universität Bayreuth, Lehrstuhl Theoretische Physik II, D-8580 Bayreuth, FRG.
KARCHER, C.	Kernforshung. Karl. GmbH, D-7500 Karlsruhe Postfach 3640, FRG.
KRAMER, L.	Universität Bayreuth, Postfach 101251, D-8580 Bayreuth, FRG.
KUBIN, L.P.	CNRS-ONERA, Laboratoire d'Etude des Microstructures, Av. de la Division Leclerc, B.P.72, F-92322 Chatillon Cedex, France.
LAJZEROWICZ, L.	Université de Grenoble 1, Laboratoire de Spectroscopie Physique, B.P.53, F-38041 Grenoble Cedex, France.
LAURE, P.	Université de Nice, Laboratoire de Mathematique, Parc Valrose, F-06034 Nice Cedex, France.
LEGA, J.	Université de Nice, Laboratoire de Physique Théorique, Parc Valrose, F-06034 Nice Cedex, France.
LEROUX, Ch.	Universiteit Antwerpen (RUCA), Elektro. Material Onder., Groenenborgerlaan 171, B-2020 Antwerpen, Belgium.
LEVINE, H.	Institute for Nonlinear Science UCSD, La Jolla, CA-92093-0402 USA.
LIAO, W.	California Institute of Technology, Sloan Annex 114-36, CA 91125 Pasadena, USA.
MASHAAL, M.	Ecole Normale Supérieure, Laboratoire de Physique Statistique, Rue Lhomond 24, F-75231 Paris Cedex 05, France.
MICHALLAND, S.	Ecole Normale Supérieure, LPS, Rue Lhomond 24, F-75005 Paris Cedex, France.
MISBAH, C.	Ecole Normale Supérieure, ENS, Place Jussieu 2, Tour 23, F-75005 Paris Cedex 05, France.

MITUS, A.C.	Universität des Saarlandes, EB Physik/Theor.Physik, Bau 88, D-6600 Saarbrücken, West Germany.
MOZOS, J.L.	Universidad de Barcelona, Facultad de Fisica, Av. Diagonal 647, E-08028 Barcelona, Spain.
MÜLLER, H.	Universität des Saarlandes, Institut für Theor. Physik, D-6600 Saarbrücken, West Germany.
MURPHY, S.	Harwell Laboratory, DIDCOT, Materials Devel. Division, Oxon OX1 0RA, Great Britain.
MUTABAZI, I.	Ecole Normale de Physique et de Chimie de Paris, Rue Vauquelin 10, F-75231 Paris Cedex, France.
NEUHAUSER, H.	Univervität Carolo-Wihelmina, Mendelssohnstraße 3, D-3300 Braunschweig, FRG.
ORTOLEVA, P.J.	Indiana University, Department of Chemistry, IN-47405 Bloomington, USA.
PLESSING, J.	Universität Carolo-Wilhelmina, Mendelssohnstraße 3, D-3300 Braunschweil, FRG.
POMEAU, Y.	ENS, Groupe de Physique des Solides, Rue Lhomond 24, F-75231 Paris Cedex 05, France.
PURWINS, H.G.	Westräl. Wilhelmina Univ. Münster, Correnstr. 2/4, D-4400 Münster, FRG.
RABAUD, M.	Ecole Normale Supérieure, LPS, Rue Lhomond 24, F-75005 Paris Cedex, France.
RIBOTTA, R.	Université de Paris Sud, Laboratoire de Physique des Solides, Bât.510, F-91405 Orsay, France.
ROCHWERGER, D.	Centre St Jérôme, Laboratoire de Recherche en Combustion, F-13397 Marseille Cedex 13, France.
ROSSI, M.	University of Southern California, Dept of Aerospace Eng., LA CA 9089-1191, USA.
ROTH, D.	Universität des Saarlandes, Institüt für Theoretische Physik, D-6600 Saarbrücken,

	West Germany.
SAINTY, J.P.	Avenue Moutier 43, F-93190 Livry-Gargan, France.
SALAN-SANTOS, J.	Universidad de Barcelona, Dept Estr. y Const. Materia, Av. Diagonal 647, E-08028 Barcelona, Spain.
SALAZAR-CRUZ, M.	Université Libre de Bruxelles, Service de Chimie Physique, C.P.231, Bd du Triomphe, B-1050 Bruxelles, Belgium.
SAN MIGUEL, M.	Universitat de les Illes Baleares, Dept de Fisica, E-07071 Palma de Mallorca, Spain.
STONE, C.	University of California, 6291 Boelter Hall, UCLA, Los Angeles, CA 90024-1587, USA.
TRAINOFF, S.	UCSB Physics Department, CA 93106 Santa Barbara, USA.
TSATIS, D.E.	University of Patras, Dept of Physics, GR 26110 Patras, Greece.
VAN DRIEL, H.M.	University of Toronto, Dept of Physics, M5S 1A7 Ontario, Canada.
WALGRAEF, D.	Université Libre de Bruxelles, Service de Chimie Physique, C.P.231, Bd du Triomphe, B-1050 Bruxelles, Belgium.
WEBER, A.	Universität Bayreuth, Institut für Theoretische Physik II, Postfach 101251, D-8580 Bayreuth, FRG.